Analytical Geometry: Two Dimensions

Prof. Sibdas Karmakar
Retd. Associate Professor of Mathematics
Ramananda College, Bishnupur, Bankura

Dr. Samiran Karmakar
Assistant Professor of Mathematics
Bankura Sammilani College, Bankura

Levant Books
India

First published 2022
by CRC Press
4 Park Square, Milton Park, Abingdon, Oxon, OX14 4RN

and by CRC Press
6000 Broken Sound Parkway NW, Suite 300, Boca Raton, FL 33487-2742

© 2022 Sibdas Karmakar, Samiran Karmakar and Levant Books

CRC Press is an imprint of Informa UK Limited

The right of Sibdas Karmakar and Samiran Karmakar to be identified as the authors of this work has been asserted in accordance with sections 77 and 78 of the Copyright, Designs and Patents Act 1988.

All rights reserved. No part of this book may be reprinted or reproduced or utilised in any form or by any electronic, mechanical, or other means, now known or hereafter invented, including photocopying and recording, or in any information storage or retrieval system, without permission in writing from the publishers.

For permission to photocopy or use material electronically from this work, access www.copyright.com or contact the Copyright Clearance Center, Inc. (CCC), 222 Rosewood Drive, Danvers, MA 01923, 978-750-8400. For works that are not available on CCC please contact mpkbookspermissions@tandf.co.uk

Trademark notice: Product or corporate names may be trademarks or registered trademarks, and are used only for identification and explanation without intent to infringe.

Print edition not for sale in South Asia (India, Sri Lanka, Nepal, Bangladesh, Pakistan or Bhutan).

British Library Cataloguing-in-Publication Data
A catalogue record for this book is available from the British Library

Library of Congress Cataloging-in-Publication Data
A catalog record has been requested

ISBN: 9781032270203 (hbk)
ISBN: 9781003293248 (ebk)

DOI: 10.4324/9781003293248

LEVANT

This book is dedicated to PARNIKA KARMAKAR, *the youngest daughter of Samiran.*

Preface

This book entitled "***Analytical Geometry: Two Dimensions***" has been written in accordance with the modern trend in the teaching of Geometry. In preparing this volume, the authors have endeavoured to write a drill book for beginners which presents the fundamental concepts of this topic. The number of propositions with formal proofs has been reduced and stress laid on the solutions of the problems. Authors believe that this text is complete for students finishing their studies of Mathematics with a course in Analytic Geometry of Two Dimensions. All new materials as well as the old one have been treated from the same standpoint.

The present work does not claim to be original. Author's aim of this book is to make the subject thoroughly intelligible. All the important properties of the conics have been discussed either in the articles or in illustrative examples. A large number of problems have been completely solved and exercises containing carefully graded and motivating examples have been incorporated at the end of each chapter.

Highlights of some specialties:

- The book is self-contained.

- Simple, comprehensive, lucid and rigorous discussions are provided.

- Basic school level algebra and geometry is sufficient to understand the topics of this book.

- Treatment of matrix theory has been shown in some topics, e.g., classification of conics, of this book.

Attention is called to the method of treatment. The subject is developed after the Euclidean method of definition and theorems. However, the formal presentation is avoided. Emphasis has been given on the analytical side everywhere.

The authors hope the book will prove useful not only to the students of Mathematics but also to the users of other branches of science and technology as well as in the applications of recently developing Business and financial sectors throughout the world.

In bringing out this book, Authors would wish to record their appreciation to the Publishers and Printers for their care and efforts. Criticisms and any constructive suggestions for the improvement of this book will be gratefully acknowledged.

We are thankful to the students who has constantly inspired us to complete this work by expressing the difficulties and problems they face frequently in this subject.

Authors

Contents

Preface		v
0	**Preliminaries**	**1**
	0.1 Introduction	1
	0.2 Terminologies and Fundamental Concepts	1
	0.2.1 Directed Lines	1
	0.2.2 Directed Line Segment and its Value	1
	0.2.3 Projection	2
	0.2.4 Half Line	3
	0.3 Some Tips and Tricks	3
	0.3.1 On Coordinate and Locus	3
	0.3.2 Straight Lines	9
	0.3.3 The Circle	12
	0.3.4 The Parabola	14
	0.3.5 The Ellipse	15
	0.3.6 The Hyperbola	16
	0.3.7 Common Properties of all Conics	18
	0.3.8 Properties of Tangents and Normals on Conics	18
	0.3.9 Conjugate Diameters	19
1	**Transformation of Rectangular Cartesian Coordinates**	**21**
	1.1 Introduction	21
	1.2 Coordinate Transformation Formulas	21
	1.2.1 Translation (or transformation by translation of axes)	21
	1.2.2 Rotation (or transformation by rotation of axes) . . .	22
	1.2.3 Rigid Motion or Orthogonal Transformation	24
	1.3 Use of Matrix Notation	24
	1.3.1 Translation	24
	1.3.2 Rotation	25

		1.3.3 Rigid Motion	25

    1.4 Definition of Orthogonal Transformation 26
    1.5 General Orthogonal Transformation 26
    1.6 Invariants 27
    1.7 Purpose of Transformation of Axes 27
    1.8 To Investigate whether there is any Fixed Point under Translation, Rotation or Rigid Motion 27
        1.8.1 Fixed Point under Translation 27
        1.8.2 Fixed Point under Rotation 28
        1.8.3 Fixed Point under Rigid Motion 28
    1.9 Invariants of Translation, Rotation and Rigid Motions 29
    1.10 Work Out Examples 34
    1.11 Exercises 48
        Answers 53

2 General Equation of Second Degree: Classification of Conics   **55**
    2.1 Introduction 55
    2.2 Some Definitions 56
        2.2.1 Centre of a second order curve 56
    2.3 Determination of Centre of a Second Order Curve 57
    2.4 Solution of Simultaneous Equations Using Matrix Notation . 60
    2.5 Canonical Form of an Equation of a Second Degree Curve . . 61
    2.6 Reduction of an Equation of a Conic to Canonical Form . . . 65
    2.7 Rank and Classification of Second Order Curves 69
    2.8 Conic through Common Points 69
    2.9 Conditions for Pair of Straight Lines and proper Conics . . . 70
    2.10 Worked Out Examples 74
    2.11 Exercises 94
        Answers 99

3 Polar Coordinates and Equations   **101**
    3.1 Polar Coordinates 101
    3.2 Discussion on Polar Coordinates 101
    3.3 Relation between Polar and Cartesian Coordinates of a Point 102
    3.4 Distance between Two Points 103
    3.5 Area of a triangle 104
    3.6 Straight Lines 104
        3.6.1 Equation of a straight line in polar coordinates in general form 104

		3.6.2	Equation of a Line Passing through the Two Given Points 105
		3.6.3	Polar equation of a line in normal form 107
	3.7	The Circle 108	
		3.7.1	Polar equation of a circle 108
	3.8	Polar Equation of a Conic 111	
		3.8.1	Definition of a conic 111
		3.8.2	Polar equation of a conic whose semi latus rectum is l and focus is the pole 111
	3.9	Polar Equation of the Chord of a Given Conic 114	
	3.10	Polar Equation of the Tangent to a Given Conic 115	
	3.11	Polar Equation of the Normal to a Given Conic 116	
	3.12	Chord of Contact 117	
		3.12.1	Equation of Chord of Contact 117
	3.13	Asymptotes 120	
		3.13.1	Equation of the asymptote 120
	3.14	Polar Equation of the Directrices of an Ellipse 121	
	3.15	A Few Properties 123	
	3.16	Worked-Out Examples 124	
	3.17	Exercises 154	
		Answers 164	

4 Pair of Straight Lines 165
    4.1 General Equations 165
    4.2 To Find a Necessary Condition that the General Equation of Second Degree Should Represent a Pair of Straight Lines . . 167
    4.3 Angle Between a Pair of Lines 174
    4.4 Equation of Bisectors of the Angle between a pair of straight lines 175
    4.5 Equation of Two Lines Joining the Origin to the Points in Which a Line Meets a Conic 176
    4.6 Worked Out Examples 177
    4.7 Exercises 199
        Answers 207

5 Tangents and Normals, Pair of Tangents, Chord of Contact & Pole and Polar 209
    5.1 Tangent to a Curve 209
        5.1.1 To Find the Equation of the Tangent to a Conic at a Given Point on the Conic 209

		5.1.2 Tangents of the Standard Equations (Conics) 212

       5.1.2 Tangents of the Standard Equations (Conics) 212
       5.1.3 Condition for Tangency of a Straight Line to a Conic . 213
       5.1.4 Condition of Tangency for Standard Conics of a Line . 214
       5.1.5 Some Important Remarks 214
  5.2 Normals 216
       5.2.1 Equation of a Normal of a Conic at a Given Point . . 216
       5.2.2 Normals of the Standard Equations of Conics 217
  5.3 Pair of Tangents 219
       5.3.1 Equations of the chord of contact of tangents to a conic from any point outside it 219
       5.3.2 Equation of the pair of tangents from an external point to a conic 220
  5.4 Director Circle 222
       5.4.1 Equation of the director circle of a conic 222
       5.4.2 Equations of Director Circles of Standard Conics . . . 223
  5.5 Chords in Terms of Middle Point 223
  5.6 Diameter of a Conic 225
       5.6.1 Equation of a Diameter 225
       5.6.2 Conjugate Diameter 225
  5.7 Pole and Polar 226
       5.7.1 Equation to the polar of a point with respect to a non-singular conic 226
       5.7.2 Deduction of the Equation of Polar from the Alternative Definition 228
       5.7.3 Particular Cases 230
  5.8 Pole of a Polar with respect to a Conic 233
       5.8.1 Pole of a Given Straight Line with respect to a Given Conic 233
  5.9 Conjugate Points and Conjugate Lines 237
  5.10 Properties of Pole and Polar 237
       5.10.1 Self-polar Triangle 238
  5.11 Conditions for Conjugate Lines 239
  5.12 Worked Out Examples 241
  5.13 Exercises 265
       Answers 271

6 Diameters and Conjugate Diameters   273

  6.1 Diameters 273
  6.2 Important Theorems 273
  6.3 Conjugate Diameter 276

x

6.4	Condition that two given Straight Lines may be Conjugate Diameters of an Ellipse	276
6.5	Properties of Conjugate Diameters of an Ellipse	277
6.6	Properties of Conjugate Diameters of a Hyperbola	280
6.7	Equi-conjugate Diameters of an Ellipse	283
6.8	Worked Out Examples	284
6.9	Exercises	297
	Answers	300

Bibliography   **301**

Index   **303**

Chapter 0

Preliminaries

0.1 Introduction

The coordination of algebra and geometry is called *Coordinate Geometry*. In 1619, the French Mathematician, R. Descartes gave an idea how the points in a plane can be located by pairs of real numbers. Using coordinates we can study the properties of figures with the help of geometric equations. In this chapter, we recapitulate, some basic ideas, important formulae and deductions which we have studied in earlier classes.

0.2 Terminologies and Fundamental Concepts

0.2.1 Directed Lines

Definition 0.2.1 *A line of a plane on which one direction is defined as positive and the opposite direction is negative is called a directed line or an axis.*

In the **Figure 0.1**, the line g is a directed line. If the direction of $\overrightarrow{OB}$ is considered as positive direction then obviously the opposite direction $\overrightarrow{OB'}$ is negative direction.

0.2.2 Directed Line Segment and its Value

Definition 0.2.2 *The segment between any two points is called a directed line segment.*

In the same **Figure 0.1**, if A, B be two points on a directed line g, then $\overrightarrow{AB}$ is the directed line segment of g. Its direction is taken from A to B i.e., A is the initial point and B is the terminal point.

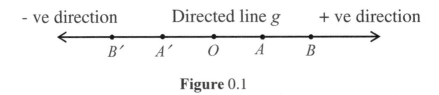

Figure 0.1

The value (or the length) of the directed line segment $\overrightarrow{AB}$ is denoted by AB which is defined as follows:

$$AB = |\overrightarrow{AB}| = -|\overrightarrow{A'B'}|, \text{ where the direction of the segment is taken along the +ve direction of } g$$

$$= -|\overrightarrow{AB}| = |\overrightarrow{A'B'}|, \text{ where the direction of the segment is taken along the -ve direction of } g$$

and $\quad = |\overrightarrow{AB}| \quad = \text{length of } \overrightarrow{AB} = \text{length of } \overrightarrow{A'B'}.$

0.2.3 Projection

(i) Projection of a point on a directed line

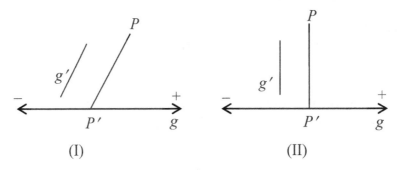

Figure 0.2

The projection of a point on an axis g in a particular direction is the point of intersection of the axis with the line through that point drawn

Preliminaries

parallel to the given direction. In **Figure 0.2** (I), P' is the projection of P on g along g'.

If the given direction is perpendicular to the axis g, then it is called an *orthogonal projection*. In **Figure 0.2** (II), P' is the orthogonal projection of P on g along the perpendicular direction g'.

Note 0.2.1 *By projection we always mean an orthogonal projection unless and otherwise stated.*

(ii) Projection of a directed line segment on a directed line

Let PQ is an arbitrary line segment and g be the axis (directed line). From P and Q perpendiculars are drawn to meet the axis g at P', Q' respectively. Then the values of the segment $\overline{P'Q'}$ is called the projection of $\overline{PQ}$ on g and P', Q' are called the *projections* of the points P and Q respectively on g [*vide* **Figure 0.3**].

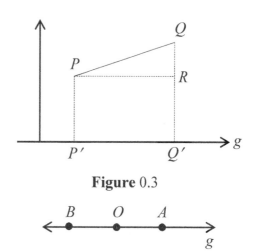

Figure 0.3

Figure 0.4

Note 0.2.2 *From Figure 0.3, it is evident that $|P'Q'| = |PR|$, where R is the foot of perpendicular from P on QQ'.*

0.2.4 Half Line

Any point on a directed line divide it into two parts, each is called a half line. In **Figure 0.4**, $\overrightarrow{OA}$ and $\overrightarrow{OB}$ are half lines generated by a point O on the directed line.

0.3 Some Tips and Tricks

0.3.1 On Coordinate and Locus

1. Location of a point

A) The point $P(x,y)$ in a plane is in

(i) The first quadrant $\Leftrightarrow x > 0, y > 0$.
(ii) The second quadrant $\Leftrightarrow x < 0, y > 0$.
(iii) The third quadrant $\Leftrightarrow x < 0, y < 0$.
(iv) The fourth quadrant $\Leftrightarrow x > 0, y < 0$.

B) The point $P(x, y)$ is on
(i) The x-axis $\Leftrightarrow y = 0$. (ii) The y-axis $\Leftrightarrow x = 0$.

C) If the Cartesian coordinates of a point are (x, y), by taking the origin as pole and the polar axis as the positive direction of the x-axis, the polar axis coordinates (r, θ) of the same point are given by $x = r\cos\theta$, $y = r\sin\theta$ and hence $r = \sqrt{x^2 + y^2}$, $\tan\theta = \dfrac{y}{x}$.

2. Distance formula

The distance between two points $P(x_1, y_1)$ and $Q(x_2, y_2)$ is given by

$$PQ = \sqrt{(x_2 - x_1)^2 + (y_2 - y_1)^2}.$$

Distance of $P(x_1, y_1)$ from the origin $= OP = \sqrt{x_1^2 + y_1^2}$.

3. Section Formula

i) If $R(x, y)$ divides the line joining two points $P(x_1, y_1)$ and $Q(x_2, y_2)$ in the ratio $m_1 : m_2$ $(m_1, m_2 > 0)$, then

$$x = \frac{m_1 x_2 + m_2 x_1}{m_1 + m_2}; \quad y = \frac{m_1 y_2 + m_2 y_1}{m_1 + m_2} \quad \text{(divided internally)}$$

and $\quad x = \dfrac{m_1 x_2 - m_2 x_1}{m_1 - m_2}; \quad y = \dfrac{m_1 y_2 - m_2 y_1}{m_1 - m_2} \quad$ (divided externally).

ii) If $R(x, y)$ divides the line joining two points $P(x_1, y_1)$ and $Q(x_2, y_2)$ in the ratio $\lambda : 1$, then

$$x = \frac{\lambda x_2 + x_1}{\lambda + 1}; \quad y = \frac{\lambda y_2 + y_1}{\lambda + 1} \quad \text{(for internal division)}$$

$$x = \frac{\lambda x_2 - x_1}{\lambda - 1}; \quad y = \frac{\lambda y_2 - y_1}{\lambda - 1} \quad \text{(for external division)}.$$

iii) The coordinates of the mid point of PQ are given by $\left(\dfrac{x_1 + x_2}{2}, \dfrac{y_1 + y_2}{2}\right)$.

Preliminaries

4. Area of a triangle

(i) The area of the $\triangle ABC$ with vertices $A(x_1, y_1)$, $B(x_2, y_2)$ and $C(x_3, y_3)$ is given by

$$\triangle = \frac{1}{2}[x_1(y_2 - y_3) + x_2(y_3 - y_1) + x_3(y_1 - y_2)] = \frac{1}{2}\begin{vmatrix} x_1 & y_1 & 1 \\ x_2 & y_2 & 1 \\ x_3 & y_3 & 1 \end{vmatrix},$$

where $\triangle$ denotes the area of the $\triangle ABC$.

(ii) The area of the triangle whose sides are $a_1 x + b_1 y + c_1 = 0$, $a_2 x + b_2 y + c_2 = 0$ and $a_3 x + b_3 y + c_3 = 0$ is given by $\triangle = \dfrac{1}{2C_1 C_2 C_3} \begin{vmatrix} a_1 & b_1 & c_1 \\ a_2 & b_2 & c_2 \\ a_3 & b_3 & c_3 \end{vmatrix}$, where C_1, C_2, C_3 are the cofactors of c_1, c_2, c_3 respectively in the determinant.

5. Area of a polygon

The area of the polygon whose vertices are $(x_1, y_1), (x_2, y_2), \ldots, (x_n, y_n)$ is

$$A = \frac{1}{2}[(x_1 y_2 - x_2 y_1) + (x_2 y_3 - x_3 y_2) + \cdots + (x_n y_1 - y_n x_1)].$$

Note 0.3.1 *The area of a triangle can also be found by easy method, viz.,* ***Stair Method*** *as follows:*

$$\triangle = \frac{1}{2}\begin{vmatrix} x_1 & y_1 \\ x_2 & y_2 \\ x_3 & y_3 \\ x_1 & y_1 \end{vmatrix} = \frac{1}{2}|\{(x_1 y_2 + x_2 y_3 + x_3 y_1) - (y_1 x_2 + y_2 x_3 + y_3 x_1)\}|.$$

Note 0.3.2 *The area of the triangle with vertices $O(0,0)$, $A(x_1, y_1)$ and $B(x_2, y_2)$ is given by $\triangle OAB = \frac{1}{2}|x_1 y_2 - y_1 x_2|$.*

Note 0.3.3 *If three points A, B and C are collinear, then the area of the $\triangle ABC$ formed by them will vanish i.e., $\triangle = 0$.*

Note 0.3.4 ***Sign of the area:*** *If the points A, B and C are plotted in a two-dimensional plane and taken in a anti-clockwise sense, then the area calculated of the $\triangle ABC$ will be positive, while if the points are taken in the clockwise sense, then the area calculated will be negative. But if the points taken arbitrary, then the area calculated may be positive or negative, the numerical value being the same in both cases. However, in case, the calculated area becomes negative, it should be considered as positive.*

Note 0.3.5 Area of a polygon by Stair Method: The area of the polygon with vertices $(x_1, y_1), (x_2, y_2), \ldots, (x_n, y_n)$ obtained by this method is given by

$$\frac{1}{2} \begin{vmatrix} x_1 & y_1 \\ & \times & \\ x_2 & y_2 \\ & \times & \\ x_3 & y_3 \\ \vdots & & \vdots \\ x_n & y_n \\ & \times & \\ x_1 & y_1 \end{vmatrix} = \frac{1}{2} |\{(x_1 y_2 + x_2 y_3 + \cdots + x_n y_1) - (y_1 x_2 + y_2 x_3 + \cdots + y_n x_1)\}|.$$

in which the first coordinates are repeated one time in last and where the down arrows are taken positive sign and for up arrows, the negative sign is taken. Also the points should be taken in cyclic order in the coordinate plane.

Particular Cases:

(a) If the two vertices be taken on the x-axis, say $(a, 0)$ and $(b, 0)$ and the third vertex be (h, k), then $\triangle = \frac{1}{2}$ base $\times$ altitude $= \frac{1}{2}|(a-b)k|$.

(b) Similarly if the two vertices be on the y-axis whose coordinates are $(0, c)$ and $(0, d)$ and third vertex in (h, k), then $\triangle = \frac{1}{2}|(c-d)h|$.

(c) The area of the $\triangle OAB$, where $O = (0,0), A = (a, 0)$ and $B = (0, b)$ then $\triangle OAB = \frac{1}{2}|ab|$.

(d) $\triangle ABC = 50$ sq. units $\Rightarrow |\triangle| = 50 \Rightarrow \triangle = \pm 50$.

6. Standard Centers of a Triangle

i) **Centroid:** The point of concurrency of the *medians* of a triangle is called the *centroid*. This point divides each median in the ratio $2:1$. The centroid of a triangle with vertices $(x_1, y_1), (x_2, y_2)$ and (x_3, y_3) is

$$\left(\frac{x_1 + x_2 + x_3}{3}, \frac{y_1 + y_2 + y_3}{3} \right).$$

Centroid is generally denoted by G.

ii) **Incentre:** The point of concurrency of the internal bisectors of the angles of a triangle is called the *incentre* of the triangle. This is the centre of the circle which touches the sides of the given triangle. Its coordinates are given by
$$\left(\frac{ax_1 + bx_2 + cx_3}{a+b+c}, \frac{ay_1 + by_2 + cy_3}{a+b+c} \right),$$
where a, b and c are the lengths of the sides of the triangle. Incentre is generally denoted by I.

iii) **Circum-Centre of a triangle:** The point of concurrency of the *perpendicular bisectors* of the three sides of a triangle is called the *Circumcentre* of the triangle. This is a point which is equidistant from the vertices of the triangle. This is generally denoted by O.

The circumcentre of a triangle ABC with vertices $(x_1, y_1), (x_2, y_2)$ and (x_3, y_3) is given by
$$\left(\frac{x_1 \sin 2A + x_2 \sin 2B + x_3 \sin 2C}{\sin 2A + \sin 2B + \sin 2C}, \frac{y_1 \sin 2A + y_2 \sin 2B + y_3 \sin 2C}{\sin 2A + \sin 2B + \sin 2C} \right),$$
where A, B and C have their usual meaning.

Alternatively,

If $P(x, y)$ be the circumcentre of the triangle ABC, then by its property, we have $PA = PB = PC$, which gives two equations in x and y. Solving these we get circumcentre.

iv) **Ex-centre of a triangle:** The *point of intersection I_1 of the external bisectors* of the angles B and C of the $\triangle ABC$ (*vide* **Figure 0.5**) is called the *Ex-centre* of the *excribed circle* opposite to an angle A.

Similarly, the bisectors of the external angles C and A and that of A and B give the other two ex-centers I_2 and I_3 respectively of the triangle ABC. The coordinates of the three Ex-centers I_1, I_2 and I_3 are given by

$$I_1 \equiv \left(\frac{-ax_1 + bx_2 + cx_3}{-a+b+c}, \frac{-ay_1 + by_2 + cy_3}{-a+b+c} \right),$$

$$I_2 \equiv \left(\frac{ax_1 - bx_2 + cx_3}{a-b+c}, \frac{ay_1 - by_2 + cy_3}{a-b+c} \right),$$

and $I_3 \equiv \left(\frac{ax_1 + bx_2 - cx_3}{a+b-c}, \frac{ay_1 + by_2 - cy_3}{a+b-c} \right).$

(v) **Orthocentre of a triangle:** The point of concurrency of the altitudes of a triangle is called the *orthocentre* of the triangle. The orthocentre

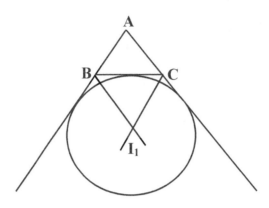

Figure 0.5

of the triangle ABC, denoted by H, is given by

$$H \equiv \left(\frac{x_1 \tan A + x_2 \tan B + x_3 \tan C}{\tan A + \tan B + \tan C}, \frac{y_1 \tan A + y_2 \tan B + y_3 \tan C}{\tan A + \tan B + \tan C} \right),$$

where A, B and C have their usual meaning.

7. Relation between the centroid G, orthocentre H, incentre I and circumcentre O of a triangle ABC

i) The points O, G, H are collinear and G divides the line OH in the ratio $1:2$, i.e., $OG:GH = 1:2$.

ii) In an isosceles triangle ABC, all the four points O, G, H and I are collinear and in an equilateral triangle, all these four points coincide.

8. Locus and its equation

Definition 0.3.1 *The curve described by a variable point P obeying some geometrical conditions is called its locus.*

As for example, let the particle moves so that its distance form the origin is constant. Then we get $\sqrt{x^2 + y^2}$ = constant or $x^2 + y^2 = k^2$ = a constant. So the point moves so as to form a circle of radius k.

Thus a geometrical condition defining a locus leads to an equation involving the coordinates of any point on it.

Preliminaries

0.3.2 Straight Lines

1. Various forms of equations of straight lines

i) **Point-slope form:** The equation of a straight line which passes through the point (x_1, y_1) and makes an angle of θ with the positive direction of the x-axis is
$$y - y_1 = m(x - x_1),$$
where $m = \tan\theta$ is called the *slope* or *gradient* of the line.

ii) **Slope-intercept form:** The equation of the line whose slope is m and which cuts off an intercept c on the y-axis is
$$y = mx + c.$$

iii) **Two point form:** The equation of the line passing through two given points (x_1, y_1) and (x_2, y_2) is
$$y - y_1 = \frac{y_2 - y_1}{x_2 - x_1}(x - x_1) \text{ or, } \begin{vmatrix} x_1 & y_1 & 1 \\ x_2 & y_2 & 1 \\ x_3 & y_3 & 1 \end{vmatrix} = 0.$$

iv) **Intercept form:** The equation of the straight line which cuts off intercepts a and b units on the coordinate axes is
$$\frac{x}{a} + \frac{y}{b} = 1.$$

v) **Normal form:** The equation of the straight line of which length of perpendicular from the origin is p and this normal makes and angle α with the positive direction of the x-axis is
$$x\cos\alpha + y\sin\alpha = p.$$

vi) **Parametric form:** The equation of a line passing through (x_1, y_1) and making an angle θ with the positive direction of the x-axis is
$$\frac{x - x_1}{\cos\theta} = \frac{y - y_1}{\sin\theta} = r \text{ (say)}.$$
At any point on this line is given by $(x_1 + r\cos\theta, y_1 + r\sin\theta)$. Here r denotes the distance of any point (x, y) from (x_1, y_1). For positive values of r, it will lie on the positive side of (x_1, y_1) and for negative values of r, it is on the left side of (x_1, y_1).

vii) **General form:** The equation of the form $Ax + By + C = 0$ (first degree equation in two variables x and y) will be considered as General form of the equation of a straight line.

2. Angle between two straight lines

Angle between two straight lines having slopes m_1 and m_2 is given by

$$\theta = \tan^{-1}\left|\frac{m_1 \sim m_2}{1 + m_1 m_2}\right|.$$

Important Notes:

i) If the lines are parallel then $m_1 = m_2$ and conversely.
ii) The two lines are perpendicular iff $m_1 m_2 = -1$.
iii) A line parallel to $ax + by + c = 0$ is given by $ax + by + c' = 0$.
iv) A line perpendicular to $ax + by + c = 0$ is given by $bx - ay + c' = 0$.
v) A line parallel to $ax + by + c = 0$ and passing through (x_1, y_1) is given by $a(x - x_1) + b(y - y_1) = 0$.
vi) A line perpendicular to $ax + by + c = 0$ and passing through (x_1, y_1) is given by $b(x - x_1) - a(y - y_1) = 0$.

3. Length of perpendicular form a point to a line

The length of perpendicular from the point (x_1, y_1) to a given straight line $ax + by + c = 0$ is given by $\left|\dfrac{ax_1 + by_1 + c}{\sqrt{a^2 + b^2}}\right|.$

4. Distance between two parallel lines

The distance between two parallel lines $ax + by + c_1 = 0$ and $ax + by + c_2 = 0$ is given by $\left|\dfrac{c_1 - c_2}{\sqrt{a^2 + b^2}}\right|.$

Alternatively,

We can find the coordinates of a point on any one of the two given lines and then find the perpendicular distances from this point to the other line.

5. A line equally inclined with two lines

If two given lines with slopes m_1 and m_2 be equally inclined to a line with slope m, then

$$\frac{m_1 - m}{1 + mm_1} = \frac{m - m_2}{1 + mm_2}.$$

6. Equations of two equally inclined straight lines with a line and which passes through a given point

The equations of two straight lines which passes though the point (x_1, y_1) and make an angle α with the line $y = mx + c$ are given by

$$\left. \begin{array}{l} y - y_1 = \tan(\theta - \alpha)(x - x_1) \\ y - y_1 = \tan(\theta + \alpha)(x - x_1) \end{array} \right\}.$$

7. Equations of bisectors

Equations of the bisectors of the angles between the two given lines $a_1x + b_1y + c_1 = 0$ and $a_2x + b_2y + c_2 = 0$ are

$$\frac{a_1x + b_1y + c_1}{\sqrt{a_1^2 + b_1^2}} = \pm \frac{a_2x + b_2y + c_2}{\sqrt{a_2^2 + b_2^2}}.$$

Important Notes:

i) If $a_1a_2 + b_1b_2 > 0$, then negative sign gives the acute angle bisector and positive sign gives the obtuse angle bisector.

ii) If $a_1a_2 + b_1b_2 < 0$, then negative sign gives the obtuse angle bisector and positive sign gives the acute angle bisector.

8. Concurrent lines

Three lines are said to be *Concurrent*, if they meet in a point. To show it we proceed as follows:

First Method: First we find the point of intersection of the first two equations (by solving these) and then we show that it satisfies the third also.

Second Method: The three given lines $L_1 : a_1x + b_1y + c_1 = 0$; $L_2 : a_2x + b_2y + c_2 = 0$ and $L_3 : a_3x + b_3y + c_3 = 0$ are concurrent if $\begin{vmatrix} a_1 & b_1 & c_1 \\ a_2 & b_2 & c_2 \\ a_3 & b_3 & c_3 \end{vmatrix} = 0$.

Third Method: The lines $L_1 = 0, L_2 = 0$ and $L_3 = 0$ will be concurrent if there exists three constants λ_1, λ_2 and λ_3, not all zero at a time such that $\lambda_1 L_1 + \lambda_2 L_2 + \lambda_3 L_3 = 0$.

9. Family of straight lines

Any line passing through the point of intersection of the lines $L_1 = 0$ and $L_2 = 0$ is given by $L_1 + \lambda L_2 = 0$ where $\lambda \in \mathbb{R}$, the set of reals.

These lines form a family of straight lines from the point of intersection.

0.3.3 The Circle

1. Equations of Circles

i) $(x - \alpha)^2 + (y - \beta)^2 = a^2$; Centre at (α, β); Radius $= a$.

ii) $x^2 + y^2 = a^2$; Centre at $(0, 0)$; Radius $= a$.

iii) General equation: $x^2 + y^2 + 2gx + 2fy + c = 0$; Centre at $(-g, -f)$; Radius $= \sqrt{g^2 + f^2 - c}$.

iv) $(x - x_1)(x - x_2) + (y - y_1)(y - y_2) = 0$ is the equation of a circle, whose one diameter is the line segment joining the points (x_1, y_1) and (x_2, y_2).

v) Parametric equations of a circle: The parametric coordinates of any point on the circle $(x - \alpha)^2 + (y - \beta)^2 = a^2$ are $(\alpha + a\cos\theta, \beta + a\sin\theta)$. So it is the parametric equation of the circle. In particular $(a\cos\theta, a\sin\theta)$ is the parametric equation of $x^2 + y^2 = a^2$.

2. Tangents and Normals

i) Equation of tangent to the circle $x^2 + y^2 = a^2$ at (x_1, y_1) is $xx_1 + yy_1 = a^2$.

ii) Equation of tangent to the circle $x^2 + y^2 + 2gx + 2fy + c = 0$ at (x_1, y_1) is $xx_1 + yy_1 + g(x + x_1) + f(y + y_1) + c = 0$.

iii) Equation of normal to the circle $x^2 + y^2 = a^2$ at (x_1, y_1) is $\dfrac{x}{x_1} = \dfrac{y}{y_1}$.

iv) Equation of normal to the circle $x^2 + y^2 + 2gx + 2fy + c = 0$ at (x_1, y_1) is $\dfrac{x - x_1}{x_1 + g} = \dfrac{y - y_1}{y_1 + f}$.

v) **Condition of tangency:** Condition that the line $y = mx + c$ may touch the circle $x^2 + y^2 = a^2$ is $c = \pm a\sqrt{1 + m^2}$, $(m \neq 0)$ and so $y = mx \pm a\sqrt{1 + m^2}$ is always a tangent to the circle $x^2 + y^2 = a^2$ for all values of m.

Preliminaries 13

vi) **Point of contact:** If $y = mx+c$ is a tangent to the circle $x^2+y^2 = a^2$, then the point of contact will be $\left(-\dfrac{am}{\sqrt{1+m^2}}, \dfrac{a}{\sqrt{1+m^2}}\right)$.

vii) **Pair of tangents:** The equation of the pair of tangents drawn form (x_1, y_1) to the circle $x^2 + y^2 = a^2$ is $SS_1 = T^2$ where $S \equiv x^2 + y^2 - a^2$, $S_1 \equiv x_1^2 + y_1^2 - a^2$ and $T \equiv xx_1 + yy_1 - a^2$.

viii) **Length of tangent:** Length of the tangent form an external point (x_1, y_1) to the circle $x^2 + y^2 + 2gx + 2fy + c = 0$ is

$$\sqrt{x_1^2 + y_1^2 + 2gx_1 + 2fy_1 + c} = \sqrt{S_1}.$$

3. Length of chord of circle intercepted by a line

The lenght of the chord of the circle $x^2 + y^2 = a^2$ intercepted by the line $y = mx + c$ is $2\sqrt{\dfrac{a^2(1+m^2) - c^2}{1+m^2}}$.

4. Equation of chord of circle in terms of its middle point

The equation of the chord of the circle $S = 0$ in terms of the coordinates of its middle point (i.e., bisected) (x_1, y_1) is given by $T = S_1$ where
$S = x^2 + y^2 + 2gx + 2fy + c,$
$T = xx_1 + yy_1 + g(x + x_1) + f(y + y_1) + c,$
$S_1 = x_1^2 + y_1^2 + 2gx_1 + 2fy_1 + c.$

5. Orthogonal Circles

Two circles $x^2 + y^2 + 2g_1x + 2f_1y + c_1 = 0$ and $x^2 + y^2 + 2g_2x + 2f_2y + c_2 = 0$ cut orthogonally if $2g_1g_2 + 2f_1f_2 = c_1 + c_2$.

6. Radical Axis

The radical axis of two circles is the locus of the point which moves such that the lengths of the tangents drawn form it to the two circles are equal. If $S_1 \equiv x^2 + y^2 + 2g_1x + 2f_1y + c_1 = 0$ and $S_2 \equiv x^2 + y^2 + 2g_2x + 2f_2y + c_2 = 0$, then the equation of their radical axis is

$$S_1 - S_2 = 0, \text{ i.e., } 2(g_1 - g_2) + 2(f_1 - f_2) + (c_1 - c_2) = 0$$

which is clearly a straight line.

0.3.4 The Parabola

1. Equation of parabola in different forms

i) Standard forms: $y^2 = 4ax, x^2 = 4ay, y^2 = -4ax, x^2 = -4ay$ all of whose vertices are $(0,0)$.

ii) Equation of the parabola with axis parallel to the x-axis is of the form $x = ay^2 + by + c$.

iii) Equation of the parabola with axis parallel to the y-axis is of the form $y = ax^2 + bx + c$.

iv) Equation of the parabola whose focus is (α, β) and directrix is $ax + by + c = 0$ is given by

$$(x-\alpha)^2 + (y-\beta)^2 = \frac{(ax+by+c)^2}{a^2+b^2}.$$

This will be of the form $(\alpha x - \beta y)^2 + 2gx + 2fy + c = 0$.

This is known as the general equation of a parabola. It should be seen that *second degree terms in the general equation of a parabola forms a perfect square*.

v) $(y-\beta)^2 = 4a(x-\alpha)$ is the equation of a parabola whose vertex is at the point (α, β).

vi) Parametric form of $y^2 = 4ax$ is $x = at^2, y = 2at$.

vii) Parametric form of $(y-\beta)^2 = 4a(x-\alpha)$ is $x = \alpha + at^2, y = \beta + 2at$.

2. Important terms regarding the parabola $y^2 = 4ax$; tangents, normals, chord of contact and length of chord etc.

i) Latus rectum $= 4a$.

ii) Vertex is $(0,0)$.

iii) Focus is $(a,0)$.

iv) Axis is $y = 0$.

v) Directrix is $x = -a$ i.e., $x + a = 0$.

vi) Coordinates of the end points of the latus rectum are $(a, \pm 2a)$.

vii) Equation of the tangent at vertex $(0,0)$ is $x = 0$.

viii) Equation of the tangent at any point (x_1, y_1) on the parabola is $yy_1 = 2a(x + x_1)$.

Preliminaries

ix) Equation of the normal at the point (x_1, y_1) is $(y - y_1) = -\dfrac{y_1}{2a}(x - x_1)$.

x) Equation of the chord in terms of the coordinates of its mid point is $(y - y_1)y_1 = 2a(x - x_1)$ [or use $T = S_1$].

xi) Equation of the chord intercepted by the straight lines $y = mx + c$ is $\dfrac{4}{m^2}\sqrt{a(a - mc)(1 + m^2)}$.

xii) Condition that $y = mx + c$ may touch the parabola $y^2 = 4ax$ is $c = \dfrac{a}{m}$ $(m \neq 0)$ i.e., the line $y = mx + \dfrac{a}{m}$ is always a tangent to the parabola for all values of m (except zero), the point of contact being $\left(\dfrac{a}{m^2}, \dfrac{2a}{m}\right)$.

xiii) Equation of the diameter is $y = \dfrac{2a}{m}$.

Definition 0.3.2 *Diameter of a parabola: The locus of the middle points of any system of parallel chords of a parabola is a straight line parallel to its axis which is known as the diameter of the parabola.*

0.3.5 The Ellipse

1. Equation of ellipse in different forms

i) Standard form: $\dfrac{x^2}{a^2} + \dfrac{y^2}{b^2} = 1$.

ii) Latus rectum: $2a(1 - e^2) = 2\dfrac{b^2}{a}$.

iii) Centre: $(0, 0)$.

iv) Vertices: $(\pm ae, 0)$.

v) Eccentricity: $b^2 = a^2(1 - e^2)$ or $e^2 = \dfrac{a^2 - b^2}{a^2}$.

vi) Equation of the directrixes: $x = \pm\dfrac{a}{e}$.

vii) Equations of the latus rectums are $x = \pm ae$.

viii) Coordinates of the feet of directrixes are $\left(\pm\dfrac{a}{e}, 0\right)$.

ix) Ends of the latus rectum are $\left(\pm ae, \pm\dfrac{b^2}{a}\right)$.

x) Focal distances of $P(x_1, y_1)$: $SP = a - ex_1$, $S'P : a + ex_1$, $SP + S'P = 2a$, where S and S' are the two foci of the ellipse.

2. Tangent, Normal, Chord of contact and Length of chord etc.

i) Tangent at (x_1, y_1): $\dfrac{xx_1}{a^2} + \dfrac{yy_1}{b^2} = 1$.

ii) Normal at (x_1, y_1): $\dfrac{x - x_1}{\frac{x_1}{a^2}} = \dfrac{y - y_1}{\frac{y_1}{b^2}}$.

iii) Equation of the chord in terms of the coordinates of its mid point is
$$(x - x_1)\dfrac{x_1}{a^2} + (y - y_1)\dfrac{y_1}{b^2} = 0 \text{ [or use } T = S_1\text{]}.$$

vi) Length of the chord intercepted by the line $y = mx + c$ on the ellipse $\dfrac{x^2}{a^2} + \dfrac{y^2}{b^2} = 1$ is $\dfrac{2ab\sqrt{1 + m^2}\sqrt{a^2m^2 + b^2 - c^2}}{a^2m^2 + b^2}$.

v) Condition of tangency: The line $y = mx + c$ is a tangent to the ellipse if $c = \pm\sqrt{a^2m^2 + b^2}$ so the lines $y = mx \pm \sqrt{a^2m^2 + b^2}$ always touch the ellipse $\dfrac{x^2}{a^2} + \dfrac{y^2}{b^2} = 1$ at $\left(\pm\dfrac{a^2m}{\sqrt{a^2m^2 + b^2}}, \pm\dfrac{b^2}{\sqrt{a^2m^2 + b^2}}\right)$.

vi) Auxiliary circle: $x^2 + y^2 = a^2$.

vii) Parametric representation: $x = a\cos\theta$, $y = b\sin\theta$, θ is the parameter.

viii) Diameter: $y = -\dfrac{b^2}{a^2 m}x$.

ix) Director circle: $x^2 + y^2 = a^2 + b^2$.

0.3.6 The Hyperbola

1. Equation of hyperbola in different forms

i) Standard form: $\dfrac{x^2}{a^2} - \dfrac{y^2}{b^2} = 1$.

ii) Latus rectum: $2a(e^2 - 1) = 2\dfrac{b^2}{a}$.

iii) Centre: $(0, 0)$.

iv) Vertices: $(\pm ae, 0)$.

v) Eccentricity: $b^2 = a^2(e^2 - 1)$ or $e^2 = \dfrac{a^2 + b^2}{a^2}$.

vi) Equation of the directrixes: $x = \pm\dfrac{a}{e}$.

vii) Equations of the latus rectums are $x = \pm ae$.

Preliminaries 17

viii) Length of latus rectum $= \dfrac{2b^2}{a}$.

ix) Coordinates of the feet of directrixes are $\left(\pm\dfrac{a}{e}, 0\right)$.

x) Ends of the latus rectum are $\left(\pm ae, \pm\dfrac{b^2}{a}\right)$.

xi) Focal radii: $SP = ex - a$ and $S'P : a + ex$, $\therefore S'P - SP = 2a =$ transverse axis, where S and S' are the two foci of the hyperbola.

xii) Parametric representation: $x = a\sec\phi$, $y = b\tan\phi$, ϕ is the parameter.

xiii) For rectangular hyperbola or equilateral hyperbola: $a = b, e = \sqrt{2}$, equation : $x^2 - y^2 = a^2$.

2. Tangent, Normal, Chord of contact and Length of chord etc.

i) Tangent at (x_1, y_1): $\dfrac{xx_1}{a^2} - \dfrac{yy_1}{b^2} = 1$.

ii) Normal at (x_1, y_1): $\dfrac{x - x_1}{\frac{x_1}{a^2}} = \dfrac{y - y_1}{-\frac{y_1}{b^2}}$.

iii) Equation of the chord in terms of the coordinates of its mid point is

$$(x - x_1)\dfrac{x_1}{a^2} - (y - y_1)\dfrac{y_1}{b^2} = 0 \text{ [or use } T = S_1].$$

vi) Length of the chord intercepted by the line $y = mx + c$ on the hyperbola $\dfrac{x^2}{a^2} - \dfrac{y^2}{b^2} = 1$ is $\dfrac{2ab\sqrt{1 + m^2}\sqrt{c^2 - a^2m^2 + b^2}}{a^2m^2 - b^2}$.

v) Condition of tangency: The line $y = mx + c$ is a tangent to the hyperbola if $c = \pm\sqrt{a^2m^2 - b^2}$ so the lines $y = mx \pm \sqrt{a^2m^2 - b^2}$ always touch the hyperbola $\dfrac{x^2}{a^2} - \dfrac{y^2}{b^2} = 1$, the point of contact being

$$\left(\pm\dfrac{a^2m}{\sqrt{a^2m^2 - b^2}}, \mp\dfrac{b^2}{\sqrt{a^2m^2 - b^2}}\right).$$

vi) Equations of the asymptotes: $y = \pm\dfrac{b}{a}x$.

vii) Equation of the Diameter: $y = \dfrac{b^2}{a^2m}x$.

viii) Director circle: $x^2 + y^2 = a^2 - b^2$.

0.3.7 Common Properties of all Conics

1. The *length of any tangent* from an external point (x_1, y_1) to the conic $S = 0$ is $\sqrt{S}$.

2. The *equation of the chord* of the conic $S = 0$ in *terms of the coordinates of the middle point* (x_1, y_1) is given by $T = S_1$.

3. The equation of the *chord of contact of the tangents* drawn form an external point to a conic is given by $T = 0$.

4. The equation of the *pair of tangents* drawn form an external point (x_1, y_1) to a conic is given by $SS_1 = T^2$.

Note 0.3.6 *All the notations used above have their usual meaning.*

0.3.8 Properties of Tangents and Normals on Conics

A. On Parabola

1. The tangents at the extremities of a focal chord of a parabola intersect at right angles on the directrix.

2. Locus of foot of perpendicular from focus upon any tangent is tangent at vertex.

3. The tangent at any point of a parabola bisects the angle between the focal distance of the point and the perpendicular on the directrix from the focus.

4. Length of tangent between the point of contact and the point where it meets the directrix, subtends a right angle at the focus.

5. Normal other than the axis of the parabola never passes through the focus.

B. On Ellipse

1. The locus of feet of perpendicular from the foci upon any tangent is an auxiliary circle $x^2 + y^2 = a^2$.

2. Normal other than the major axis never passes through the focus.

3. The normal at any point P on the ellipse $\dfrac{x^2}{a^2} + \dfrac{y^2}{b^2} = 1$ bisects the angle SPS' where S and S' are the foci of the ellipse.

C. On Hyperbola

1. Normal other than transverse axis, never passes through the focus.

2. The portion of the tangent between the point of contact and the directrix subtends a right angle at the corresponding focus.

3. The locus of feet of perpendiculars from the foci upon any tangent is its auxiliary circle $x^2 + y^2 = a^2$.

4. The normal at any point P on the hyperbola $\dfrac{x^2}{a^2} - \dfrac{y^2}{b^2} = 1$ bisects the angle SPS' where S and S' are the foci of the hyperbola.

0.3.9 Conjugate Diameters

Definition 0.3.3 *Two diameters are said to be conjugate when each bisects all chords parallel to the other.*

If $y = m_1 x$ and $y = m_2 x$ be two conjugate diameters of an *ellipse* then $m_1 m_2 = -\dfrac{b^2}{a^2}$ and if these be two conjugate diameters of a *hyperbola* then $m_1 m_2 = \dfrac{b^2}{a^2}$.

Properties of Conjugate Diameters

1. The eccentric angle of the extremities of a pair of conjugate diameters differ by a right angle.

2. If CP and CD are two conjugate semi-diameters of the ellipse $\dfrac{x^2}{a^2} + \dfrac{y^2}{b^2} = 1$, then $CP^2 + CD^2 = a^2 + b^2$.

3. If CP and CD are two conjugate semi-diameters of the hyperbola $\dfrac{x^2}{a^2} - \dfrac{y^2}{b^2} = 1$, then $CP^2 - CD^2 = a^2 - b^2$.

4. The area of the parallelogram with CP and CD as adjacent sides $= ab$ for both ellipse and hyperbola.

Chapter 1

Transformation of Rectangular Cartesian Coordinates

1.1 Introduction

In analytical geometry the position of a point is determined with the help of a coordinate system, for which we require an arbitrarily chosen origin and two straight lines intersecting at the origin. This follows that the coordinates of a point are different for different sets of axes. That is the coordinates of a point are relative. This may happen when the origin is shifted to a point keeping the directions of the axes unchanged or when the axes are rotated through an angle keeping the origin unaltered. The former is called ***translation*** and the later is called ***rotation***. The coordinates of any point may also be affected by a combination of these two, called a ***rigid motion*** or ***orthogonal transformation***.

In this chapter, we shall establish formulas corresponding to the above transformations.

1.2 Coordinate Transformation Formulas

1.2.1 Translation (or transformation by translation of axes)

If the origin be shifted to any other point without changing the directions of coordinate axes, the result is known as ***translation***.

Let (x, y) be the coordinates of P with respect to a rectangular axes OX

and OY and (x', y') be its coordinates with respect to a new set of axes $O'X'$ and $O'Y'$ which are parallel to the original axes $O'X'$ and $O'Y'$ respectively.

Let (a, b) be the coordinates of the new origin O' with respect to the old axes OX and OY. PM is perpendicular to OX. It meets $O'X'$ at N. $O'L$ is perpendicular to OX. Therefore, from the **Figure 1.1**,

$$OM = x, \ PM = y, \ O'N = x', \ NP = y', \ OL = a, \ O'L = b.$$
$$\therefore \quad x = OM = OL + LM = O'N + OL = x' + a$$
$$\text{and} \quad y = PM = PN + NM = PN + O'L = y' + b.$$

Hence the required translation formulas are given by

$$\left.\begin{array}{l} x = x' + a \\ y = y' + b \end{array}\right\} \Rightarrow \left.\begin{array}{l} x' = x - a \\ y' = y - b. \end{array}\right\}$$

Thus the origin will be transferred to the point (a, b), the axes remaining parallel to the original axes, if we substitute $x' + a$ for x and $y' + b$ for y and inversely, in the equation of a locus referred to new system of axes (x', y') will be replaced by $(x - a, y - b)$, when the equation is referred to old pair of axes.

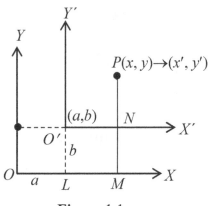

Figure 1.1

1.2.2 Rotation (or transformation by rotation of axes)

When both the axes are turned in the same sense through the same angle without shifting the origin, then the result thus obtained is known as *rotation*.

Formula for Rotation:

First Method: Let the coordinate axes be turned through an angle α. Let (x, y) and (x', y') be the Cartesian coordinates of any point P with respect to old and the new coordinate systems respectively (*vide.* **Figure 1.2**). Also let us consider that (r, θ) and (r, θ') be their coordinates in polar coordinates. Then we have

$$x = r\cos\theta, \ x' = r\cos\theta', \ y = r\sin\theta, \ y' = r\sin\theta'$$
$$\text{and} \quad \theta = \theta' + \alpha \ \text{i.e.,} \ \theta - \theta' = \alpha.$$

Transformation of Coordinates

So we get

$$\begin{aligned}
x = r\cos\theta = r\cos(\theta' + \alpha) &= r\cos\theta'\cos\alpha - r\sin\theta'\sin\alpha \\
&= x'\cos\alpha - y'\sin\alpha \\
y = r\sin\theta = r\sin(\theta' + \alpha) &= r\sin\theta'\cos\alpha + r\cos\theta'\sin\alpha \\
&= y'\cos\alpha + x'\sin\alpha.
\end{aligned}$$

So the required formulae are

$$\left.\begin{aligned} x &= x'\cos\alpha - y'\sin\alpha \\ y &= x'\sin\alpha + y'\cos\alpha. \end{aligned}\right\}$$

Again since the new coordinate system is obtained by a rotation of the old coordinate system through an angle α, the old system will be obtained by a rotation of the new system through an angle $-\alpha$.

Therefore, interchanging the old and new coordinates in the above formulae and replacing α by $-\alpha$ we get

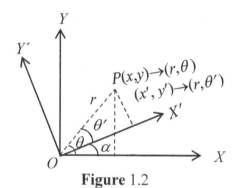

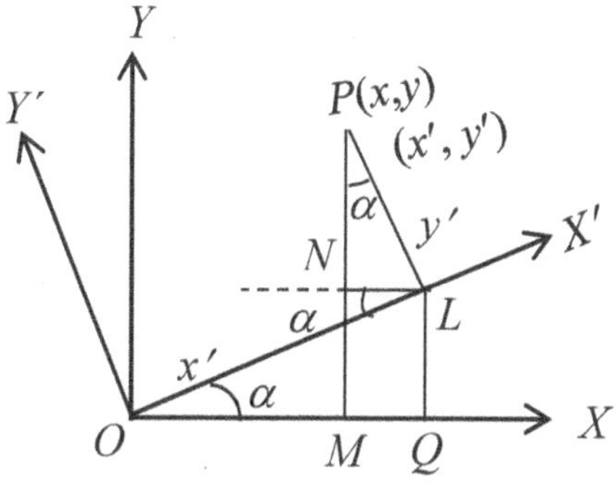

Figure 1.2

$$\begin{aligned}
x' &= x\cos(-\alpha) - y\sin(-\alpha) \\
&= x\cos\alpha + y\sin\alpha \\
y' &= x\sin(-\alpha) + y\cos(-\alpha) \\
&= -x\sin\alpha + y\cos\alpha.
\end{aligned}$$

Hence the new coordinates (x', y') of P are expressed in terms of its old coordinates (x, y) as follows:

$$\left.\begin{aligned} x' &= x\cos\alpha + y\sin\alpha \\ y' &= -x\sin\alpha + y\cos\alpha. \end{aligned}\right\}$$

Figure 1.3

Second Method: Here the position of origin remains unchanged. Let the axes be rotated through an angle α. From the **Figure 1.3**, we get

$$\left.\begin{aligned} x &= OM = OQ - MQ = OQ - LN = x'\cos\alpha - y'\sin\alpha \\ y &= PM = PN + NM = PN + LQ = y'\cos\alpha + x'\sin\alpha. \end{aligned}\right\}$$

So we get the rotational formula as

$$\left.\begin{array}{l} x = x'\cos\alpha - y'\sin\alpha \\ y = x'\sin\alpha + y'\cos\alpha \end{array}\right\} \Rightarrow \left.\begin{array}{l} x' = x\cos\alpha + y\sin\alpha \\ y' = -x\sin\alpha + y\cos\alpha. \end{array}\right\}$$

[by interchanging $(x,y) \to (x',y')$ and by replacing α by $(-\alpha)$ as before.]

1.2.3 Rigid Motion or Orthogonal Transformation

This is the combination of translation and rotation both. Let us first consider that the origin O is shifted to the point $O'(a,b)$ and let the coordinates (x,y) be changed to (X,Y) by translation. So we get by translation formulae

$$\left.\begin{array}{l} x = X + a \\ y = Y + b. \end{array}\right\}$$

Next we rotate the axes through an angle α by which let (X,Y) be changed into (x',y'). That is we get

$$\left.\begin{array}{l} X = x'\cos\alpha - y'\sin\alpha \\ Y = x'\sin\alpha + y'\cos\alpha \end{array}\right\}$$

and so finally we get

$$\left.\begin{array}{l} x = x'\cos\alpha - y'\sin\alpha + a \\ y = x'\sin\alpha + y'\cos\alpha + b \end{array}\right\} \Rightarrow \left.\begin{array}{l} x' = (x-a)\cos\alpha + (y-b)\sin\alpha \\ y' = -(x-a)\sin\alpha + (y-b)\cos\alpha \end{array}\right\}$$

which is the rigid motion or orthogonal transformation formulae.

1.3 Use of Matrix Notation

1.3.1 Translation

The translation formulae are

$$\left.\begin{array}{l} x = x' + a \\ y = y' + b \end{array}\right\} \text{ or, } \left.\begin{array}{l} x' = x - a \\ y' = y - b. \end{array}\right\}$$

In matrix notation, these can be written as

$$\begin{pmatrix} x \\ y \end{pmatrix} = \begin{pmatrix} 1 & 0 \\ 0 & 1 \end{pmatrix}\begin{pmatrix} x' \\ y' \end{pmatrix} + \begin{pmatrix} a \\ b \end{pmatrix}$$

and $$\begin{pmatrix} x' \\ y' \end{pmatrix} = \begin{pmatrix} 1 & 0 \\ 0 & 1 \end{pmatrix}\begin{pmatrix} x \\ y \end{pmatrix} - \begin{pmatrix} a \\ b \end{pmatrix}.$$

As $\begin{pmatrix} 1 & 0 \\ 0 & 1 \end{pmatrix}$ is an orthogonal matrix, so translation is an orthogonal transformation.

1.3.2 Rotation

The rotational formulae are
$$\left.\begin{array}{l}x = x'\cos\alpha - y'\sin\alpha \\ y = x'\sin\alpha + y'\cos\alpha\end{array}\right\} \text{ or } \left.\begin{array}{l}x' = x\cos\alpha + y\sin\alpha \\ y' = -x\sin\alpha + y\cos\alpha\end{array}\right\}$$

which can be written in matrix notation as
$$\begin{pmatrix}x\\y\end{pmatrix} = \begin{pmatrix}\cos\alpha & -\sin\alpha \\ \sin\alpha & \cos\alpha\end{pmatrix}\begin{pmatrix}x'\\y'\end{pmatrix} \text{ and } \begin{pmatrix}x'\\y'\end{pmatrix} = \begin{pmatrix}\cos\alpha & \sin\alpha \\ -\sin\alpha & \cos\alpha\end{pmatrix}\begin{pmatrix}x\\y\end{pmatrix}.$$

Clearly the matrices $\begin{pmatrix}\cos\alpha & -\sin\alpha \\ \sin\alpha & \cos\alpha\end{pmatrix}$ and $\begin{pmatrix}\cos\alpha & \sin\alpha \\ -\sin\alpha & \cos\alpha\end{pmatrix}$ are orthogonal matrices. So the transformation of rotation of the axes is an orthogonal transformation.

Note 1.3.1 *These formulae can be remembered by the scheme:*

	x'	y'
x	$\cos\alpha$	$-\sin\alpha$
y	$\sin\alpha$	$\cos\alpha$

Note 1.3.2 *The rotation of axes may be taken as*
$$\left.\begin{array}{l}x = px' - qy' \\ y = qx' + py'\end{array}\right\} \text{ where } \begin{vmatrix}p & q \\ -q & p\end{vmatrix} = p^2 + q^2 = 1.$$

1.3.3 Rigid Motion

The rigid motion formulae are
$$\left.\begin{array}{l}x = x'\cos\alpha - y'\sin\alpha + a \\ y = x'\sin\alpha + y'\cos\alpha + b\end{array}\right\}$$

which can be written as
$$\begin{pmatrix}x\\y\end{pmatrix} = \begin{pmatrix}\cos\alpha & \sin\alpha \\ -\sin\alpha & \cos\alpha\end{pmatrix}\begin{pmatrix}x'\\y'\end{pmatrix} + \begin{pmatrix}a\\b\end{pmatrix}.$$

This is the orthogonal transformation formula in matrix form.

Note 1.3.3 *The rigid motion formulae may also be taken as*
$$\left.\begin{array}{l}x = px' - qy' + a \\ y = qx' + py' + b\end{array}\right\} \text{ where } \begin{vmatrix}p & q \\ -q & p\end{vmatrix} = p^2 + q^2 = 1.$$

Note 1.3.4 *The inverse formulae are given by*
$$\left.\begin{array}{l}x' = (x-a)\cos\alpha + (y-b)\sin\alpha = p(x-a) + q(y-b) \\ y' = -(x-a)\sin\alpha + (y-b)\cos\alpha = -q(x-a) + p(y-b).\end{array}\right\}$$

1.4 Definition of Orthogonal Transformation

The transformation in a plane, determined by

$$\left. \begin{array}{l} x' = a_1 x + b_1 y + c_1 \\ y' = a_2 x + b_2 y + c_2 \end{array} \right\} \text{ i.e., } \begin{pmatrix} x' \\ y' \end{pmatrix} = \begin{pmatrix} a_1 & b_1 \\ a_2 & b_2 \end{pmatrix} \begin{pmatrix} x \\ y \end{pmatrix} + \begin{pmatrix} c_1 \\ c_2 \end{pmatrix}$$

is called orthogonal if the matrix $\begin{pmatrix} a_1 & b_1 \\ a_2 & b_2 \end{pmatrix}$ is orthogonal.

Examples of orthogonal transformation of axes are Translation, Rotation and their combination (called the rigid motion).

1.5 General Orthogonal Transformation

Let OX, OY be the old and $O'X'$, $O'Y'$ be the new set of axes. Also let the equation of the new axes $O'X'$ and $O'Y'$ referred to old set OX and OY are respectively given as

$$\left. \begin{array}{l} lx + my + n = 0 \\ \text{and} \quad mx - ly + n' = 0. \end{array} \right\}$$

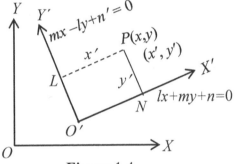

Figure 1.4

If the coordinates of any point P referred to the first system be (x, y) and referred to the second system be (x', y'), then $O'N = PL = x'$ and $PN = y'$.

Now PL and PN are the lengths of perpendiculars from P on $O'X'$ and $O'Y'$, i.e., on the lines $lx + my + n = 0$ and $mx - ly + n' = 0$. So we get

$$\left. \begin{array}{l} x' = \frac{mx - ly + n'}{\sqrt{l^2 + m^2}} \\ \text{and} \quad y' = \frac{lx + my + n}{\sqrt{l^2 + m^2}}. \end{array} \right\}$$

These are called the general orthogonal transformation.
Solving for x and y, we get

$$\left. \begin{array}{l} x = \frac{mx' + ly'}{\sqrt{l^2 + m^2}} - \frac{ln + mn'}{\sqrt{l^2 + m^2}} \\ y = \frac{-lx' + my'}{\sqrt{l^2 + m^2}} - \frac{mn - ln'}{\sqrt{l^2 + m^2}}. \end{array} \right\}$$

1.6 Invariants

Under orthogonal transformation, some expressions remains unchanged which are known as invariants. These are
1. The degree of the equation.
2. The distance between two points.
3. Area of a triangle.
4. The nature of the curve.

1.7 Purpose of Transformation of Axes

It is already known that there are three types of *standard conics*, namely parabola, ellipse, and hyperbola whose equations in rectangular Cartesian coordinates x, y are equations of the second degree in x and y. It is to be noted that, although every general equation of first degree in x and y always represents a straight line, every general equation of second degree in x and y does not always represent a *conic*, for it may represent a *circle* or a *pair of straight lines* or it *may not represent any geometric object at all*.

For the purpose of study of a second degree curve, we have to choose suitable coordinate system by applying *translation* or *rotation* or *both* in which the equation will reduce to the *simplest form*, known as *normal form* or *canonical form*. So the purpose of transformation of axes is to determine the type of a conic represented by a second degree equation, for by such transformation, the nature of the curve remains invariant. Since the standard equation of a conic does not contain the term xy, so to identify a curve we require to *remove the term containing xy which can be done by applying rotational formula only* i.e., by turning the axes through a suitable angle.

Also, *the purpose of translation is to remove the terms containing x or y or both*. These will be discussed in the next chapter, viz., **general equation of second degree and classification of conics**.

1.8 To Investigate whether there is any Fixed Point under Translation, Rotation or Rigid Motion

1.8.1 Fixed Point under Translation

We consider the formulae for translation given by

$$\left. \begin{array}{l} x = x' + a \\ y = y' + b \end{array} \right\}, (a,b) \neq (0,0).$$

Let (α, β) be a fixed point. Then it will satisfy the above equation, i.e.,

$$\left. \begin{array}{l} \alpha = \alpha + a \\ \beta = \beta + b \end{array} \right\} \Rightarrow a = 0, \ b = 0$$

which is not possible, so there is no fixed point under translation.

1.8.2 Fixed Point under Rotation

The rotational formulae are

$$\left. \begin{array}{l} x = x' \cos \theta - y' \sin \theta \\ y = x' \sin \theta + y' \cos \theta \end{array} \right\}, \theta \neq 0.$$

Let (α, β) be a fixed point. Then it will satisfy the above equation, i.e.,

$$\left. \begin{array}{l} \alpha = \alpha \cos \theta - \beta \sin \theta \\ \beta = \alpha \sin \theta + \beta \cos \theta \end{array} \right\} \Rightarrow \left. \begin{array}{l} (\cos \theta - 1)\alpha - \beta \sin \theta = 0 \\ \alpha \sin \theta + (\cos \theta - 1)\beta = 0 \end{array} \right\} \quad (1.1)$$

The coefficient determinant

$$\left| \begin{array}{cc} \cos \theta - 1 & -\sin \theta \\ \sin \theta & \cos \theta - 1 \end{array} \right| = (\cos \theta - 1)^2 + \sin^2 \theta = 2(1 - \cos \theta) \neq 0 \text{ as } \theta \neq 0.$$

So the homogeneous system of equations (1.1) has unique solution and the solution is $\alpha = 0$ and $\beta = 0$.

Hence $(0, 0)$ is the only fixed point under rotation.

1.8.3 Fixed Point under Rigid Motion

Let us consider the rigid motion given by

$$\left. \begin{array}{l} x = x' \cos \theta - y' \sin \theta + a \\ y = x' \sin \theta + y' \cos \theta + b \end{array} \right\}, \theta \neq 0.$$

Let (α, β) be a fixed point. Then it will satisfy the above equations, i.e.,

$$\left. \begin{array}{l} \alpha = \alpha \cos \theta - \beta \sin \theta + a \\ \beta = \alpha \sin \theta + \beta \cos \theta + b \end{array} \right\} \Rightarrow \left. \begin{array}{l} (1 - \cos \theta)\alpha + \beta \sin \theta = a \\ -\alpha \sin \theta + (1 - \cos \theta)\beta = b \end{array} \right\} \quad (1.2)$$

This is a non homogeneous system of equations and its coefficient determinant =

$$\left| \begin{array}{cc} 1 - \cos \theta & \sin \theta \\ -\sin \theta & 1 - \cos \theta \end{array} \right| = (1 - \cos \theta)^2 + \sin^2 \theta = 2(1 - \cos \theta) \neq 0 \text{ as } \theta \neq 0.$$

Transformation of Coordinates

So, there is always one fixed point under rigid motion which is given by solving the system of equations (1.2) as

$$\frac{\alpha}{\begin{vmatrix} a & \sin\theta \\ b & 1-\cos\theta \end{vmatrix}} = \frac{\beta}{\begin{vmatrix} 1-\cos\theta & a \\ -\sin\theta & b \end{vmatrix}} = \frac{1}{\begin{vmatrix} 1-\cos\theta & \sin\theta \\ -\sin\theta & 1-\cos\theta \end{vmatrix}}$$

$$\Rightarrow \frac{\alpha}{a(1-\cos\theta) - b\sin\theta} = \frac{\beta}{b(1-\cos\theta) + a\sin\theta} = \frac{1}{2(1-\cos\theta)}$$

$$\therefore \quad \alpha = \frac{a(1-\cos\theta) - b\sin\theta}{2(1-\cos\theta)}, \quad \beta = \frac{b(1-\cos\theta) + a\sin\theta}{2(1-\cos\theta)}.$$

1.9 Invariants of Translation, Rotation and Rigid Motions

Theorem 1.9.1 *The distance between two points remains invariant under orthogonal transformation.*

Proof: Let the orthogonal transformation formulae be

$$\left.\begin{array}{l} x = x'\cos\alpha - y'\sin\alpha + a \\ y = x'\sin\alpha + y'\cos\alpha + b \end{array}\right\} \qquad (1.3)$$

Let, by transformation (1.3) the coordinates of two points $P(x_1, y_1)$ and $Q(x_2, y_2)$ becomes (x'_1, y'_1) and (x'_2, y'_2) respectively. Then we get

$$\begin{aligned} x_1 &= x'_1\cos\alpha - y'_1\sin\alpha + a, \\ y_1 &= x'_1\sin\alpha + y'_1\cos\alpha + b; \\ x_2 &= x'_2\cos\alpha - y'_2\sin\alpha + a, \\ y_2 &= x'_2\sin\alpha + y'_2\cos\alpha + b. \end{aligned}$$

From these we get, on subtracting,

$$\left.\begin{array}{l} x_2 - x_1 = (x'_2 - x'_1)\cos\alpha - (y'_2 - y'_1)\sin\alpha \\ y_2 - y_1 = (x'_2 - x'_1)\sin\alpha + (y'_2 - y'_1)\cos\alpha. \end{array}\right\}$$

$$\begin{aligned} \therefore (x_2-x_1)^2 + (y_2-y_1)^2 &= [(x'_2-x'_1)\cos\alpha - (y'_2-y'_1)\sin\alpha]^2 \\ &\quad + [(x'_2-x'_1)\sin\alpha + (y'_2-y'_1)\cos\alpha]^2 \\ &= (x'_2-x'_1)^2(\cos^2\alpha + \sin^2\alpha) + (y'_2-y'_1)^2(\cos^2\alpha + \sin^2\alpha) \\ &= (x'_2-x'_1)^2 + (y'_2-y'_1)^2. \end{aligned}$$

Therefore $\sqrt{(x_2-x_1)^2 + (y_2-y_1)^2} = \sqrt{(x'_2-x'_1)^2 + (y'_2-y'_1)^2}$.

So the distance between two points remains invariant. $\square$

Theorem 1.9.2 *The area of a triangle remains invariant under orthogonal transformation.*

Proof: Let the orthogonal transformation formulae be

$$\left.\begin{array}{l} x = x' \cos \alpha - y' \sin \alpha + a \\ y = x' \sin \alpha + y' \cos \alpha + b \end{array}\right\} \qquad (1.4)$$

Let the coordinates of the vertices $A(x_1, y_1), B(x_2, y_2)$ and $C(x_3, y_3)$ becomes $(x'_1, y'_1), (x'_2, y'_2)$ and (x'_3, y'_3) respectively. So by formula (1.4) we get

$$\begin{array}{l} x_1 = x'_1 \cos \alpha - y'_1 \sin \alpha + a, \\ y_1 = x'_1 \sin \alpha + y'_1 \cos \alpha + b; \\ x_2 = x'_2 \cos \alpha - y'_2 \sin \alpha + a, \\ y_2 = x'_2 \sin \alpha + y'_2 \cos \alpha + b; \\ x_3 = x'_3 \cos \alpha - y'_3 \sin \alpha + a, \\ y_3 = x'_3 \sin \alpha + y'_3 \cos \alpha + b. \end{array}$$

Now the area of $\triangle ABC = \dfrac{1}{2} \begin{vmatrix} x_1 & x_2 & x_3 \\ y_1 & y_2 & y_3 \\ 1 & 1 & 1 \end{vmatrix}$

$$= \frac{1}{2} \begin{vmatrix} x'_1 \cos \alpha - y'_1 \sin \alpha + a & x'_2 \cos \alpha - y'_2 \sin \alpha + a & x'_3 \cos \alpha - y'_3 \sin \alpha + a \\ x'_1 \sin \alpha + y'_1 \cos \alpha + b & x'_2 \sin \alpha + y'_2 \cos \alpha + b & x'_3 \sin \alpha + y'_3 \cos \alpha + b \\ 1 & 1 & 1 \end{vmatrix}$$

$$= \frac{1}{2} \begin{vmatrix} x'_1 & x'_2 & x'_3 \\ y'_1 & y'_2 & y'_3 \\ 1 & 1 & 1 \end{vmatrix} \times \begin{vmatrix} \cos \alpha & -\sin \alpha & a \\ \sin \alpha & \cos \alpha & b \\ 0 & 0 & 1 \end{vmatrix} \quad \text{(taking column-row product.)}$$

$$= \frac{1}{2} \begin{vmatrix} x'_1 & x'_2 & x'_3 \\ y'_1 & y'_2 & y'_3 \\ 1 & 1 & 1 \end{vmatrix} = \text{area obtained by new system.}$$

Hence the area remains invariant. □

Theorem 1.9.3 *The coefficients of x^2, xy, y^2 and Δ obtained from the expression $ax^2 + 2hxy + by^2 + 2gx + 2fy + c$, where $\Delta = \begin{vmatrix} a & h & g \\ h & b & f \\ g & f & c \end{vmatrix}$ remains invariant under translation.*

Proof: Let us consider the translation formula

$$\left.\begin{array}{l} x = x' + \alpha \\ y = y' + \beta. \end{array}\right\}$$

Transformation of Coordinates

Then the given expression $ax^2 + 2hxy + by^2 + 2gx + 2fy + c$ becomes

$$a(x' + \alpha)^2 + 2h(x' + \alpha)(y' + \beta) + b(y' + \beta)^2 + 2g(x' + \alpha)$$
$$+ 2f(y' + \beta) + c$$

$$= ax'^2 + 2hx'y' + by'^2 + 2(a\alpha + h\beta + g)x' + 2(h\alpha + b\beta + f)y'$$
$$+ (a\alpha^2 + 2h\alpha\beta + b\beta^2 + 2g\alpha + 2f\beta + c)$$

$$= a'x'^2 + 2h'x'y' + b'y'^2 + 2g'x' + 2f'y' + c' \text{ (say)}$$

where $a' = a$, $b' = b$, $h' = h$, $g' = a\alpha + h\beta + g$, $f' = h\alpha + b\beta + f$ and $c' = a\alpha^2 + 2h\alpha\beta + b\beta^2 + 2g\alpha + 2f\beta + c$ which shows that

$$\text{coefficient of } x'^2 = \text{coefficient of } x^2,$$
$$\text{coefficient of } y'^2 = \text{coefficient of } y^2 \text{ and}$$
$$\text{coefficient of } x'y' = \text{coefficient of } xy.$$

Again $\Delta' = \begin{vmatrix} a' & h' & g' \\ h' & b' & f' \\ g' & f' & c' \end{vmatrix}$

$$= \begin{vmatrix} a & h & a\alpha + h\beta + g \\ h & b & h\alpha + b\beta + f \\ a\alpha + h\beta + g & h\alpha + b\beta + f & a\alpha^2 + 2h\alpha\beta + b\beta^2 + 2g\alpha + 2f\beta + c \end{vmatrix}$$

$$= \begin{vmatrix} a & h & g \\ h & b & f \\ a\alpha + h\beta + g & h\alpha + b\beta + f & g\alpha + f\beta + c \end{vmatrix} \quad [C'_3 \to C_3 - (\alpha C_1 + \beta C_2)].$$

$$= \begin{vmatrix} a & h & g \\ h & b & f \\ g & f & c \end{vmatrix} = \Delta. \quad [R'_3 \to R_3 - (\alpha R_1 + \beta R_2)]$$

Hence Δ is also an invariant. $\square$

Theorem 1.9.4 *If the expression* $ax^2 + 2hxy + by^2 + 2gx + 2fy + c$ *becomes* $ax'^2 + 2hx'y' + by'^2 + 2gx' + 2fy' + c$ *under rotation then*

(i) $a' + b' = a + b$ (ii) $c' = c$ (iii) $a' + b' + c' = a + b + c$

(iv) $a'b' - h'^2 = ab - h^2$ *and* (v) $\Delta' = \Delta$.

Proof: Let the axes be rotated through an angle α. Then we have the rotational formulae

$$\left. \begin{array}{l} x = x' \cos\alpha - y' \sin\alpha \\ y = x' \sin\alpha + y' \cos\alpha. \end{array} \right\}$$

Using these rotational formulae the given expression $ax^2 + 2hxy + by^2 + 2gx + 2fy + c$ becomes

$$a(x'\cos\alpha - y'\sin\alpha)^2 + 2h(x'\cos\alpha - y'\sin\alpha)(x'\sin\alpha + y'\cos\alpha)$$
$$+ b(x'\sin\alpha + y'\cos\alpha)^2 + 2g(x'\cos\alpha - y'\sin\alpha)$$
$$+ 2f(x'\sin\alpha + y'\cos\alpha) + c$$
$$= (a\cos^2\alpha + 2h\cos\alpha\sin\alpha + b\sin^2\alpha)x'^2 + 2(-a\cos\alpha\sin\alpha$$
$$+ h\cos^2\alpha - h\sin^2\alpha + b\cos\alpha\sin\alpha)x'y' + (a\sin^2\alpha - 2h\sin\alpha\cos\alpha$$
$$+ b\cos^2\alpha)y'^2 + 2(g\cos\alpha + f\sin\alpha)x' + 2(-g\sin\alpha + f\cos\alpha)y' + c$$
$$= a'x'^2 + 2h'x'y' + b'y'^2 + 2g'x' + 2f'y' + c' \text{ (say)}$$

where, we get by equating the coefficients

$$a' = a\cos^2\alpha + 2h\cos\alpha\sin\alpha + b\sin^2\alpha$$
$$b' = a\sin^2\alpha - 2h\sin\alpha\cos\alpha + b\cos^2\alpha$$
$$h' = -a\cos\alpha\sin\alpha + h\cos^2\alpha - h\sin^2\alpha + b\cos\alpha\sin\alpha$$
$$g' = g\cos\alpha + f\sin\alpha, \quad f' = -g\sin\alpha + f\cos\alpha, \quad c' = c.$$

So we get

(i) $a' + b' = a(\cos^2\alpha + \sin^2\alpha) + b(\cos^2\alpha + \sin^2\alpha) = a + b.$

(ii) $c' = c$ is already proved.

(iii) Since $c' = c$, it is obvious from (i) that $a' + b' + c' = a + b + c$.

To prove (iv) and (v), let us take

$$a_1 = a\cos\alpha + h\sin\alpha \quad a_2 = -a\sin\alpha + h\cos\alpha$$
$$b_1 = h\cos\alpha + b\sin\alpha \quad b_2 = -h\sin\alpha + b\cos\alpha$$
$$\therefore \quad a' = a_1\cos\alpha + b_1\sin\alpha \quad b' = -a_2\sin\alpha + b_2\cos\alpha$$
$$\text{and} \quad h' = -a_1\sin\alpha + b_1\sin\alpha = a_2\cos\alpha + b_2\sin\alpha.$$

Now (iv) $a'b' - h'^2 = \begin{vmatrix} a' & h' \\ h' & b' \end{vmatrix}$

$$= \begin{vmatrix} a_1\cos\alpha + b_1\sin\alpha & -a_1\sin\alpha + b_1\cos\alpha \\ a_2\cos\alpha + b_2\sin\alpha & -a_2\sin\alpha + b_2\cos\alpha \end{vmatrix}$$

$$= \begin{vmatrix} a_1 & b_1 \\ a_2 & b_2 \end{vmatrix} \times \begin{vmatrix} \cos\alpha & \sin\alpha \\ -\sin\alpha & \cos\alpha \end{vmatrix} = \begin{vmatrix} a_1 & b_1 \\ a_2 & b_2 \end{vmatrix}$$

$$= \begin{vmatrix} a\cos\alpha + h\sin\alpha & h\cos\alpha + b\sin\alpha \\ -a\sin\alpha + h\cos\alpha & -h\sin\alpha + b\cos\alpha \end{vmatrix}$$

$$= \begin{vmatrix} a & h \\ h & b \end{vmatrix} \times \begin{vmatrix} \cos\alpha & \sin\alpha \\ -\sin\alpha & \cos\alpha \end{vmatrix} = \begin{vmatrix} a & h \\ h & b \end{vmatrix} = ab - h^2.$$

Transformation of Coordinates

(v) $\Delta' = \begin{vmatrix} a' & h' & g' \\ h' & b' & f' \\ g' & f' & c' \end{vmatrix} = g' \begin{vmatrix} h' & g' \\ b' & f' \end{vmatrix} - f' \begin{vmatrix} a' & g' \\ h' & f' \end{vmatrix} + c' \begin{vmatrix} a' & h' \\ h' & b' \end{vmatrix}$

(expanding through third row or third column)

$= g' \begin{vmatrix} a_2 \cos\alpha + b_2 \sin\alpha & g \cos\alpha + f \sin\alpha \\ -a_2 \sin\alpha + b_2 \cos\alpha & -g \sin\alpha + f \cos\alpha \end{vmatrix}$

$\quad - f' \begin{vmatrix} a_1 \cos\alpha + b_1 \sin\alpha & g \cos\alpha + f \sin\alpha \\ -a_1 \sin\alpha + b_1 \sin\alpha & -g \sin\alpha + f \cos\alpha \end{vmatrix} + c \begin{vmatrix} a & h \\ h & b \end{vmatrix}$

$\left[\because c' = c \text{ and } \begin{vmatrix} a' & h' \\ h' & b' \end{vmatrix} = \begin{vmatrix} a & h \\ h & b \end{vmatrix} \right]$

$= g' \begin{vmatrix} a_2 & b_2 \\ g & f \end{vmatrix} \times \begin{vmatrix} \cos\alpha & \sin\alpha \\ -\sin\alpha & \cos\alpha \end{vmatrix}$

$\quad - f' \begin{vmatrix} a_1 & b_1 \\ g & f \end{vmatrix} \times \begin{vmatrix} \cos\alpha & \sin\alpha \\ -\sin\alpha & \cos\alpha \end{vmatrix} + c \begin{vmatrix} a & h \\ h & b \end{vmatrix}$

$= g' \begin{vmatrix} a_2 & b_2 \\ g & f \end{vmatrix} - f' \begin{vmatrix} a_1 & b_1 \\ g & f \end{vmatrix} + c \begin{vmatrix} a & h \\ h & b \end{vmatrix}$

$= g'(a_2 f - b_2 g) - f'(a_1 f - b_1 g) + c \begin{vmatrix} a & h \\ h & b \end{vmatrix}$

$= g(b_1 f' - b_2 g') - f(a_1 f' - a_2 g') + c \begin{vmatrix} a & h \\ h & b \end{vmatrix}$

$= g \begin{vmatrix} b_1 & b_2 \\ g' & f' \end{vmatrix} - f \begin{vmatrix} a_1 & a_2 \\ g' & f' \end{vmatrix} + c \begin{vmatrix} a & h \\ h & b \end{vmatrix}$

$= g \begin{vmatrix} h \cos\alpha + b \sin\alpha & -h \sin\alpha + b \cos\alpha \\ g \cos\alpha + f \sin\alpha & -g \sin\alpha + f \cos\alpha \end{vmatrix}$

$\quad - f \begin{vmatrix} a \cos\alpha + h \sin\alpha & -a \sin\alpha + h \cos\alpha \\ g \cos\alpha + f \sin\alpha & -g \sin\alpha + f \cos\alpha \end{vmatrix} + c \begin{vmatrix} a & h \\ h & b \end{vmatrix}$

$= g \begin{vmatrix} h & b \\ g & f \end{vmatrix} \times \begin{vmatrix} \cos\alpha & \sin\alpha \\ -\sin\alpha & \cos\alpha \end{vmatrix}$

$\quad - f \begin{vmatrix} a & h \\ g & f \end{vmatrix} \times \begin{vmatrix} \cos\alpha & \sin\alpha \\ -\sin\alpha & \cos\alpha \end{vmatrix} + c \begin{vmatrix} a & h \\ h & b \end{vmatrix}$

$= g \begin{vmatrix} h & b \\ g & f \end{vmatrix} - f \begin{vmatrix} a & h \\ g & f \end{vmatrix} + c \begin{vmatrix} a & h \\ h & b \end{vmatrix} = \begin{vmatrix} a & h & g \\ h & b & f \\ g & f & c \end{vmatrix} = \Delta.$

$\therefore \quad \Delta' = \Delta.$

Hence the results. □

1.10 Work Out Examples

Example 1.10.1 *(i) Find the form of the equation $\frac{x}{a} + \frac{y}{b} = 1$ when the origin is shifted to the point (a,b).*

(ii) What will be the form of the equation $x\cos\alpha + y\sin\alpha = p$ when the axes are rotated through an angle α?

Solution: (i) Here the translation formulae are $x = x'+a$ and $y = y'+b$ and so the transformed equation becomes $\frac{x'+a}{a} + \frac{y'+b}{b} = 1 \Rightarrow \frac{x'}{a} + \frac{y'}{b} + 1 = 0.$

(ii) The rotational formulae are $x = x'\cos\alpha - y'\sin\alpha$ and $y = x'\sin\alpha + y'\cos\alpha$ and the transformed equation becomes

$$(x'\cos\alpha - y'\sin\alpha)\cos\alpha + (x'\sin\alpha + y'\cos\alpha)\sin\alpha = p$$

or, $x'(\cos^2\alpha + \sin^2\alpha) = p \Rightarrow x' = p.$ □

Example 1.10.2 *(i) Find the angle by which the axes should be turned so that the equation $ax^2 + 2hxy + by^2 = 0$ becomes another equation in which the term xy is absent.*

(ii) Show that the angle through which the axes are to be turned so that the equation $x^2 + 2\sqrt{3}xy - y^2 = 2$ may be reduced to the form $x'^2 - y'^2 = 1$ is $30°$.

(iii) Find the rotation so that the equation $3x^2 + 10xy + 3y^2 = 0$ will want the product term $x'y'$.

Solution: (i) Let the axes be rotated through an angle θ. Then the rotational formulae are

$$\left.\begin{array}{l} x = x'\cos\theta - y'\sin\theta \\ y = x'\sin\theta + y'\cos\theta. \end{array}\right\}$$

So the given equation is transformed to

$$a(x'\cos\theta - y'\sin\theta)^2 + 2h(x'\cos\theta - y'\sin\theta)(x'\sin\theta + y'\cos\theta)$$
$$+ b(x'\sin\theta + y'\cos\theta)^2 = 0$$
$$\Rightarrow (a\cos^2\theta + h\sin 2\theta + b\sin^2\theta)x'^2 + (-a\sin 2\theta + 2h\cos 2\theta$$
$$+ b\sin 2\theta)x'y' + (a\sin^2\theta + h\sin 2\theta + b\cos^2\theta)y'^2 = 0.$$

Now, we find θ, so that the coefficient of $x'y'$ vanishes, i.e.,

$$(b-a)\sin 2\theta + 2h\cos 2\theta = 0 \Rightarrow \tan 2\theta = \frac{2h}{a-b} \Rightarrow \theta = \frac{1}{2}\tan^{-1}\frac{2h}{a-b}.$$

Transformation of Coordinates

Note 1.10.1 *In particular, when* $a = b$, $\theta = \dfrac{\pi}{4}$.

(ii) Let the original axes be rotated through an angle θ, i.e., we have the rotational formulae

$$\left.\begin{array}{l} x = x' \cos\theta - y' \sin\theta \\ y = x' \sin\theta + y' \cos\theta. \end{array}\right\}$$

Then the given equation becomes

$$(x' \cos\theta - y' \sin\theta)^2 + 2\sqrt{3}(x' \cos\theta - y' \sin\theta)(x' \sin\theta + y' \cos\theta)$$
$$-(x' \sin\theta + y' \cos\theta)^2 = 0$$
$$\Rightarrow (\cos 2\theta + \sqrt{3}\sin 2\theta)x'^2 + 2(\sqrt{3}\cos 2\theta - \sin 2\theta)x'y'$$
$$-(\cos 2\theta + \sqrt{3}\sin 2\theta)y'^2 = 0.$$

If this equation is reduced to the form $x'^2 - y'^2 = 1$, then the coefficient of $x'y'$ must vanishes, i.e.,

$$\sqrt{3}\cos 2\theta - \sin 2\theta = 0 \Rightarrow \tan 2\theta = \sqrt{3} = \tan 60° \Rightarrow 2\theta = 60° \therefore \theta = 30°$$

and in that case the transformed equation becomes

$$(\cos 60° + \sqrt{3}\sin 60°)x'^2 - (\cos 60° + \sqrt{3}\sin 60°)y'^2 = 2$$
$$\Rightarrow \left(\frac{1}{2} + \frac{3}{2}\right)x'^2 - \left(\frac{1}{2} + \frac{3}{2}\right)y'^2 = 2 \Rightarrow x'^2 - y'^2 = 1.$$

An alternative method:
Refer to **Example 1.10.2 (i)** above, here $a = 1, h = \sqrt{3}$ and $b = -1$

$$\therefore \theta = \frac{1}{2}\tan^{-1}\frac{2\sqrt{3}}{1+1} = \frac{1}{2}\tan^{-1}\sqrt{3}, \text{ i.e., } \tan 2\theta = \sqrt{3} \Rightarrow \theta = 60°.$$

Now substituting this value in the given equation we get $x'^2 - y'^2 = 1$ (as above). Hence the required angle of rotation is 30°.

(iii) The given equation is $3x^2 + 10xy + 3y^2 = 0$. Here the coefficients of x^2 and y^2 are equal and so the required angle of rotation is $\dfrac{\pi}{4}$ and so the formulae for rotation in this case are

$$x = \frac{1}{\sqrt{2}}x' - \frac{1}{\sqrt{2}}y', \quad y = \frac{1}{\sqrt{2}}x' + \frac{1}{\sqrt{2}}y'$$

and the transformed equation is

$$3\left(\frac{x'-y'}{\sqrt{2}}\right)^2 + 10\left(\frac{x'-y'}{\sqrt{2}}\right)\left(\frac{x'+y'}{\sqrt{2}}\right) + 3\left(\frac{x'+y'}{\sqrt{2}}\right)^2 = 0$$

or, $\frac{3}{2}\{(x'-y')^2 + (x'+y')^2\} + 5(x'^2 - y'^2) = 0$

$\Rightarrow 3(x'^2 + y'^2) + 5(x'^2 - y'^2) = 0 \Rightarrow 8x'^2 - 2y'^2 = 0 \Rightarrow 4x'^2 - y'^2 = 0.$

which is of the form $ax^2 + by^2 = 0$. □

Example 1.10.3 *Two coordinate systems have a common origin O, the transformation from one system to another being accomplished by a rotation through a certain angle. The coordinates of the point $P(4, -3)$ are given with reference to the first system. Find the coordinate transformation formulas if the positive direction of the new x-axis is determined by the segment OP.*

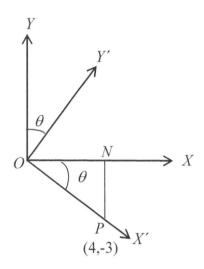

Solution: The point $P(4, -3)$ is referred to the old system (OX, OY) and the point lies on the new x-axis, i.e., on OX'. So if the angle of rotation be θ, we get $\tan\theta = \frac{NP}{ON} = \frac{-3}{4}$ which gives $\sin\theta = -\frac{3}{5}$ and $\cos\theta = \frac{4}{5}$ and hence the transformation formulae for rotation are obtained as

$$\left.\begin{array}{l} x = x'\cos\theta - y'\sin\theta = \frac{4}{5}x' + \frac{3}{5}y' \\ \\ y = x'\sin\theta + y'\cos\theta = -\frac{3}{5}x' + \frac{4}{5}y'. \end{array}\right\}$$

□

Transformation of Coordinates

Example 1.10.4 *The origin is shifted to the point $(-1, 2)$ and the axes are then rotated through an angle $\tan^{-1}\dfrac{5}{12}$. If the coordinates of a point in the new system be $(13, -13)$, find the coordinates of the point in the old system.*

Solution: Let (x, y) and (x', y') be the coordinates of a point in the old and the new systems respectively. Then the rigid motion formulae are

$$\left.\begin{array}{l} x = x'\cos\alpha - y'\sin\alpha + a \\ y = x'\sin\alpha + y'\cos\alpha + b. \end{array}\right\}$$

Here $a = -1, b = 2$ and $\tan\alpha = \dfrac{5}{12} \Rightarrow \sin\alpha = \dfrac{5}{13}, \cos\alpha = \dfrac{12}{13}$. Putting these values in the above formulae we get

$$\left.\begin{array}{l} x = x'\frac{12}{13} - y'\frac{5}{13} - 1 \\ y = x'\frac{5}{13} + y'\frac{12}{13} + 2 \end{array}\right\} \Rightarrow \left.\begin{array}{l} x = 13 \times \frac{12}{13} - (-13) \times \frac{5}{13} - 1 = 16 \\ y = 13 \times \frac{5}{13} + (-13) \times \frac{12}{13} + 2 = -5. \end{array}\right\}$$

Hence the coordinates in the old system are $(16, -5)$. □

Example 1.10.5 *(i) Show that there is only one point whose coordinates do not alter due to a rigid motion.*

(ii) Show that there is only one point whose coordinates remains the same by changing the axes of coordinates by the rigid motion $x' = \dfrac{4}{5}x - \dfrac{3}{5}y + 2$ and $y' = \dfrac{3}{5}x + \dfrac{4}{5}y - 2$ and show that the coordinates of the point are $(4, 2)$.

Solution: (i) See article **1.8.3**.

(ii) Let (α, β) be the point. Since the point remains same after rigid motion, i.e., (α, β) is a fixed point. So it will satisfy the given equation and we have

$$\left.\begin{array}{l} \alpha = \frac{4}{5}\alpha - \frac{3}{5}\beta + 2 \Rightarrow \alpha + 3\beta - 10 = 0 \\ \beta = \frac{3}{5}\alpha + \frac{4}{5}\beta - 2 \Rightarrow 3\alpha - \beta - 10 = 0. \end{array}\right\}$$

The coefficient determinant $= \begin{vmatrix} 1 & 3 \\ 3 & -1 \end{vmatrix} = -1 - 9 = -10 \neq 0$.

Therefore the point is unique. Hence proved the first part.

Now by solving the above system of equations we get $\alpha = 4, \beta = 2$. Hence the required coordinates of the point are $(4, 2)$. □

Example 1.10.6 *$P(9, -1)$ and $Q(-3, 4)$ are two points. If the origin is shifted to P and the coordinate axes are rotated so that the positive direction of the new x-axis agrees with the direction of the segment $\overline{PQ}$, find the coordinate transformation formulae.*

Solution: The coordinate transformation formulae can be taken as

$$\left. \begin{array}{l} x = x' \cos\alpha - y' \sin\alpha + a \\ y = x' \sin\alpha + y' \cos\alpha + b \end{array} \right\} \quad (1.5)$$

By the condition of the problem, new origin is $P(9, -1)$, i.e., we get $a = 9$,

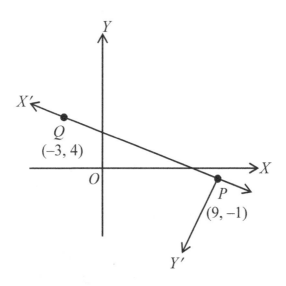

$b = -1$. So equation (1.5) takes the form

$$\left. \begin{array}{l} x = x' \cos\alpha - y' \sin\alpha + 9 \\ y = x' \sin\alpha + y' \cos\alpha - 1 \end{array} \right\} \quad (1.6)$$

Let (r, s) be the coordinates of the point $Q(-3, 4)$ in the new coordinate system. Then clearly $s = 0$, but $r \neq 0$ and the corresponding to the point $(-3, 4)$ we get the new coordinates $(x', y') = (r, 0)$. So from equation (1.6) we get

$$\left. \begin{array}{ll} -3 = r\cos\alpha + 9 & \Rightarrow \quad r\cos\alpha = -12 \\ \text{and} \quad 4 = r\sin\alpha - 1 & \Rightarrow \quad r\sin\alpha = 5 \end{array} \right\} \Rightarrow \frac{\cos\alpha}{-12} = \frac{\sin\alpha}{5}.$$

Hence $\dfrac{\sin\alpha}{5} = \dfrac{\cos\alpha}{-12} = \dfrac{1}{13}$.

Therefore $\sin\alpha = \dfrac{5}{13}$ and $\cos\alpha = \dfrac{-12}{13}$ and we get the required formulae as under

$$\left. \begin{array}{l} x = -\frac{12}{13}x' - \frac{5}{13}y' + 9 \\ y = \frac{5}{13}x' - \frac{12}{13}y' - 1. \end{array} \right\}$$

Transformation of Coordinates 39

An Alternative Method:

Gradient of the line joining $P(9, -1)$ and $Q(-3, 4)$ is

$$\tan \alpha = \frac{4+1}{-3-9} = \frac{5}{-12}, \quad \therefore \sin \alpha = \frac{5}{13} \text{ and } \cos \alpha = \frac{-12}{13}.$$

So the required formulae for rigid motion are
$$\left. \begin{array}{l} x = -\frac{12}{13}x' - \frac{5}{13}y' + 9 \\ y = \frac{5}{13}x' - \frac{12}{13}y' - 1 \end{array} \right\} \qquad \square$$

Example 1.10.7 *When the axes are turned through an angle if the expression (i) $ax + by$ becomes $a'x' + b'y'$ referred to new axes; then show that $a'^2 + b'^2 = a^2 + b^2$*

and if (ii) $ax+by$ and $cx+dy$ are changed into $a'x'+b'y'$ and $c'x'+d'y'$, then show that $a'd' - b'c' = ad - bc$.

Solution: Applying the rotational formulae

$$\left. \begin{array}{l} x = x' \cos \alpha - y' \sin \alpha \\ y = x' \sin \alpha + y' \cos \alpha \end{array} \right\}$$

we get (i)
$$\begin{aligned} ax + by &= a(x' \cos \alpha - y' \sin \alpha) + b(x' \sin \alpha + y' \cos \alpha) \\ &= (a \cos \alpha + b \sin \alpha)x' + (-a \sin \alpha + b \cos \alpha)y' \\ &= a'x' + b'y' \end{aligned}$$

which imply $a' = a \cos \alpha + b \sin \alpha$ and $b' = -a \sin \alpha + b \cos \alpha$. Therefore we get

$$\begin{aligned} a'^2 + b'^2 &= (a \cos \alpha + b \sin \alpha)^2 + (-a \sin \alpha + b \cos \alpha)^2 \\ &= a^2(\cos^2 \alpha + \sin^2 \alpha) + b^2(\cos^2 \alpha + \sin^2 \alpha) \\ &= a^2 + b^2 \end{aligned}$$

(ii) Again
$$\begin{aligned} cx + dy &= c(x' \cos \alpha - y' \sin \alpha) + d(x' \sin \alpha + y' \cos \alpha) \\ &= (c \cos \alpha + d \sin \alpha)x' + (-c \sin \alpha + d \cos \alpha)y' \\ &= c'x' + d'y' \end{aligned}$$

which imply $c' = c \cos \alpha + d \sin \alpha$ and $d' = -c \sin \alpha + d \cos \alpha$. Thus

$$\begin{aligned} a'd' - b'c' &= \begin{vmatrix} a' & b' \\ c' & d' \end{vmatrix} = \begin{vmatrix} a \cos \alpha + b \sin \alpha & -a \sin \alpha + b \cos \alpha \\ c \cos \alpha + d \sin \alpha & -c \sin \alpha + d \cos \alpha \end{vmatrix} \\ &= \begin{vmatrix} a & b \\ c & d \end{vmatrix} \times \begin{vmatrix} \cos \alpha & \sin \alpha \\ -\sin \alpha & \cos \alpha \end{vmatrix} = \begin{vmatrix} a & b \\ c & d \end{vmatrix} = ad - bc. \end{aligned}$$

An Alternative Method:

(i) Let $A = \begin{bmatrix} a & b \end{bmatrix}, X = \begin{bmatrix} x \\ y \end{bmatrix}$. Then $ax + by = AX$. Let us consider the rotation given by $X = SX'$ where $S = \begin{bmatrix} \cos\theta & \sin\theta \\ -\sin\theta & \cos\theta \end{bmatrix}$.

Now, $AX = ASX' = \begin{bmatrix} a & b \end{bmatrix} \times \begin{bmatrix} \cos\theta & \sin\theta \\ -\sin\theta & \cos\theta \end{bmatrix} \times \begin{bmatrix} x' \\ y' \end{bmatrix}$

$= (a\cos\theta + b\sin\theta)x' + (-a\sin\theta + b\cos\theta)y' = a'x' + b'y'$

$\Rightarrow\quad a' = a\cos\theta + b\sin\theta, b' = -a\sin\theta + b\cos\theta$

$\therefore\quad a'^2 + b'^2 = (a\cos\theta + b\sin\theta)^2 + (-a\sin\theta + b\cos\theta)^2$

$= a^2 + b^2$ [as before].

(ii) Applying the same method we can get $c' = c\cos\theta + d\sin\theta$ and $d' = -c\sin\theta + d\cos\theta$ and hence $a'd' - b'c' = ad - bc$. [as before] □

Example 1.10.8 *Prove that the transformation of rectangular axes which converts $\dfrac{x^2}{p} + \dfrac{y^2}{q}$ into $ax'^2 + 2hx'y' + by'^2$ will convert $\dfrac{x^2}{p - \lambda} + \dfrac{y^2}{q - \lambda}$ into*

$$\dfrac{ax'^2 + 2hx'y' + by'^2 - \lambda(ab - h^2)(x'^2 + y'^2)}{1 - (a - b)\lambda + (ab - h^2)\lambda^2}.$$

Solution: Here the transformation is from one set of rectangular axes to another set of rectangular axes, the origin remaining same, therefore $x^2 + y^2 = x'^2 + y'^2$, for each of them are square of the distance of the same point from the origin.

Again by the theorem of invariants [ref. **Theorem 1.9.4**], we have $\dfrac{1}{p} + \dfrac{1}{q} = a + b$ and $\dfrac{1}{pq} = ab - h^2$. Hence

$$\dfrac{x^2}{p - \lambda} + \dfrac{y^2}{q - \lambda} = \dfrac{qx^2 + py^2 - \lambda(x^2 + y^2)}{pq - \lambda(p + q) + \lambda^2} = \dfrac{\frac{x^2}{p} + \frac{y^2}{q} - \frac{\lambda}{pq}(x^2 + y^2)}{1 - \lambda(\frac{1}{p} + \frac{1}{q}) + \frac{\lambda^2}{pq}}$$

$$= \dfrac{ax'^2 + 2hx'y' + by'^2 - \lambda(ab - h^2)(x'^2 + y'^2)}{1 - (a - b)\lambda + (ab - h^2)\lambda^2}.$$ □

Example 1.10.9 *Show that the transformed equation of the curve $(4x + 3y + 1)(3x - 4y + 2) = 75$ when the axes are $4x + 3y + 1 = 0$ and $3x - 4y + 2 = 0$ is $x'y' = 3$ where (x', y') be the new coordinates.*

Transformation of Coordinates

Solution: (x, y) and (x', y') are the coordinates in the old and new system. Then $x' =$ perpendicular distance of the point (x, y) from the new y-axis $3x - 4y + 2 = 0$ [see § **1.5**],

i.e., $x' = \dfrac{3x - 4y + 2}{\sqrt{3^2 + (-4)^2}} = \dfrac{3x - 4y + 2}{5} \Rightarrow 3x - 4y + 2 = 5x'$

and $y' =$ perpendicular distance of the point (x, y) from the new axis $4x + 3y + 1 = 0$ [see § **1.5**],

i.e., $y' = \dfrac{4x + 3y + 1}{\sqrt{4^2 + 3^2}} = \dfrac{4x + 3y + 1}{5} \Rightarrow 4x + 3y + 1 = 5y'$.

Therefore the given equation becomes $5x' 5y' = 75$ or, $x'y' = 3$. □

Example 1.10.10 *Transform the equation* $(a^2 + b^2)(x^2 + y^2) + 2(ac + bd)x + 2(bc - ad)y + (c^2 + d^2) = (a^2 + b^2)r^2$ *to a new axes of x and y whose equations are the mutually perpendicular lines $ax + by + c = 0$ and $bx - ay + d = 0$ respectively.*

Solution: The transformation by § **1.5** are

$$x' = \dfrac{bx - ay + d}{\sqrt{a^2 + b^2}} \text{ and } y' = \dfrac{ax + by + c}{\sqrt{a^2 + b^2}}.$$

The given equation can be written as

$$a^2 x^2 + a^2 y^2 + b^2 x^2 + b^2 y^2 + 2acx + 2bdx + 2bcy - 2ady +$$
$$c^2 + d^2 = (a^2 + b^2)r^2$$
$$\Rightarrow (a^2 x^2 + b^2 y^2 + c^2 + 2abxy + 2acx + 2bcy) + (a^2 y^2 + b^2 x^2 + d^2$$
$$- 2abxy - 2ady + 2bdx) = (a^2 + b^2)r^2$$
$$\Rightarrow (ax + by + c)^2 + (bx - ay + d)^2 = (a^2 + b^2)r^2.$$

Hence on substitution the values of x', y', the equation becomes

$$(a^2 + b^2)x'^2 + (a^2 + b^2)y'^2 = (a^2 + b^2)r^2.$$

Suppressing the dashes, the equation becomes $x^2 + y^2 = r^2$. □

Example 1.10.11 *By orthogonal transformation without change of origin if the equation $ax^2 + 2hxy + by^2 = c$ is changed into one in which there is no term involving xy, show that the transformed equation is $(a + b + \lambda)x^2 + (a + b - \lambda)y^2 = 2c$ where $\lambda = \sqrt{(a-b)^2 + 4h^2}$.*

Solution: Since the origin remains unchanged we apply only rotational formulae

$$\left.\begin{array}{l}x = x'\cos\theta - y'\sin\theta \\ y = x'\sin\theta + y'\cos\theta\end{array}\right\}$$

where the angle of rotation is θ, which is obtained as

$$\theta = \frac{1}{2}\tan^{-1}\frac{2h}{a-b} \quad \text{[See \textbf{Example 1.10.2} before]}$$

$$\Rightarrow \quad 2\theta = \tan^{-1}\frac{2h}{a-b} \quad \text{or} \quad \tan 2\theta = \frac{2h}{a-b}$$

$$\therefore \quad \sec^2 2\theta = 1 + \tan^2 2\theta = 1 + \frac{4h^2}{(a-b)^2} = \frac{(a-b)^2 + 4h^2}{(a-b)^2}$$

$$\therefore \quad (a-b)^2 + 4h^2 = (a-b)^2 \sec^2 2\theta$$

i.e., $(a-b)\sec 2\theta = \sqrt{(a-b)^2 + 4h^2} = \lambda.$

Now we propose to apply the method of invariants.

Let the transformed equation be $a'x'^2 + b'y'^2 = c$, then we have $a' + b' = a + b$ and $a'b' = ab - h^2$.

$$\therefore a' - b' = \sqrt{(a'+b')^2 - 4a'b'} = \sqrt{(a+b)^2 - 4ab + 4h^2}$$
$$= \sqrt{(a-b)^2 + 4h^2} = \lambda$$

$$\therefore \quad a' = \frac{a+b+\lambda}{2} \quad \text{and} \quad b' = \frac{a+b-\lambda}{2}.$$

Hence suppressing the dashes, the transformed equation is

$$(a+b+\lambda)x^2 + (a+b-\lambda)y^2 = 2c. \qquad \square$$

Example 1.10.12 *Find the rigid motion for which the point $(4,2)$ remains $(4,2)$.* *[a converse problem of **Example 1.10.5 (ii)**]*

Solution: Let (x,y) and (x',y') be the coordinates in the old and new system. Let the origin be shifted to the point (α, β) and θ be the angle of rotation. Then we get the transformation as

$$\left.\begin{array}{l}x = x'\cos\theta - y'\sin\theta + \alpha \\ y = x'\sin\theta + y'\cos\theta + \beta.\end{array}\right\}$$

For unalteration of coordinates, we have

$$\left.\begin{array}{l}x = x\cos\theta - y\sin\theta + \alpha \\ y = x\sin\theta + y\cos\theta + \beta.\end{array}\right\}$$

Transformation of Coordinates

Now for the point $(4, 2)$, $x = 4, y = 2$, so the above transformation becomes

$$\left.\begin{array}{l} 4 = 4\cos\theta - 2\sin\theta + \alpha \\ 2 = 4\sin\theta + 2\cos\theta + \beta \end{array}\right\}$$

Now, $(4-\alpha)^2 + (2-\beta)^2 = (4\cos\theta - 2\sin\theta)^2$
$$+ (4\sin\theta + 2\cos\theta)^2 = 4^2 + 2^2$$

or, $\alpha^2 - 8\alpha + 4^2 + \beta^2 - 4\beta + 2^2 = 4^2 + 2^2$

or, $\alpha(\alpha - 8) + \beta(\beta - 4) = 0$

or, $\alpha(\alpha - 8) = -\beta(\beta - 4)$.

For rigid motion, $\alpha \neq 0, \beta \neq 0$. By trial $\alpha = 6, \beta = 6$ or $\alpha = 2, \beta = -2$.

Case 1: When $\alpha = 6, \beta = 6$,

$$4 = 4\cos\theta - 2\sin\theta + 6 \Rightarrow 2\cos\theta + 1 = \sin\theta$$
$$\Rightarrow (2\cos\theta + 1)^2 = 1 - \cos^2\theta$$

or, $5\cos^2\theta + 4\cos\theta = 0$ or, $\cos\theta(5\cos\theta + 4) = 0$ or, $\cos\theta = 0, -\dfrac{4}{5}$.

If $\cos\theta = 0, \sin\theta = 1$, these values do not satisfy $2 = 6 + 4\sin\theta + 2\cos\theta$.
To satisfy $4 = 6 + 4\cos\theta - 2\sin\theta$ and $2 = 6 + 4\sin\theta + 2\cos\theta$ we get $\cos\theta = -\dfrac{4}{5}$ and $\sin\theta = -\dfrac{3}{5}$. So the rigid motion is

$$\left.\begin{array}{l} x = -\frac{4}{5}x' + \frac{3}{5}y' + 6 \\ y = -\frac{3}{5}x' - \frac{4}{5}y' + 6. \end{array}\right\}$$

Case 2: When $\alpha = 2, \beta = -2$,

$$4 = 4\cos\theta - 2\sin\theta + 2 \Rightarrow 2\cos\theta - 1 = \sin\theta$$
$$\Rightarrow (2\cos\theta - 1)^2 = 1 - \cos^2\theta$$

or, $5\cos^2\theta - 4\cos\theta = 0$ or, $\cos\theta(5\cos\theta - 4) = 0$.

If $\cos\theta = 0, \sin\theta = 1$, these values satisfy $2 = 4\sin\theta + 2\cos\theta - 2$.
Also $\cos\theta = \dfrac{4}{5}$ and $\sin\theta = \dfrac{3}{5}$ satisfy $4 = 2 + 4\cos\theta - 2\sin\theta$ and $2 = -2 + 4\sin\theta + 2\cos\theta$. Thus the rigid motions are

$$\left.\begin{array}{l} x = -2y' + 2 \\ y = 4x' - 2 \end{array}\right\} \text{ and } \left.\begin{array}{l} x = \frac{4}{5}x' - \frac{3}{5}y' + 2 \\ y = \frac{3}{5}x' + \frac{4}{5}y' - 2. \end{array}\right\} \qquad \square$$

Example 1.10.13 *Find the old coordinates of the new origin if the point $(2, -3)$ lies on the new x-axis and the point $(1, 2)$ lies on the new y-axis and the corresponding axes of the old and new coordinate systems have the same directions.*

Solution: This problem is a translation only as the two systems have the same directions. Since the point $P(1, 2)$ lies on the new y-axis $O'Y'$, we get

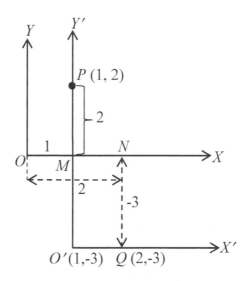

from the figure $OM = 1$. Again the point $Q(2, -3)$ lies on the new x-axis $O'X'$. Figure shows that $NQ = MO' = -3$.

From these we see that coordinates of new origin $O' \equiv (OM, MO') \equiv (1, -3)$. □

Example 1.10.14 *Find the acute angle through which the axes must be turned so that the equation $ax + by + c = 0$ may be reduced to the form $x' = $ constant and determine the value of this constant.*

Solution: Let the axes be turned through an angle θ, then

$$\left. \begin{array}{l} x = x' \cos \theta - y' \sin \theta \\ y = x' \sin \theta + y' \cos \theta. \end{array} \right\}$$

So the given equation becomes

$$a(x' \cos \theta - y' \sin \theta) + b(x' \sin \theta + y' \cos \theta) + c = 0$$
or, $\quad (a \cos \theta + b \sin \theta)x' + (-a \sin \theta + b \cos \theta)y' + c = 0.$

Transformation of Coordinates

This is of the form $x' = $ constant, if $-a\sin\theta + b\cos\theta = 0 \Rightarrow \tan\theta = \dfrac{a}{b}$.

$\Rightarrow \dfrac{\sin\theta}{b} = \dfrac{\cos\theta}{a} = \dfrac{1}{\sqrt{a^2+b^2}}$, $\therefore \sin\theta = \dfrac{b}{\sqrt{a^2+b^2}}$, $\cos\theta = \dfrac{a}{\sqrt{a^2+b^2}}$.

Therefore the transformed equation is

$$\left(\dfrac{a.a}{\sqrt{a^2+b^2}} + \dfrac{b.b}{\sqrt{a^2+b^2}}\right)x' + c = 0 \text{ or } \sqrt{a^2+b^2}\,x' = -c \text{ i.e., } x = \dfrac{-c}{\sqrt{a^2+b^2}}.$$

Hence the required value of the constant is $\dfrac{-c}{\sqrt{a^2+b^2}}$. □

Example 1.10.15 *Find the coordinates of the point where the origin is to be shifted so that the equation $3x^2 + 8xy + 3y^2 - 2x + 2y - 2 = 0$ can be reduced to one which is free from linear terms.*

Solution: Let the origin be shifted to the point (α, β), i.e., we have the translation formulae $x = x' + \alpha, y = y' + \beta$. Then the transformed equation becomes

$$3(x'+\alpha)^2 + 8(x'+\alpha)(y'+\beta) + 3(y'+\beta)^2 - 2(x'+\alpha)$$
$$+ 2(y'+\beta) - 2 = 0$$

or, $3x'^2 + 8x'y' + 3y'^2 + 2(3\alpha + 4\beta - 1)x' + 2(4\alpha + 3\beta + 1)y'$
$$+ 3\alpha^2 + 8\alpha\beta + 3\beta^2 - 2\alpha + 2\beta - 2 = 0.$$

This equation will be from first degree terms iff

$$\left.\begin{array}{l}3\alpha + 4\beta - 1 = 0 \\ 4\alpha + 3\beta + 1 = 0.\end{array}\right\}$$

On solving we get $\alpha = -1, \beta = 1$.

Hence the origin is to be shifted to the point $(-1, 1)$. □

Example 1.10.16 *By the rotation of axes after transferring the origin to the point $(2,3)$ an equation $3x^2 + 2xy + 3y^2 - 18x - 22y + 50 = 0$ is transformed to $4x^2 + 2y^2 = 1$. Find the inclination of the later axes to the former.*

Solution: Transferring the origin to $(2,3)$ without changing the directions of the axes, we have

$$3(x'+2)^2 + 2(x'+2)(y'+3) + 3(y'+3)^2 - 18(x'+2)$$
$$-22(y'+3) + 50 = 0$$

or, $3x'^2 + 2x'y' + 3y'^2 + 2(6 + 3 - 9)x' + 2(2 + 9 - 11)y'$
$$+ 12 + 12 + 27 - 36 - 66 + 50 = 0$$

$\Rightarrow 3x'^2 + 2x'y' + 3y'^2 - 1 = 0.$

Suppressing the dashes the new equation is

$$3x^2 + 2xy + 3y^2 - 1 = 0 \tag{1.7}$$

Let the axes now be turned through an angle θ. Since the resulting equation does not contain the product term xy, so $\tan 2\theta = \dfrac{2}{3-3} = \infty \therefore \theta = \dfrac{\pi}{4}$.
So the transformation gives

$$\left. \begin{array}{l} x = x' \cos \frac{\pi}{4} - y' \sin \frac{\pi}{4} = \frac{1}{\sqrt{2}}(x' - y') \\ y = x' \sin \frac{\pi}{4} + y' \cos \frac{\pi}{4} = \frac{1}{\sqrt{2}}(x' + y'). \end{array} \right\}$$

Substituting in equation (1.7), we get

$$\frac{3}{2}(x' - y')^2 + \frac{3}{2}(x' + y')^2 + (x' - y')(x' + y') - 1 = 0$$

or, $3(x'^2 + y'^2) + (x'^2 - y'^2) - 1 = 0$ or, $4x'^2 + 2y'^2 = 1$.

Suppressing the dashes again, as above, we get $4x^2 + 2y^2 = 1$. Hence the inclination of the later axes to the former is $\dfrac{\pi}{4}$. □

Example 1.10.17 *Find the coordinate transformation formulae if the point $(3, -4)$ lies on the new x-axis and the point $(1, -6)$ lies on the new y-axis and the corresponding axes of old and new coordinate systems have the same directions.*

Solution: Proceeding as in **Example 1.10.13**, the coordinates of the new origin are obtained as $(1, -4)$ and therefore the transformation formulae are

$$\left. \begin{array}{l} x = x' + 1 \\ y = y' - 4 \end{array} \right\}, \text{ as it is a simple translation.} \qquad \square$$

Example 1.10.18 *Find out the transformation under which the circle $x^2 + y^2 - 2x - 4 = 0$ is transformed to one with centre at $(0, 1)$.*

Solution: The equation of the given circle is written as

$$(x - 1)^2 + y^2 = 5 \tag{1.8}$$

The equation of the circle whose centre is $(0, 1)$ will be of the form

$$x^2 + (y - 1)^2 = \text{constant}.$$

Writing, by interchanging the values of x and y in the form $x - 1 = Y$, $y =$

Transformation of Coordinates

$X - 1$ and then applying the rotational formulae $\left.\begin{array}{l} X = x' \cos\theta - y' \sin\theta \\ Y = x' \sin\theta + y' \cos\theta \end{array}\right\}$

equation (1.8) gives

$$Y^2 + (X-1)^2 = 5$$
or, $(x' \sin\theta + y' \cos\theta)^2 + (x' \cos\theta - y' \sin\theta - 1)^2 = 5$
or, $x'^2 + y'^2 - 2x' \cos\theta + 2y' \sin\theta - 4 = 0.$

The centre of the circle is $(\cos\theta, -\sin\theta) \equiv (0, 1)$. Hence

$$\cos\theta = 0 \text{ and } \sin\theta = -1 \Rightarrow \theta = \frac{3\pi}{2}.$$

Therefore the coordinate transformation formulae are

$$\left.\begin{array}{l} x = Y + 1 = x' \sin\theta + y' \cos\theta + 1 = x' \sin\dfrac{3\pi}{2} + y' \cos\dfrac{3\pi}{2} + 1 \\ \qquad = -x' + 1 \\ y = X - 1 = x' \cos\theta - y' \sin\theta - 1 = x' \cos\dfrac{3\pi}{2} - y' \sin\dfrac{3\pi}{2} - 1 \\ \qquad = y' - 1. \end{array}\right\} \quad \Box$$

Remark: To check the answer put $x = -x' + 1, y = y' - 1$ in the given equation you will be satisfied.

1.11 Exercises

Section A: Objective Type Questions

1. Find the form of the equation $3x + 4y = 5$ due to change of origin to the point $(3, -2)$ only.

2. Find the equation of the line $y = \sqrt{3}x$ when the axes are rotated through an angle $\dfrac{\pi}{3}$.

3. Is the following transformation $x' = -x + a$, $y' = -y + b$ a rigid motion?

4. To what point the origin is to be transferred to get rid of the first degree terms from the equation $8x^2 + 10xy + 3y^2 + 26x + 16y + 21 = 0$?

5. By shifting the origin to the point (α, β) without changing the direction of axes, each of the equation $x - y + 3 = 0$ and $2x - y + 1 = 0$ is reduced to the form $ax' + by' = 0$. Find α and β.

6. Find the transformed form of the equation $x^2 - y^2 = 4$ when the axes are rotated through an angle of $45°$ keeping the origin fixed.

7. Transform the equation $3(12x - 5y + 39)^2 + 2(5x - 12y - 26)^2 = 169$ taking the lines $12x - 5y + 39 = 0$ and $5x - 12y - 26 = 0$ as the new axes of x and y respectively.

8. The coordinates of the axes are rotated through an angle of $\dfrac{\pi}{3}$. If the transformed coordinates of a point are $(2\sqrt{3}, -6)$, find the original coordinates.

9. Find the coordinate transformation formulae if the origin be shifted to the point $(-1, 2)$ without changing the direction of the axes.

10. The origin is shifted to the point $(1, -2)$ without changing the directions of the axes. Find the coordinates in the old coordinate system if its coordinates with respect to the new coordinate system are given as $(2, 4)$.

11. If keeping the origin fixed, the axes are rotated through an angle of $30°$, then show that the new coordinates of the point whose old coordinates were $(2, 4)$ are $(\sqrt{3} + 2, 2\sqrt{3} - 1)$.

12. If the origin be shifted to the point $(1, -2)$ without the rotation of the axes what do the following equations becomes?
    (i) $2x^2 + y^2 - 4x + 4y = 0$, (ii) $y^2 - 4x + 4y + 8 = 0$.

13. At what point should the origin be shifted if the coordinates of a point $(4, 5)$ become $(-3, 9)$?

14. Given the equation $4x^2 + 2\sqrt{3}xy + 2y^2 = 1$. Through what angle should the axes be rotated so that the term xy is removed form the transformed equation?

Transformation of Coordinates

15. The point $(2,3)$ is transformed to $(3,2)$ under a reflection about a line. Find the equation of the line.

[Hints: The required line is perpendicular to the line joining the points and it passes through the mid-point of the line joining the given points. The equation is $x = y$.]

16. Find the equations of the rigid motion that transforms $x^2 + 2x + y^2 - 10y + 25 = 0$ to a circle about the new origin as centre.

$\left[Hints: \text{The centre is } (-1, 5), \text{ so } \begin{array}{l} x = x' \cos\theta - y' \sin\theta - 1 \\ y = x' \sin\theta + y' \cos\theta + 5. \end{array} \right\}\right].$

17. Transform the point $(1, 2)$ to the point $(3, -2)$ by the repeated application of two transformations of which one is a translation and the other is a reflection.

[Hints: The line of reflection is $y = 2(x-2)$.]

18. Show that $x^2 + y^2$ remains invariant under rotation of the xy-plane.

19. Transform the point $(3, 4)$ to the point $(0, 5)$ by a rotation.

[Hints: If θ be the angle of rotation then $\left. \begin{array}{l} 0 = 3\cos\theta + 4\sin\theta \\ 5 = -3\sin\theta + 4\cos\theta \end{array} \right\}$. From these equations we get $\sin\theta = \dfrac{3}{5}$, $\cos\theta = \dfrac{4}{5}$.]

20. Is the transformation $x' = -x - a$, $y' = -y + b$ a rigid motion.

21. For what values of a, the transformation $x' = -x + 2$, $y' = ay + 3$ is a translation.

22. Is the transformation $x' = x, y' = y$ a rigid motion?

[Hints: No, it is the identity transformation.]

23. Transform the equation $y^2 - 2y = x$ with respect to parallel axes through $(-1, 1)$.

24. Determine the angle through which the axes must be rotated so that the equation $\sqrt{2}x + y + 6 = 0$ may reduce to the form $x = c$.

Section B: Broad Answer Type Questions

1. Three points $P(2, 4), Q(-1, 7)$ and $R(-2, -4)$ are given. Find their new coordinates when the axes are rotated through an angle of $-45°$.

2. The origin is shifted to the point $(3, -1)$ and the coordinate axes are rotated through an angle $\alpha = \tan^{-1}\dfrac{3}{4}$. Find the coordinates of the point $(5, -2)$ in the new coordinate system.

3. (i) Shift the origin to a suitable point so that the equation $y^2 + 4y + 8x - 2 = 0$ will not contain a term in y and the constant term.

(ii) Find the equation to which the equation $x^2 + 7xy - 2y^2 + 17x - 26y - 60 = 0$ is transformed if the origin is shifted to the point $(2, -3)$, the axes remaining parallel to the original axes.

(iii) What does the equation $2x^2 + 4xy - 5y^2 + 20x - 22y - 14 = 0$ become when referred to the rectangular axes through the point $(-2, -3)$, the new axes being inclined at an angle of $45°$ with the old axes?

4. (i) Find the angle through which the axes must be turned, so that the equation $lx + my + n = 0 (l \neq 0)$ may reduce to the form $ax + b = 0$.

(ii) If (x, y) and (x', y') be the coordinates of the same point referred to two sets of rectangular axes with same origin and if $x = kx' + ly'$; $y = k'y' + l'y'$, then prove that $kl' - k'l = 1$.

5. (i) Transform the equation $(ax + by + c)(bx - ay + d) = a^2 + b^2$ to new axes of x and y whose equations are $ax + by + c = 0$ and $bx - ay + d = 0$ respectively.

(ii) Transform the equation $11x^2 - 4xy + 14y^2 - 58x - 44y + 126 = 0$ to the new axes of x and y whose equations are $x - 2y + 1 = 0$ and $2x + y - 8 = 0$ respectively.

6. (i) Show that the transformed equation of the curve $(4x + 3y + 1)(3x - 4y + 2) = 75$ when the axes are $4x + 3y + 1 = 0$ and $3x - 4y + 2 = 0$ is $x'y' = 3$, where (x', y') are the new coordinates.

(ii) If the perpendicular straight lines $ax + by + c = 0$ and $bx - ay + c' = 0$ be taken as the axes of x and y respectively, then show that the equation $(ax + by + c)^2 - 2(bx - ay + c')^2 = 1$ will be transformed into $y'^2 - 2x'^2 = \dfrac{1}{a^2 + b^2}$.

7. (i) Transform the equation $17x^2 + 18xy - 7y^2 - 16x - 32y - 18 = 0$ to one in which there is no term involving x, y and xy, both sets of axes being rectangular.

(ii) If the origin is shifted to the point $(0, 1)$ and the axes be rotated through an angle $\dfrac{\pi}{4}$, then find the transformed form of the equation $5x^2 - 2xy + 5y^2 - 10y - 7 = 0$.

8. (i) If the origin is shifted to the point $\left(\dfrac{ab}{a-b}, 0\right)$ without rotation then prove that the equation $(a - b)(x^2 + y^2) - 2abx = 0$ becomes $(a - b)^2(x^2 + y^2) = a^2 b^2$.

(ii) Find out the transformation under which the circle $x^2 + y^2 - 2x - 4 = 0$ is transformed to one with centre at $(0, 1)$.

[*Hints:* See Worked Out **Example 1.10.18**]

9. (i) Give the definition of a rigid motion in a plane. Show that

$\begin{vmatrix} x_1 & y_1 & 1 \\ x_2 & y_2 & 1 \\ x_3 & y_3 & 1 \end{vmatrix}$ is an invariant in respect of three points $(x_i, y_i), (i = 1, 2, 3)$ under rigid motion in the plane.

Give a geometrical significance of the invariant.

(ii) If by an orthogonal transformation of coordinate axes the coordinates of two points $P(x_1, y_1)$ and $Q(x_2, y_2)$ become (x'_1, y'_1) and (x'_2, y'_2) respectively, then prove that $(x'_1 - x'_2)^2 + (y'_1 - y'_2)^2 = (x_1 - x_2)^2 + (y_1 - y_2)^2$.

Give its geometrical interpretation.

10. (i) If the origin is shifted to $P(7,3)$ and the coordinate axes are rotated so that the positive direction of the new x-axis coincides with that of vector $\overline{PQ}$, Q being the point $(10, 7)$, find the formulae for coordinate transformation.

[*Hints:* See Worked Out **Example 1.10.6**.]

(ii) Determine a rigid motion to transform the circle $x^2 + y^2 + 2y - 8 = 0$ into one with centre $(0,0)$.

[*Hints:* $x^2 + (y+1)^2 = 9$ where $\left.\begin{array}{l} x = x'\cos\theta - y'\sin\theta \\ y = -1 + x'\sin\theta + y'\cos\theta \end{array}\right\}$. Also See Worked Out **Example 1.10.18**.]

11. (i) If the quadratic expression $ax^2 + 2hxy + by^2 + 2gx + 2fy + c$ is changed to $a'x'^2 + 2h'x'y' + b'y'^2 + 2g'x' + 2f'y' + c'$ only by the transformation of rotation of coordinate axes, then prove the following invariants:

(a) $c' = c$ \hspace{2em} (b) $a' + b' = a + b$
(c) $a'b' - h'^2 = ab - h^2$ \hspace{2em} (d) $f'^2 + g'^2 = f^2 + g^2$
(e) $a'b'c' + 2f'g'h' - a'f'^2 - b'g'^2 - c'h'^2 = abc + 2fgh - af^2 - bg^2 - ch^2$
(f) $b'c' + c'a' + a'b' - f'^2 - g'^2 - h'^2 = bc + ca + ab - f^2 - g^2 - h^2$
(g) $2f'g'h' - a'f'^2 - b'g'^2 = 2fgh - af^2 - bg^2$ \hspace{1em} (h) $a' + b' + c' = a + b + c$.

(ii) When the axes are turned through angle, the expression $ax + by$ becomes $AX + BY$ referred to new axes, show that $a^2 + b^2 = A^2 + B^2$.

(iii) If, by a rotation of rectangular axes about the origin, the expression $ax^2 + 2hxy + by^2$ changes to $Ax^2 + 2Hxy + By^2$, then prove that $a+b = A+B$ and $ab - h^2 = AB - H^2$.

12. Show that the radius of a circle remains unchanged due to a rigid body motion.

13. (i) If by a rotation of axes about the origin, $(ax + by)$ and $(cx + dy)$ be changed to $(a'x' + b'y')$ and $(c'x' + d'y')$ respectively, then show that $ad - bc = a'd' - b'c'$.

(ii) If the expressions $ax^2 + 2hxy + by^2$ and $Ax^2 + 2Hxy + By^2$ are changed into $a'x'^2 + 2h'x'y' + b'y'^2$ and $A'x'^2 + 2H'x'y' + B'y'^2$ respectively by a rotation of coordinate axes then show that the two expressions $aA + bB + 2hH$ and $aA + bB - 2hH$ are invariants of rotation.

14. If the expressions $ax + by + c$ and $Ax + By + C$ are changed into $a'x' + b'y' + c'$ and $A'x' + B'y' + C'$ respectively by a rigid motion, show that $aA + bB = 0$ if and only if $a'A' + b'B' = 0$ and that when $aA + bB \neq 0$,
$$\frac{a'B' - A'b'}{a'A' + b'B'} = \frac{aB - Ab}{aA + bB}.$$

15. If a point (α, β) be changed into (α', β') and the expression $ax+by+c$ be changed to $a'x' + b'y' + c'$ by a rigid motion, show that
$$\left| \frac{a'\alpha' + b'\beta' + c'}{\sqrt{a'^2 + b'^2}} \right| = \left| \frac{a\alpha + b\beta + c}{\sqrt{a^2 + b^2}} \right|.$$

16. (i) If by a rotation of axes, without change of origin, the equation $ax^2 + 2hxy + by^2$ becomes $a'x'^2 + 2h'x'y' + b'y'^2$, prove that
$$(a - b)^2 + 4h^2 = (a' - b')^2 + 4h'^2.$$

(ii) By orthogonal transformation, without change of origin, if the equation $ax^2 + 2hxy + by^2 = c$ be changed into one in which there is no term involving xy, show that the transformed equation is
$$(a + b + \lambda)x^2 + (a + b - \lambda)y^2 = 2c, \text{ where } \lambda = \sqrt{(a-b)^2 + 4h^2}.$$

17. (i) Show that there is one and only one point whose coordinates do not alter due to a rigid motion.

(ii) Find the fixed point of the rigid motion
$$\left. \begin{array}{l} x = \frac{3}{5}x' - \frac{4}{5}y' - 1 \\ y = \frac{4}{5}x' + \frac{3}{5}y' - 1 \end{array} \right\}.$$

18. Show by rotating the axes through an angle $\frac{\pi}{4}$, the expression $ax + by + c = 0$ can be transformed to the form $x = $ constant, if $a = b$ and to the form $y = $ constant, if $a = -b$.

19. Find the angle of rotation about the origin which will transform the equation $\sqrt{3}(x^2 - y^2) - 2xy = 8$ into $x'y' = 2$.

Transformation of Coordinates

ANSWERS

Section A: 1. $3x' + 4y' = 4$; **2.** $y' = 0$; **3.** Yes, for the given equation can be written as $\left.\begin{array}{l} x = -x' + a = x'\cos\pi - y'\sin\pi + a \\ y = -y' + b = x'\sin\pi + y'\cos\pi + b \end{array}\right\}$ and so it is a rigid motion, the angle of rotation being π and the new origin is (a,b). More specific, it is a translation; **4.** $(-1,-1)$; **5.** $\alpha = 2, \beta = 5$; **6.** $x'y' = -2$; **7.** $2x'^2 + 3y'^2 = 169$; **8.** $(4\sqrt{3}, 0)$; **9.** $x = x' - 1, y = y' + 2$; **10.** $(3,2)$; **12. (i)** $2x'^2 + y'^2 = 6$; **(ii)** $y'^2 = 4x'$; **13.** $(7,-4)$; **14.** $\theta = \frac{\pi}{6}, \frac{2\pi}{3}$; **15.** $y = x$; **16.** $\left.\begin{array}{l} x = x'\cos\theta - y'\sin\theta - 1 \\ y = x'\sin\theta + y'\cos\theta + 5 \end{array}\right\}$; **20.** Yes, proceed for justification similarly as **Exercise 3** of this section; **21.** $a = \pm 1$; **22.** No.; **23.** $y^2 = x$; **24.** $\frac{\pi}{6}$.

Section B: 1. $P(-\sqrt{2}, 3\sqrt{2}), Q(-4\sqrt{2}, 3\sqrt{2}), R(\sqrt{2}, -3\sqrt{2})$; **2.** $(1,-2)$; **3. (i)** $(\frac{3}{4}, -2)$; **(ii)** $x'^2 + 7x'y' - 2y'^2 = 4$; **(iii)** $x'^2 - 14x'y' - 7y'^2 = 2$; **4.** $\tan^{-1}(\frac{m}{7})$; **5. (i)** $x'y' = 1$; **(ii)** $2x'^2 + 3y'^2 = 1$; **7. (i)** If the equation is transformed to rectangular axes through the point $(1,-1)$ inclined at an angle $\frac{1}{2}\tan^{-1}\frac{3}{4}$ to the original axes, the equation becomes $2x'^2 - y'^2 = 1$; **(ii)** $2x'^2 + 3y'^2 = 6$; **8. (ii)** Axes will be rotated through an angle $\frac{3\pi}{2}$, $x = -x' + 1, y = y' - 1$ [See Worked out **Example 1.10.18**]; **10. (i)** $\left.\begin{array}{l} x = \frac{3}{5}x' - \frac{4}{5}y' + 7 \\ y = \frac{4}{5}x' + \frac{3}{5}y' + 3 \end{array}\right\}$; **(ii)** $x^2 + (y+1)^2 = 9$, where $\left.\begin{array}{l} x = x'\cos\theta - y'\sin\theta \\ y = -1 + x'\sin\theta + y'\cos\theta, \end{array}\right\}$ θ can take any value; **17. (ii)** $(\frac{1}{2}, -\frac{3}{2})$.

Chapter 2

General Equation of Second Degree: Classification of Conics

2.1 Introduction

The general equation of second degree in x and y is of the form

$$ax^2 + 2hxy + by^2 + 2gx + 2fy + c = 0 \qquad (2.1)$$

has really five independent arbitrary constants, because it can be divided by any one of these constants. So five conditions are necessary to obtain the five constants uniquely. Curves satisfying this equation are curves of the second degree, called a *second-order curve* usually known as *conics*. Hence one conic can always be found to pass through five points, and, in general, only one such conic can be found. We shall see later in this chapter that in case of a parabola, $ab = h^2$, in case of a rectangular hyperbola $a+b = 0$ etc. So there are only four constants in these cases. Moreover, an exceptional case occurs when as many as four points are collinear. This chapter is mainly devoted to find the nature of the locus of a point where coordinates satisfy equation (2.1).

2.2 Some Definitions

2.2.1 Centre of a second order curve

A point C is called centre of the second order curve (2.1), if for any point P on the curve, there exists a point P' on the curve as well as well on the line CP, such that $|CP'| = |CP|$. This means that a curve is symmetrical about its centre.

Accordingly the curve (2.1) has a single centre or it has infinitely many centres or even it has no centres and we classify (2.1) into two categories, viz., **(a)** *a central curve* and **(b)** *a non central curve* defined as under.

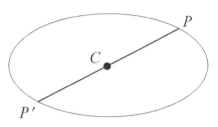

Figure 2.1

(a) Central Curve

A curve is said to be central, if it has only one centre.

(b) Non Central Curve

A curve which is not central is called non-central. Therefore, non-central curve has either no centre or infinitely many centers.

Example 2.2.1 *Give an example of a central conic.*

Solution: Let us consider the curve

$$3x^2 - 4xy + 2y^2 + 4x - 2y - 7 = 0 \tag{2.2}$$

Shifting the origin to the point $O'\left(-1, -\dfrac{1}{2}\right)$, i.e., taking the translation $x = x' - 1, y = y' - \dfrac{1}{2}$ we get

$$3(x'-1)^2 - 4(x'-1)\left(y' - \dfrac{1}{2}\right) + 2\left(y' - \dfrac{1}{2}\right)^2 + 4(x'-1)$$
$$-2\left(y' - \dfrac{1}{2}\right) - 7 = 0$$

$$\Rightarrow \quad 3x'^2 - 4x'y' + 2y'^2 + 2(-3+1+2)x' + 2(2-1-1)y' +$$
$$3.(-1)^2 - 4(-1)\left(-\dfrac{1}{2}\right) + 2\left(-\dfrac{1}{2}\right)^2 + 4(-1) - 2\left(-\dfrac{1}{2}\right) - 7 = 0$$

General Equation of Second Degree 57

$$\Rightarrow 3x'^2 - 4x'y' + 2y'^2 - 8\frac{1}{2} = 0$$
$$\Rightarrow 6x'^2 - 8x'y' + 4y'^2 - 17 = 0 \qquad (2.3)$$

Now, let $P(x_1, y_1)$ be a point on the curve and P' be the point $(-x_1, -y_1)$ lying on the line $O'P$, then $|O'P| = \sqrt{x_1^2 + y_1^2}$ and $|O'P'| = \sqrt{x_1^2 + y_1^2}$. So $|O'P'| = |O'P|$. Further P' is a point on (2.3) because its coordinates satisfy the equation (2.3). Thus O' is a point such that when P is a point on (2.3), another point $P'(\neq P)$ lying on $O'P$ and satisfying $|O'P'| = |O'P|$ is also a point on (2.3). Hence O' is a centre of the curve (2.3). In other words, $\left(-1, -\frac{1}{2}\right)$ is the centre of the curve (2.2). □

2.3 Determination of Centre of a Second Order Curve

The following theorem will provide the determination of centre of a second order curve.

Theorem 2.3.1 *A point (α, β) is a centre of the curve represented by $ax^2 + 2hxy + by^2 + 2gx + 2fy + c = 0$, if and only if it satisfies the two equations $ax + hy + g = 0$ and $hx + by + f = 0$.*

Proof: Condition Necessary:

Let us first suppose that the coordinates (α, β) is a centre of the given second order curve. Shifting the origin to this point, i.e., applying the translation $x = x' + \alpha, y = y' + \beta$ we get

$$a(x' + \alpha)^2 + 2h(x' + \alpha)(y' + \beta) + b(y' + \beta)^2 + 2g(x' + \alpha)$$
$$+ 2f(y' + \beta) + c = 0$$
$$\Rightarrow ax'^2 + 2hx'y' + by'^2 + 2(a\alpha + h\beta + g)x' + 2(h\alpha + b\beta + f)y'$$
$$+ a\alpha^2 + 2h\alpha\beta + b\beta^2 + 2g\alpha + 2f\beta + c = 0$$

which can again be written as

$$\Rightarrow ax'^2 + 2hx'y' + by'^2 + 2(a\alpha + h\beta + g)x' + 2(h\alpha + b\beta + f)y'$$
$$+ (a\alpha + h\beta + g)\alpha + (h\alpha + b\beta + f)\beta + (g\alpha + f\beta + c) = 0 \quad (2.4)$$

In new coordinates $(0, 0)$ is a centre. So (2.4) has a centre $(0, 0)$, i.e., equation (2.4) will be symmetric about $(0, 0)$. So, if (x'_1, y'_1) is a point on (2.4),

$(-x'_1, -y'_1)$ is also a point on (2.4) i.e., we get

$$ax'^2_1 + 2hx'_1y'_2 + by'^2_1 + 2(a\alpha + h\beta + g)x'_1 + 2(h\alpha + b\beta + f)y'_1$$
$$+(a\alpha + h\beta + g)\alpha + (h\alpha + b\beta + f)\beta + (g\alpha + f\beta + c) = 0 \quad (2.5)$$
and $ax'^2_1 + 2hx'_1y'_2 + by'^2_1 - 2(a\alpha + h\beta + g)x'_1 - 2(h\alpha + b\beta + f)y'_1$
$$+(a\alpha + h\beta + g)\alpha + (h\alpha + b\beta + f)\beta + (g\alpha + f\beta + c) = 0 \quad (2.6)$$

Subtracting (2.6) from (2.5) we get

$$4(a\alpha + h\beta + g)x'_1 + 4(h\alpha + b\beta + g)y'_1 = 0$$
$$\Rightarrow (a\alpha + h\beta + g)x'_1 + (h\alpha + b\beta + g)y'_1 = 0.$$

Since this is true for any point (x'_1, y'_1) on the curve, so we get

$$a\alpha + h\beta + g = 0 \text{ and } h\alpha + b\beta + g = 0.$$

Thus (α, β) satisfies the equations $ax + hy + g = 0$ and $hx + by + f = 0$.

Condition Sufficient:

Let (α, β) satisfy the equations $ax + hy + g = 0$ and $hx + by + f = 0$. Therefore, $a\alpha + h\beta + g = 0$ and $h\alpha + b\beta + g = 0$. Thus equation (2.4) becomes

$$ax'^2 + 2hx'y' + by'^2 + g\alpha + f\beta + c = 0 \quad (2.7)$$

So if (x'_1, y'_1) lies on (2.7) then $(-x'_1, -y'_1)$ also lies on (2.7). Hence $(0,0)$ is a centre of (2.7). Thus (α, β) is a centre of the given conic (2.4). This completes the proof. □

Corollary 2.3.1 *If the origin be the centre of a conic represented by the general second degree curve*

$$ax^2 + 2hxy + by^2 + 2gx + 2fy + c = 0 \quad (2.8)$$

then the coefficients of the first degree terms in the equation are all zero.

Proof: Let the centre of the conic be at the origin O, then every chord through O is bisected at O. So if POQ be a chord and the coordinates of P be (x_1, y_1) and those of Q will be $(-x_1, -y_1)$. So both the points lie on (2.8). Hence, we have

$$ax_1^2 + 2hx_1y_1 + by_1^2 + 2gx_1 + 2fy_1 + c = 0 \quad (2.9)$$
$$\text{and } ax_1^2 + 2hx_1y_1 + by_1^2 - 2gx_1 - 2fy_1 + c = 0 \quad (2.10)$$

General Equation of Second Degree

Subtracting we get $4gx_1 + 4fy_1 = 0 \Rightarrow gx_1 + fy_1 = 0$ and this is true for all values of x_1 and y_1, provided the point (x_1, y_1) lies on (2.8). Therefore $g = 0$ and $f = 0$ and the equation (2.8) reduces to

$$ax^2 + 2hxy + by^2 + c = 0$$

which completes the proof. □

Note 2.3.1 *The general equation of second degree in x, y, given by*

$$ax^2 + 2hxy + by^2 + 2gx + 2fy + c = 0$$

can be written as

$$(ax + hy + g)x + (hx + by + f)y + (gx + fy + c) = 0.$$

Such writing will often be found to be useful.

Centre of the curve: *It follows from the above theorem that the curve represented by*

$$ax^2 + 2hxy + by^2 + 2gx + 2fy + c = 0$$

is a central second order curve if the system of equations $ax + hy + g = 0$ and $hx + by + f = 0$ has unique solution.

If now $ab - h^2 \neq 0$, this system of equations is consistent and determinate, that is, it has a unique solution. In that case, the coordinates of the centre can be found as $\left(\dfrac{hf - bg}{ab - h^2}, \dfrac{gh - af}{ab - h^2} \right).$

Note 2.3.2 *If $ab - h^2 = 0$, then the centre is at infinity and the conic, in fact, is a parabola. On the other hand, if $ab - h^2 \neq 0$, then the centre is at finite distance and the conic is either an ellipse or a hyperbola.*

Note 2.3.3 *For a conic having infinitely many centers, $ab - h^2 = 0$, $hf - bg = 0$ and $gh - af = 0$.*

Note 2.3.4 *We write* $\Delta = \begin{vmatrix} a & h & g \\ h & b & f \\ g & f & c \end{vmatrix}$ *and* $\delta = ab - h^2 = \begin{vmatrix} a & h \\ h & b \end{vmatrix}$. *Again using the notations of capital letters for co-factors of Δ, the coordinates of the centre can be written as* $\left(\dfrac{G}{C}, \dfrac{F}{C} \right), C \neq 0$ *where C has the same expression as δ, used earlier.*

Note 2.3.5 *For a central conic, $\Delta \neq 0$ and $\delta \neq 0$. The conic, in this case, is either an ellipse or a hyperbola.*

Note 2.3.6 *When $ab - h^2 = 0$ and $(hf - bg, gh - af) \neq (0,0)$, then the system has no solution $\Rightarrow$ there exists no centre $\Rightarrow$ the curve is non central.*
In the geometrical point of view the given equations

$$\left. \begin{array}{l} ax + hy + g = 0 \\ hx + by + f = 0 \end{array} \right\}$$

represent straight lines and so will have unique solution, infinitely many solutions or no solutions according as the lines are intersecting, coincident or parallel i.e., according as

$$\frac{a}{h} \neq \frac{h}{b}, \quad \frac{a}{h} = \frac{h}{b} = \frac{g}{f}, \quad \frac{a}{h} = \frac{h}{b} \neq \frac{g}{f}.$$

Central conics are (i) Ellipse, (ii) Hyperbola and (iii) Pair of intersecting straight lines.

Non-central conics are (i) Parabola, (ii) Pair of parallel straight lines and (iii) Pair of coincident straight lines.

2.4 Solution of Simultaneous Equations Using Matrix Notation

Let the equations be

$$\left. \begin{array}{l} ax + hy + g = 0 \\ hx + by + f = 0 \end{array} \right\} \tag{2.11}$$

The coefficient matrix is $\begin{pmatrix} a & h \\ h & b \end{pmatrix}$ and the augmented matrix is $\begin{pmatrix} a & h & g \\ h & b & f \end{pmatrix}$

Now we discuss the following cases:

Case 1: Let $ab - h^2 \neq 0$. Then rank of the coefficient matrix = rank of the augmented matrix = number of unknowns = 2. So the system (2.11) has a unique solution, namely $x = \dfrac{hf - bg}{ab - h^2}$, $y = \dfrac{hf - bg}{ab - h^2}$.

In this case, the curve represented by the general second order equation $ax^2 + 2hxy + by^2 + 2gx + 2fy + c = 0$ is a central curve and its centre is given by $\left(\dfrac{hf - bg}{ab - h^2}, \dfrac{hf - bg}{ab - h^2} \right)$.

General Equation of Second Degree

Case 2: Let $ab - h^2 = 0$, $hf - bg = 0$ and $gh - af = 0$. Thus rank of the coefficient matrix = rank of the augmented matrix = $1 <$ number of unknowns = 2. Here the system (2.11) has infinitely many solutions $\Rightarrow$ the second degree curve has infinitely many solutions, which implies the curve is non-central.

Case 3: Let $ab - h^2 = 0$, $hf - bg \neq 0$ and $gh - af \neq 0$. Then the rank of the coefficient matrix = 1 and the rank of the augmented matrix = 2. So the simultaneous system of equations (2.11) has no solutions. Therefore the given curve is non-central.

2.5 Canonical Form of an Equation of a Second Degree Curve

The general equation of a second degree can be reduced to the standard equation of a conic by choosing suitable coordinate system with the application of translation or rotation or both. The standard equation is called its *canonical equation*. The form of the canonical equation is known as its *canonical form* or sometimes called its *normal form*.

To get the canonical form from the general equation of second degree, the following transformations are required.

(i) to remove the term in xy from the equation, by rotating the axes suitably.

(ii) to remove the terms containing x, y or both by translation of axes, i.e., by choosing new origin and

(iii) to remove the constant term also, if possible, by transformation of coordinates.

The following are the canonical forms of the conics:

1. The canonical form of the equation of a parabola is $y^2 = 4ax$, x-axis is its axis, $(0,0)$, the vertex, $(a,0)$ is the focus and the length of latus rectum is $4a$.

2. The canonical form of the equation of an ellipse is $\dfrac{x^2}{a^2} + \dfrac{y^2}{b^2} = 1$. Its centre is the origin $(0,0)$, x-axis is the major axis, y-axis is the minor axis, $(\pm a, 0)$ are the vertices, $(\pm ae, 0)$ are the foci, the length of latus rectum is $\dfrac{2b^2}{a}$ and the eccentricity e is given by $b^2 = a^2(1 - e^2)$ where $0 < e < 1$.

3. The canonical form of the equation of an hyperbola is $\dfrac{x^2}{a^2} - \dfrac{y^2}{b^2} = 1$, where transverse and conjugate axes are along the x-axis and y-axis respectively, the centre is the origin $(0,0)$, the vertices are $(\pm a, 0)$, $(\pm ae, 0)$ are the foci, the length of latus rectum is $\dfrac{2b^2}{a}$ and the eccentricity e is given by $b^2 = a^2(e^2 - 1)$ where $e > 1$.

Now we shall prove the following important theorem.

Theorem 2.5.1 *Every equation of second degree in x and y represents a conic.*

Proof: Let us consider the general equation of second degree in x and y given by

$$ax^2 + 2hxy + by^2 + 2gx + 2fy + c = 0 \qquad (2.12)$$

where a, b, h are not all zero at a time.

First of all, we shall remove the term containing xy from the equation (2.12). For this, let the axes of the coordinate system be rotated through an angle θ, in anticlockwise sense, keeping the directions of the axes unaltered, i.e., we apply the rotational formulae

$$\left. \begin{array}{l} x = x' \cos\theta - y' \sin\theta \\ y = x' \sin\theta + y' \cos\theta. \end{array} \right\}$$

where (x', y') are the new coordinates. Then equation (2.12) becomes

$$a(x'\cos\theta - y'\sin\theta)^2 + 2h(x'\cos\theta - y'\sin\theta)(x'\sin\theta + y'\cos\theta)$$
$$+ b(x'\sin\theta + y'\cos\theta)^2 + 2g(x'\cos\theta - y'\sin\theta)$$
$$+ 2f(x'\sin\theta + y'\cos\theta) + c = 0$$

$$\Rightarrow (a\cos^2\theta + h\sin 2\theta + b\sin^2\theta)x'^2 + \{2h\cos 2\theta - (a-b)\sin 2\theta\}x'y'$$
$$+ (a\sin^2\theta - h\sin 2\theta + b\cos^2\theta)y'^2 + 2(g\cos\theta + f\sin\theta)x'$$
$$+ 2(f\cos\theta - g\sin\theta)y' + c = 0 \qquad (2.13)$$

Now we choose θ such that the coefficient of the term $x'y'$ vanishes, i.e.,

$$2h\cos 2\theta - (a-b)\sin 2\theta = 0 \;\Rightarrow\; \tan 2\theta = \dfrac{2h}{a-b}$$

or, $\quad \theta = \dfrac{1}{2}\tan^{-1}\left(\dfrac{2h}{a-b}\right) \qquad (2.14)$

General Equation of Second Degree

Now since an angle can be found whose tangent is equal to any real number whatsoever, the value of θ satisfying (2.14) the values of $\sin\theta$ and $\cos\theta$ are found and substituting these values of $\sin\theta$ and $\cos\theta$ in (2.13), it reduces to the following form

$$Ax'^2 + By'^2 + 2Gx' + 2Fy' + C = 0 \qquad (2.15)$$

where by the theorem of invariants, we have

$$A + B = a + b \text{ and } D = AB = ab - h^2 \ (\because \text{ here } H = 0) \text{ and}$$
$$\Delta = abc + 2fgh - af^2 - bg^2 - ch^2 = ABC - AF^2 - BG^2 \ (\because H = 0)$$

Now we have the following cases:

Case 1: Let $ab - h^2 \neq 0$, then $AB \neq 0 \Rightarrow$ then neither A nor B is zero. In that case equation (2.15) can be written as

$$A\left(x'^2 + \frac{2G}{A}x' + \frac{G^2}{A^2}\right) + B\left(y'^2 + \frac{2F}{B}y' + \frac{F^2}{B^2}\right) = \frac{G^2}{A} + \frac{F^2}{B} - C$$

$$\Rightarrow A\left(x' + \frac{G}{A}\right)^2 + B\left(y' + \frac{F}{B}\right)^2 = \frac{G^2}{A} + \frac{F^2}{B} - C = k(\text{say}).$$

Let us now shift the origin to the point $\left(-\frac{G}{A}, -\frac{F}{B}\right)$ without changing the directions of the axes, the equation becomes

$$AX^2 + BY^2 = K \qquad (2.16)$$

where $K = k = \dfrac{G^2}{A} + \dfrac{F^2}{B} - C = \dfrac{1}{AB}\{BG^2 + AF^2 - ABC\} = \dfrac{-\Delta}{AB}$. Here again following sub cases arise:

Subcase 1: Let $K \neq 0$ i.e., $\Delta \neq 0$ and $AB > 0$, the equation (2.16) takes the form

$$\frac{X^2}{\frac{K}{A}} + \frac{Y^2}{\frac{K}{B}} = 1$$

writing $a^2 = \dfrac{K}{A}, b^2 = \dfrac{K}{B}$ we have the following three forms:

(a) $\dfrac{x^2}{a^2} + \dfrac{y^2}{b^2} = 1$, where A and B are both positive which is an ellipse.

(b) $\dfrac{x^2}{a^2} + \dfrac{y^2}{b^2} = -1$, where A and B are both negative which is a conic without a real trace, sometimes called an imaginary ellipse.

(c) $x^2 + y^2 = a^2$ when $A = B$, which is a circle.

Subcase 2: Let $K \neq 0$ and $AB < 0$, the equation (2.16) takes the form

(a) $\dfrac{x^2}{a^2} - \dfrac{y^2}{b^2} = 1$, where $a^2 = \dfrac{K}{A}$ and $b^2 = \dfrac{K}{B}$ and B which is a hyperbola,

or,

(b) $-\dfrac{x^2}{a^2} + \dfrac{y^2}{b^2} = 1$, which is again a hyperbola.

Subcase 3: Let $K = 0$, so that $\Delta = 0$ and $D = AB > 0$, then (2.16) takes the form $AX^2 + BY^2 = 0$ or of the form $\dfrac{x^2}{a^2} + \dfrac{y^2}{b^2} = 0$ where $a^2 = \dfrac{1}{A}$ and $b^2 = \dfrac{1}{B}$ which is a null ellipse having only one real point $(0,0)$. Also we can say that it represents a pair of imaginary straight lines.

Subcase 4: Let $K = 0$, so that $\Delta = 0$ and $D = AB < 0$, then (2.16) takes the form $\dfrac{x^2}{a^2} - \dfrac{y^2}{b^2} = 0$ where $a^2 = \dfrac{1}{A}$ and $b^2 = \dfrac{1}{B}$ which represents a pair of straight lines.

Case 2: Let one of A or B be zero. Let us suppose that $A = 0$, so that $D = AB = 0$. Then equation (2.15) takes the form

$$By'^2 + 2Gx' + 2Fy' + C = 0$$

or, $B\left(y' + \dfrac{F}{B}\right)^2 = -2Gx' + \dfrac{F^2 - BC}{B} = -2G\left(x' - \dfrac{F^2 - BC}{2BG}\right)$ (2.17)

Now shifting the origin to the point $\left(\dfrac{F^2 - BC}{2BG}, -\dfrac{F}{B}\right)$, without changing the directions of the axes, the equation reduces to $BY^2 = -2GX$ which is a parabola with the straight line $Y = 0$ as its axis, i.e., the axis is parallel to x'-axis.

Next if G is also zero, then equation (2.15) becomes

$$By'^2 + 2Fy' + C = 0 \text{ or, } B\left(y' + \dfrac{F}{B}\right)^2 = \dfrac{F^2 - BC}{B} \text{ or, } y' + \dfrac{F}{B} = \pm\dfrac{\sqrt{F^2 - BC}}{B}$$

which represents a pair of straight lines. Thus we see that in all cases the general equation of second degree represents a conic, either proper or degenerate. □

We also mention here the following definition:

Definition 2.5.1 *The general second degree equation represented by the equation (2.12) will be a degenerate or non-degenerate conic according as*

$$\Delta = abc + 2fgh - af^2 - bg^2 - ch^2 = 0 \text{ or } \neq 0.$$

The non-degenerate conics are mainly divided into three classes: (a) elliptic if $D > 0$, (b) parabolic if $D = 0$ and (c) hyperbolic if $D < 0$, where $D = AB = ab - h^2 = \delta$. Circle is a special case of the ellipse when the major and minor axes are equal.

Note 2.5.1 *When $h^2 = ab$, the terms of the second degree in equation (2.12) form a perfect square.*

Note 2.5.2 *When $A = 0$ and $B = 0$, then $a = 0, b = 0, h = 0$ and the equation (2.12) represents a line.*

Note 2.5.3 *If $A = 0, G = 0$, but $B \neq 0$ then equation (2.15) reduces to*

$$By'^2 + 2Fy' + C = 0 \Rightarrow \left(y' + \frac{F}{B}\right)^2 = \frac{F^2 - BC}{B^2}.$$

Here the equation represents a pair of parallel lines, a pair of coincident line or no geometric locus according as $F^2 - BC >, =, < 0$.

Note 2.5.4 *If $A \neq 0, B = 0, F = 0$ then equation (2.15) reduces to*

$$Ax'^2 + 2Gx' + C = 0 \text{ or, } \left(x' + \frac{G}{A}\right)^2 = \frac{G^2 - AC}{A^2}.$$

Here, also the equation (2.12) represents a pair of parallel lines, a pair of coincident lines or no geometric locus according as $G^2 - AC >, =, < 0$.

2.6 Reduction of an Equation of a Conic to Canonical Form

A general equation of second degree in x and y represented by

$$ax^2 + 2hxy + by^2 + 2gx + 2fy + c = 0 \tag{2.18}$$

is always a conic.

We consider two different cases:

Case 1: *Let the conic be a central conic, i.e., $\delta = ab - h^2 \neq 0$.*

If (α, β) be the centre of the conic (2.18), we shift the origin to this point, i.e., we apply the translation $x = x' + \alpha, y = y' + \beta$ without rotating the axes. Then equation (2.18) becomes

$$ax'^2 + 2hx'y' + by'^2 + 2(a\alpha + h\beta + g)x' + 2(h\alpha + b\beta + f)y' +$$
$$a\alpha^2 + 2h\alpha\beta + b\beta^2 + 2g\alpha + 2f\beta + c = 0$$

or, $\quad ax'^2 + 2hx'y' + by'^2 + 2(a\alpha + h\beta + g)x' + 2(h\alpha + b\beta + f)y' +$
$\quad (a\alpha + h\beta + g)\alpha + (h\alpha + b\beta + f)\beta + (g\alpha + f\beta + c) = 0 \quad (2.19)$

The equation is now referred to the new system (x', y') and its centre is at (α, β)

$$\therefore \quad \left.\begin{array}{l} a\alpha + h\beta + g = 0 \\ h\alpha + b\beta + f = 0 \end{array}\right\} \quad (2.20)$$

Solving these we get $\alpha = \dfrac{hf - bg}{ab - h^2}, \quad \beta = \dfrac{gh - af}{ab - h^2}.$

The equation (2.19) then becomes

$$ax'^2 + 2hx'y' + by'^2 + g\alpha + f\beta + c = 0.$$

Now $\quad g\alpha + f\beta + c = g\left(\dfrac{hf - bg}{ab - h^2}\right) + f\left(\dfrac{gh - af}{ab - h^2}\right) + c$

$$= \dfrac{1}{ab - h^2}[abc + 2fgh - af^2 - bg^2 - ch^2] = \dfrac{\Delta}{\delta} = d \text{ (say)}$$

and hence the equation (2.19), i.e., reduced equation of (2.18) is

$$ax'^2 + 2hx'y' + by'^2 + d = 0 \quad (2.21)$$

Now applying rotation of the coordinate axes through an angle

$$\theta = \dfrac{1}{2}\tan^{-1}\dfrac{2h}{a - b},$$

let the equation (2.21) becomes

$$AX^2 + BY^2 + d = 0 \quad (2.22)$$

where by the rule of invariant, we have

$$A + B = a + b \text{ and } AB = ab - h^2 \neq 0 \Rightarrow A \neq 0, B \neq 0.$$

From these we can find A and B and thereafter equation (2.22) will give the final transformed equation, which is in its *canonical form* or *normal form* or *standard form*.

General Equation of Second Degree

Note 2.6.1 *We observe that the coefficients of the second degree terms in the reduced equation (2.21) are the same as in the original equation (2.18), the linear terms are removed and the constant term c has been changed to $d = \dfrac{\Delta}{\delta}$, where*

$$\Delta = abc + 2fgh - af^2 - bg^2 - ch^2 = \begin{vmatrix} a & h & g \\ h & b & f \\ g & f & c \end{vmatrix} \text{ and } \delta = ab - h^2 = \begin{vmatrix} a & h \\ h & b \end{vmatrix}.$$

Note 2.6.2 *The axes of the conic are along $X = 0$ and $Y = 0$*

i.e., along $(x - \alpha) \cos \theta + (y - \beta) \sin \theta = 0$
and along $(x - \alpha) \sin \theta - (y - \beta) \cos \theta = 0$

where $\theta = \dfrac{1}{2} \tan^{-1} \dfrac{2h}{a - b}$.

Case 2: Let the given conic be a non-central, i.e., $\delta = ab - h^2 = 0$. (Here the conic will be called a parabolic type.)

As $ab - h^2 = 0$, the second degree terms in (2.18) together reduce to a perfect square. Let us write (2.18) as

$$(lx + my)^2 + 2gx + 2fy + c = 0$$

or $(lx + my + n)^2 = -2gx - 2fy - c + 2lnx + 2mny + n^2$
$= 2(ln - g)x + 2(mn - f)y + (n^2 - c)$ (2.23)

Now we have the following subcases:

Subcase 1: Let $\dfrac{g}{l} = \dfrac{f}{m} = n$, then equation (2.23) becomes

$$(lx + my + n)^2 = n^2 - c \tag{2.24}$$

If we consider $lx + my + n = 0$ as the x'-axis and any line perpendicular to it as the y'-axis then

$$y' = \pm \dfrac{lx + my + n}{\sqrt{l^2 + m^2}}$$

and x' depends on the choice of y'-axis. So the equation (2.24) becomes

$$(l^2 + m^2) y'^2 = n^2 - c.$$

It represents a pair of parallel straight lines, coincident straight lines or will have no real trace according as $n^2 >, =$ or $< c$.

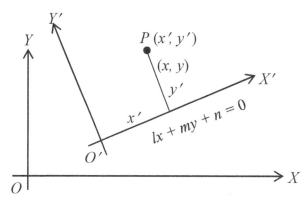

Figure 2.2

Subcase 2: Let $\dfrac{g}{l} \neq \dfrac{f}{m}$. Let us choose n such that the lines

$$lx + my + n = 0 \tag{2.25}$$
$$\text{and } 2(ln - g)x + 2(mn - f)y + n^2 - c = 0 \tag{2.26}$$

are at right angles. For this we choose n such that

$$\left(-\dfrac{l}{m}\right)\left(-\dfrac{ln-g}{mn-f}\right) = -1 \;\Rightarrow\; l^2 n - lg = -m^2 n + mf$$

or, $(l^2 + m^2)n = lg + mf$ whence $n = \dfrac{lg + mf}{l^2 + m^2}$.

If we choose now equation (2.25) as x'-axis and (2.26) as the y'-axis, then

$$x' = \dfrac{2(ln - g)x + 2(mn - f)y + n^2 - c}{2\sqrt{(ln - g)^2 + (mn - f)^2}} \quad \text{and} \quad y' = \dfrac{lx + my + n}{\sqrt{l^2 + m^2}}$$

where the new coordinates of $P(x, y)$ on the conic be (x', y'). So the equation (2.23) becomes

$$(l^2 + m^2)y'^2 = 2\sqrt{(ln-g)^2 + (mn-f)^2}\, x' \quad \text{i.e., } y'^2 = 4kx' \tag{2.27}$$

where $k = \dfrac{\sqrt{(ln-g)^2 + (mn-f)^2}}{2(l^2 + m^2)}$.

Equation (2.27) is the canonical form of the given conic (2.18) which represents a parabola of latus rectum $4|k|$ and the axis along the line $x' = 0$, i.e., along the new axis $lx + my + n = 0$.

2.7 Rank and Classification of Second Order Curves

The rank of a second order curve

$$ax^2 + 2hxy + by^2 + 2gx + 2fy + c = 0$$

is the rank of the matrix $\begin{bmatrix} a & h & g \\ h & b & f \\ g & f & c \end{bmatrix}$. The rank of the matrix may be 3, 2 or 1.

When rank = 3, the curve is *non-singular* or *non-degenerate*.

When rank = 2, the curve is *singular*.

When rank = 1, the curve is *degenerate*.

A circle, a parabola, an ellipse and a hyperbola are non-singular.
A point ellipse, a pair of intersecting or parallel lines are singular.
A pair of coincident straight lines are degenerate.

A complete classification of second order curve according to the different possible values of δ and Δ, the corresponding canonical forms and there corresponding names are given in a tabular form in **Table 2.1**.

2.8 Conic through Common Points

There are two cases:

(a) Conic through the common points of two conics

If $S = 0$ and $S' = 0$ be two conics, then the equation of a conic through their common points of intersection is given by $S + \lambda S' = 0$, where λ is an arbitrary constant. It is determined by the given condition.

(b) Conic through the common points of a conic and two lines

If $S = 0$ is the conic and the lines are $L_1 = 0$ and $L_2 = 0$, then the equation of the required conic is $S + \lambda L_1 L_2 = 0$, λ being an arbitrary constant. It is determined by the given condition.

Table 2.1: **Matrix classification of a second order curve**

δ	Δ	Canonical form	Name	Rank	Class
$\delta > 0$	$\Delta < 0$	$\frac{x^2}{a^2} + \frac{y^2}{b^2} = 1$	Ellipse	3	Non-singular
$\delta > 0$	$\Delta < 0$	$x^2 + y^2 = 1$	Circle	3	Non-singular
$\delta > 0$	$\Delta > 0$	$\frac{x^2}{a^2} + \frac{y^2}{b^2} = -1$	No geometric locus (Imaginary Ellipse)	3	Non-singular
$\delta > 0$	$\Delta = 0$	$Ax^2 + By^2 = 0$	A pair of imaginary lines or point ellipse	2	Singular
$\delta < 0$	$\Delta > 0$	$\frac{x^2}{a^2} - \frac{y^2}{b^2} = 1$	Hyperbola	3	Non-singular
$\delta < 0$	$\Delta < 0$	$\frac{x^2}{a^2} - \frac{y^2}{b^2} = -1$	Hyperbola	3	Non-singular
$\delta < 0$	$\Delta = 0$	$y^2 - k^2 x^2 = 0$	Pair of intersecting lines	2	Singular
$\delta = 0$	$\Delta \neq 0$	$y^2 = 4ax, x^2 = 4by$	Parabola	3	Non-singular
$\delta = 0$	$\Delta = 0$	$y^2 = k^2, x^2 = l^2$	Pair of parallel lines	2	Singular
$\delta = 0$	$\Delta = 0$	$y^2 = 0, x^2 = 0$	Pair of coincident lines	1	Singular

2.9 Conditions for Pair of Straight Lines and proper Conics

A. Necessary and sufficient conditions for a general equation of second degree to represent a pair of straight lines

Let the general equation of second degree be

$$ax^2 + 2hxy + by^2 + 2gx + 2fy + c = 0 \tag{2.28}$$

Let us first suppose that the equation (2.28) represents a pair of straight lines. Also let the lines be $l_1 x + m_1 y + n_1 = 0$ and $l_2 x + m_2 y + n_2 = 0$, i.e., we get

$$ax^2 + 2hxy + by^2 + 2gx + 2fy + c = (l_1 x + m_1 y + n_1)(l_2 x + m_2 y + n_2).$$

Equating the coefficients of x^2, xy, y^2, x, y and the constant terms we get

$$l_1 l_2 = a \qquad m_1 m_2 = b \qquad n_1 n_2 = c$$
$$l_1 m_2 + l_2 m_1 = 2h \quad l_1 n_2 + l_2 n_1 = 2g \quad m_1 n_2 + m_2 n_1 = 2f.$$

General Equation of Second Degree

Now we consider the determinants $\begin{vmatrix} l_1 & l_2 & 0 \\ m_1 & m_2 & 0 \\ n_1 & n_2 & 0 \end{vmatrix}$ and $\begin{vmatrix} l_2 & m_2 & n_2 \\ l_1 & m_1 & n_1 \\ 0 & 0 & 0 \end{vmatrix}$.

Since the value of each of these determinants is zero, the value of their product is also zero, i.e., we have

$$\begin{vmatrix} l_1 & l_2 & 0 \\ m_1 & m_2 & 0 \\ n_1 & n_2 & 0 \end{vmatrix} \times \begin{vmatrix} l_2 & m_2 & n_2 \\ l_1 & m_1 & n_1 \\ 0 & 0 & 0 \end{vmatrix} = 0$$

$$\Rightarrow \begin{vmatrix} 2l_1 l_2 & l_1 m_2 + l_2 m_1 & l_1 n_2 + l_2 n_1 \\ m_1 l_2 + m_2 l_1 & 2m_1 m_2 & m_1 n_2 + m_2 n_1 \\ n_1 l_2 + n_2 l_1 & n_1 m_2 + n_2 m_1 & 2n_1 n_2 \end{vmatrix} = 0$$

$$\Rightarrow \begin{vmatrix} 2a & 2h & 2g \\ 2h & 2b & 2f \\ 2g & 2f & 2c \end{vmatrix} = 0 \Rightarrow \begin{vmatrix} a & h & g \\ h & b & f \\ g & f & c \end{vmatrix} = 0 \qquad (2.29)$$

Again $4(h^2 - ab) = (l_1 m_2 + l_2 m_1)^2 - 4 l_1 l_2 m_1 m_2 = (l_1 m_2 - l_2 m_1)^2$.

So $h^2 - ab \geq 0$ \hfill (2.30)

Next we suppose that the equation (2.28) is such that (2.29) and (2.30) hold. We consider the system of equations

$$\left.\begin{array}{r} ax + hy + g = 0 \\ hx + by + f = 0 \\ gx + fy + c = 0. \end{array}\right\}$$

In virtue of (2.29), it follows that the above system has a solution. Let (x_1, y_1) be such a solution. Thus we get

$$\left.\begin{array}{r} ax_1 + hy_1 + g = 0 \\ hx_1 + by_1 + f = 0 \\ gx_1 + fy_1 + c = 0 \end{array}\right\} \qquad (2.31)$$

Now we apply the translation $x = x' + x_1, y = y' + y_1$, thus the given equation (2.28) is translated into

$$a(x' + x_1)^2 + 2h(x' + x_1)(y' + y_1) + b(y' + y_1)^2 + 2g(x' + x_1)$$
$$+ 2f(y' + y_1) + c = 0$$

$$\Rightarrow ax'^2 + 2hx'y' + by'^2 + 2(ax_1 + hy_1 + g)x' + 2(hx_1 + by_1 + f)y'$$
$$+ (ax_1 + hy_1 + g)x_1 + (hx_1 + gy_1 + f)y_1 + (gx_1 + fy_1 + c) = 0$$

$$\Rightarrow ax'^2 + 2hx'y' + by'^2 = 0 \quad \text{[by equations (2.31)]}.$$

$$\Rightarrow b\frac{y'^2}{x'^2} + 2h\frac{y'}{x'} + a = 0 \qquad (2.32)$$

Since (2.30) holds, the equation (2.32) represents a pair of real straight lines. Thus we get the required necessary and sufficient conditions for (2.28) to be a pair of straight lines are that

$$\begin{vmatrix} a & h & g \\ h & b & f \\ g & f & c \end{vmatrix} = 0 \text{ and } h^2 - ab \geq 0 \text{ i.e., } \Delta = 0 \text{ and } \delta \leq 0.$$

Note 2.9.1 *The equation (2.28) represents a pair of parallel or coincident straight lines if and only if $\Delta = 0$ and $h^2 - ab = 0$.*

Note 2.9.2 *If the two lines represented by (2.28) pass through the origin then $n_1 = 0$ and $n_2 = 0$ and consequently $g = 0, f = 0$ and $c = 0$ and the equation (2.28) reduces to $ax^2 + 2hxy + by^2 = 0$.*

Thus we see that the equation of a pair of straight lines passing through the origin is homogeneous containing only the second degree terms in x and y. However, it should be noted that any *homogeneous equation of second degree in x and y does not represent a pair of straight lines through the origin* and we may state the following:

Corollary 2.9.1 *The homogeneous equation of second degree in x and y, given by $ax^2 + 2hxy + by^2 = 0$ represents a pair of straight lines through the origin if and only if $h^2 - ab \geq 0$.*

The truth of the statement follows by taking $f = g = h = 0$.

Point of Intersection

Let us suppose that (2.28) represents a pair of intersecting straight lines having equations

$$\left. \begin{array}{r} l_1 x + m_1 y + n_1 = 0 \\ l_2 x + m_2 y + n_2 = 0. \end{array} \right\}$$

Solving we get $$\frac{x}{m_1 n_2 - m_2 n_1} = \frac{y}{n_1 l_2 - n_2 l_1} = \frac{1}{l_1 m_2 - l_2 m_1}.$$

Now $l_1 m_2 - l_2 m_1 = \sqrt{(l_1 m_2 + l_2 m_1)^2 - 4 l_1 l_2 m_1 m_2} = \sqrt{4h^2 - 4ab}$
$= 2\sqrt{h^2 - ab} \geq 0$

$$\Rightarrow x = \frac{m_1 n_2 - m_2 n_1}{l_1 m_2 - l_2 m_1}, \; y = \frac{n_1 l_2 - n_2 l_1}{l_1 m_2 - l_2 m_1}.$$

General Equation of Second Degree 73

$$
\begin{aligned}
\text{Also } 4bg - 4hf &= 2m_1m_2(l_1n_2 + l_2n_1) - (l_1m_2 + l_2m_1)(m_1n_2 + m_2n_1) \\
&= m_1m_2l_1n_2 + m_1m_2l_2n_1 - l_1n_1m_2^2 - l_2n_2m_1^2 \\
&= m_1n_2(l_1m_2 - l_2m_1) - m_2n_1(l_1m_2 - m_1l_2) \\
&= (l_1m_2 - l_2m_2)(m_1n_2 - m_2n_1).
\end{aligned}
$$

So we get $\quad x = \dfrac{(m_1n_2 - m_2n_1)(l_1m_2 - l_2m_1)}{(l_1m_2 - l_2m_1)^2} = \dfrac{4(bg - hf)}{4(h^2 - ab)} = \dfrac{hf - bg}{ab - h^2}.$

Similarly we shall get $y = \dfrac{gh - af}{ab - h^2}.$

Thus the point of intersection of the pair of lines (2.28) is given by

$$\left(\frac{hf - bg}{ab - h^2}, \frac{gh - af}{ab - h^2} \right).$$

Note 2.9.3 *We note that the point of intersection is same as the centre of a conic if (2.28) represents a central conic.*

B. Conditions for general equation of second degree to be a proper conic

Let the given equation be

$$ax^2 + 2hxy + by^2 + 2gx + 2fy + c = 0 \quad (2.33)$$

Let the equation of the directrix LM be $lx + my + n = 0$. Let $S(\alpha, \beta)$ be the focus of the conic and $P(x, y)$ be any point of the conic. Also let e be the eccentricity of the conic. Then by definition of a conic we have

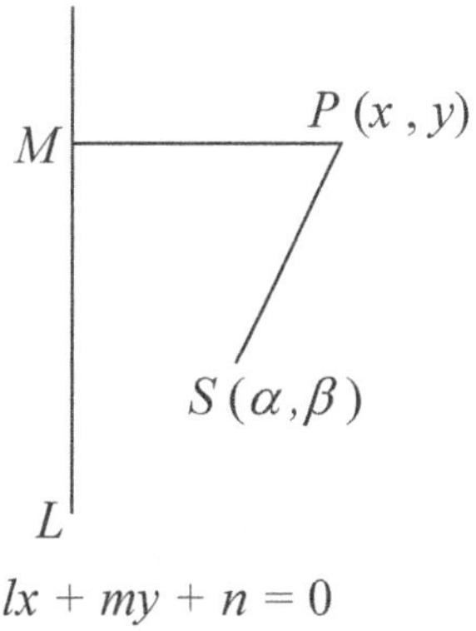

$$SP^2 = e^2 PM^2$$

$$\Rightarrow (x - \alpha)^2 + (y - \beta)^2$$
$$= e^2 \left(\frac{lx + my + n}{\sqrt{l^2 + m^2}} \right)^2 = e^2 \frac{(lx + my + n)^2}{l^2 + m^2}$$

$$\Rightarrow \{l^2(1 - e^2) + m^2\}x^2 - 2lme^2xy + \{l^2 + m^2(1 - e^2)\}y^2$$
$$- 2\{(l^2 + m^2)\alpha + lne^2\}x - 2\{(l^2 + m^2)\beta + mne^2\}y$$
$$+ (l^2 + m^2)(\alpha^2 + \beta^2) - n^2e^2 = 0 \quad (2.34)$$

Figure 2.3

Equation (2.34) will be identical to equation (2.33) if the coefficient of like powers of x, y and the constant terms are equal. Comparing equations (2.33) and (2.34) we get

$$a = l^2(1-e^2) + m^2; \quad h = -lme^2; \quad b = l^2 + m^2(1-e^2);$$
$$g = -\{(l^2+m^2)\alpha + lne^2\}; \quad f = -\{(l^2+m^2)\beta + mne^2\};$$
$$c = (l^2+m^2)(\alpha^2+\beta^2) - n^2e^2.$$

Now $\delta = ab - h^2 = \{l^2(1-e^2) + m^2\}\{l^2 + m^2(1-e^2)\} - l^2m^2e^4$
$$= (l^2+m^2)^2(1-e^2).$$

For a parabola $e = 1$, $\therefore \delta = 0$, i.e., $ab - h^2 = 0$, i.e., $h^2 = ab$.
For an ellipse $e < 1$, $\therefore \delta > 0$, i.e., $ab - h^2 > 0$, i.e., $h^2 < ab$.
For a hyperbola $e > 1$, $\therefore \delta < 0$, i.e., $ab - h^2 < 0$, i.e., $h^2 > ab$.

2.10 Worked Out Examples

Example 2.10.1 *For each of the following curves, determine whether it has a single centre, has no centre or has infinitely many centers and hence classify the equations according to the presence of centre/centers.*

(i) $5x^2 - 6xy + 2y^2 + 24x - 26y + 10 = 0$.

(ii) $x^2 - 2xy + y^2 + 2x - 4y + 3 = 0$.

(iii) $2x^2 + 4xy + 2y^2 - x - y + 1 = 0$.

(iv) $(x-2y)^2 + 3(x-2y) + k = 0, k = $ constant.

Solution: (i) Here $a = 5, b = 2, c = 10, h = -3, g = 12, f = -13$ and so $\delta = ab - h^2 = 5 \times 2 - (-3)^2 = 1 \neq 0$. Hence the conic is a central conic and the centre is given by $\left(\dfrac{hf-bg}{ab-h^2}, \dfrac{gh-af}{ab-h^2}\right)$

$$= \left(\frac{-3\times(-13) - 2\times 12}{1}, \frac{12\times(-3) - 5\times(-13)}{1}\right) = (15, 29).$$

(ii) Here $a = 1, b = 1, c = 3, h = -1, g = 1, f = -2$ and so $\delta = ab - h^2 = 0$. Hence the curve is non central. Now $hf - bg = 1 \neq 0$ and $gh - af = -1 \neq 0$. So the curve has no centre.

(iii) Here $a = 2, b = 2, c = 1, h = 2, g = -\dfrac{1}{2}, f = -\dfrac{1}{2}$ and so $\delta = ab - h^2 = 0$. Hence the curve is non central. Now $hf - bg = 0$ and $gh - af = 0$. So the curve has infinitely many centers.

(iv) The given equation can be written as $x^2 - 4xy + 4y^2 + 3x - 6y + k = 0$, from which we get, $a = 1, b = 4, c = k, h = -2, g = \frac{3}{2}, f = -3$. Therefore $\delta = ab - h^2 = 0$. Hence the curve is non central. Now $hf - bg = 0$ and $gh - af = 0$ imply that the curve has infinitely many centers. □

Note 2.10.1 *In Example 2.10.1 (iv), it should be noticed that the existence or non existence of centre does not depend upon k.*

Example 2.10.2 *Find the values of a and f for which the curve $ax^2 - 20xy + 25y^2 - 14x + 2fy - 15 = 0$ represents a conic having (i) no centre and (ii) infinitely many centers.*

Solution: For the given curve we have $a = a, b = 25, c = -15, h = -10, g = -7, f = f$ and so $\delta = ab - h^2 = 25a - 100$. Also $hf - bg = -10f + 175$ and $gh - af = 70 - af$.

(i) For a conic having no centre, we get $ab - h^2 = 0, hf - bg \neq 0, gh - af \neq 0$. Therefore $25a - 100 = 0 \Rightarrow a = 4$ and $-10f + 175 \neq 0 \Rightarrow f \neq \frac{35}{2}$. Hence $a = 4$ and $f \neq \frac{35}{2}$.

(ii) For a conic with infinitely many centers we get $ab - h^2 = 0, hf - bg = 0, gh - af = 0 \Rightarrow a = 4$ and $f = \frac{35}{2}$. □

Example 2.10.3 *Determine the nature of the locus represented by each of the following equations:*

(i) $x^2 + y^2 + 10x - 12y + 16 = 0$.
(ii) $x^2 + 6xy + 9y^2 - 5x - 15y + 6 = 0$.
(iii) $x^2 + 4xy + 4y^2 + 4x + y - 15 = 0$.
(iv) $x^2 - 2xy + 2y^2 - 4x - 6y + 3 = 0$.

Solution: (i) Here $a = 1, b = 1, h = 0, g = 5, f = -6, c = 16, \therefore a = b$ and $h = 0$ and $g^2 + f^2 - c = 5^2 + (-6)^2 - 16 = 45 > 0$. Hence the locus is a circle.

(ii) Here $a = 1, b = 9, c = 6, h = 3, g = -\frac{5}{2}, f = -\frac{15}{2}$.

$$\therefore \Delta = \begin{vmatrix} a & h & g \\ h & b & f \\ g & f & c \end{vmatrix} = \begin{vmatrix} 1 & 3 & -\frac{5}{2} \\ 3 & 9 & -\frac{15}{2} \\ -\frac{5}{2} & -\frac{15}{2} & 6 \end{vmatrix} = \frac{1}{8} \begin{vmatrix} 2 & 6 & -5 \\ 6 & 18 & -15 \\ -5 & -15 & 12 \end{vmatrix}$$

$$= \frac{1}{8} \begin{vmatrix} 2 & 6 & -5 \\ 0 & 0 & 0 \\ -5 & -15 & 12 \end{vmatrix} = 0 \text{ and } \delta = ab - h^2 = 9 - 9 = 0.$$

Hence the locus is a parallel straight lines.

(iii) Here $a = 1, b = 4, c = -15, h = 2, g = 2, f = \frac{1}{2}$.

$$\therefore \Delta = \begin{vmatrix} a & h & g \\ h & b & f \\ g & f & c \end{vmatrix} = \begin{vmatrix} 1 & 2 & 2 \\ 2 & 4 & \frac{1}{2} \\ 2 & \frac{1}{2} & -15 \end{vmatrix} = \frac{1}{4} \begin{vmatrix} 1 & 2 & 2 \\ 4 & 8 & 1 \\ 4 & 1 & -30 \end{vmatrix}$$

$$= \frac{1}{4} \begin{vmatrix} 1 & 2 & 2 \\ 0 & 0 & -7 \\ 4 & 1 & -30 \end{vmatrix} = \frac{1}{4} \times (-1)(-7)(1-8) = -\frac{49}{4} \neq 0$$

and $\delta = ab - h^2 = 1 \times 4 - 2^2 = 0$. Hence the locus is a parabola.

(iv) Here $a = 1, b = 2, c = 3, h = -1, g = -2, f = -3$

$$\therefore \Delta = \begin{vmatrix} 1 & -1 & -2 \\ -1 & 2 & -3 \\ -2 & -3 & 3 \end{vmatrix} = 1(6-9) + 1(-3-6) - 2(3+4) = -26 \neq 0.$$

and $\delta = ab - h^2 = 1.2 - (-1)^2 = 1 > 0$.

Therefore the locus is an ellipse. □

Example 2.10.4 Reduce the equation $11x^2 - 4xy + 14y^2 - 58x - 44y + 71 = 0$ to its canonical form. Name the conic, find the eccentricity of the conic.

Solution: From the equation we have $a = 11, b = 14, c = 71, h = -2, g = -29, f = -22$.

$$\therefore \Delta = \begin{vmatrix} 11 & -2 & -29 \\ -2 & 14 & -22 \\ -29 & -22 & 71 \end{vmatrix} = -9000 \neq 0.$$

So the equation does not represent a pair of straight lines.

$$\delta = ab - h^2 = 11 \times 14 - (-2)^2 = 154 - 4 = 150 > 0.$$

Since $\Delta < 0$ and $\delta > 0$, so the equation represents an ellipse. If (α, β) be the centre of the conic then we have

$$(\alpha, \beta) = \left(\frac{hf - bg}{ab - h^2}, \frac{gh - af}{ab - h^2} \right)$$

$$= \left(\frac{-2 \times (-22) - 14 \times (-29)}{150}, \frac{-29 \times (-2) - 11 \times (-22)}{150} \right) = (3, 2).$$

General Equation of Second Degree

Shifting the origin to this point $(3, 2)$, the given equation reduces to

$$11x^2 - 4xy + 14y^2 + \frac{(-9000)}{150} = 0 \Rightarrow 11x^2 - 4xy + 14y^2 = 60 \quad (2.35)$$

Now, keeping the origin fixed, let the coordinate axes be rotated about the origin through an angle θ in anticlockwise sense, so that the term involving xy is removed.

Here θ will be obtained as

$$\theta = \frac{1}{2}\tan^{-1}\frac{2h}{a-b} = \frac{1}{2}\tan^{-1}\frac{2\times(-2)}{11-14} = \frac{1}{2}\tan^{-1}\frac{4}{3}$$

and equation (2.35) reduces to

$$AX^2 + BY^2 = 60 \quad (2.36)$$

where $A+B = a+b = 11+14 = 25$ and $AB = ab - h^2 = 11\times 14 - (-2)^2 = 150$.

$$\therefore A - B = \pm\sqrt{(A+B)^2 - 4AB} = \pm\sqrt{25^2 - 4\times 150} = \pm\sqrt{25} = \pm 5,$$

i.e., $A + B = 25$ and $A - B = \pm 5$ which gives $A = 15, B = 10$ or $A = 10, B = 15$.

We take $A = 10$, $B = 15$ and the equation (2.36) becomes

$$10X^2 + 15Y^2 = 60 \text{ or, } \frac{X^2}{6} + \frac{Y^2}{4} = 1.$$

This is the canonical form of the given equation which represents an ellipse. Its eccentricity is given by

$$b^2 = a^2(1-e^2) \text{ i.e., } 4 = 6(1-e^2) \text{ or, } 1 - e^2 = \frac{4}{6} = \frac{2}{3}$$

$$\therefore e^2 = 1 - \frac{2}{3} = \frac{1}{3} \quad \therefore e = \frac{1}{\sqrt{3}}. \qquad \square$$

Note 2.10.2 *The equation of the axes are*

$$\left.\begin{array}{l}(x-3)\cos\theta + (y-2)\sin\theta = 0 \\ (x-3)\sin\theta - (y-2)\cos\theta = 0.\end{array}\right\} \text{ [§ } \textbf{2.6} \textit{ case 1 } \textbf{Note 2.6.2]}$$

Note 2.10.3 *When $A > 0$, $B > 0$, it is customary to take the smaller values of A. In case A and B are of opposite signs, the +ve value is taken for A.*

Example 2.10.5 Show that the equation $4x^2 - 4xy + y^2 + 2x - 26y + 9 = 0$ represents a parabola. Find its latus-rectum.

Solution: Here $a = 4, b = 1, c = 9, h = -2, g = 1, f = -13$.

$$\begin{aligned}\therefore \Delta &= abc + 2fgh - af^2 - bg^2 - ch^2 \\ &= 4.1.9 + 2.(-13).1.(-2) - 4.(-13)^2 - 1.1^2 - 9.(-2)^2 = -625 \neq 0 \\ \text{and } \delta &= ab - h^2 = 4.1 - (-2)^2 = 0.\end{aligned}$$

So the given equation represents a non-central conic, i.e., a parabola.

Now the given equation can be written as

$$(2x - y)^2 + 2x - 26y + 9 = 0$$

or, $\quad (2x - y + \lambda)^2 = -2x + 26y - 9 + \lambda^2 + 4\lambda x - 2\lambda y$
$$= 2(2\lambda - 1)x + 2(13 - \lambda)y + (\lambda^2 - 9) \quad (2.37)$$

where λ is any constant.

Now we choose λ such that the two lines $2x - y + \lambda = 0$ and $2(2\lambda - 1) + 2(13 - \lambda)y + (\lambda^2 - 9) = 0$ be perpendicular to each other. For this we have

$$2\frac{-(2\lambda - 1)}{13 - \lambda} = -1 \Rightarrow \lambda = 3 \text{ [Product of the gradients = -1]}$$

For this value of λ, equation (2.37) becomes

$$(2x - y + 3)^2 = 10(x + 2y) \quad (2.38)$$

If we now take the two perpendicular lines $2x - y + 3 = 0$ and $x + 2y = 0$ as the new axes of x and y respectively, then we get the following transformation formulas:

$$\begin{aligned}X &= \text{perpendicular distance of the point } (x, y) \text{ form the new } y\text{-axis} \\ &= \frac{x + 2y}{\sqrt{1^2 + 2^2}} = \frac{x + 2y}{\sqrt{5}} \Rightarrow x + 2y = \sqrt{5}X \\ \text{and } Y &= \text{perpendicular distance of the point } (x, y) \text{ form the new } x\text{-axis} \\ &= \frac{2x - y + 3}{\sqrt{2^2 + (-1)^2}} = \frac{2x - y + 3}{\sqrt{5}} \Rightarrow 2x - y + 3 = \sqrt{5}Y.\end{aligned}$$

With these transformations, equation (2.38) becomes

$$(\sqrt{5}Y)^2 = 10(x + 2y) = \sqrt{5}X \text{ or, } Y^2 = \frac{10}{\sqrt{5}}X \Rightarrow Y^2 = 2\sqrt{5}X.$$

This is the canonical form of equation of parabola. The axis of it is along the line $Y = 0$, i.e., along $2x - y + 3 = 0$, the equation of the tangent at the vertex is $X = 0$, i.e., $x + 2y = 0$ and its length of latus rectum $= 2\sqrt{5}$. □

Note 2.10.4 *The canonical form will be of the form $Y^2 = \lambda X$ or $X^2 = \mu Y$ according as the coefficient of $xy < 0$ or > 0.*

Example 2.10.6 *Reduce the equation $4x^2 + 4xy + y^2 - 4x - 2y + a = 0$ to the canonical form and determine the conic represented by it for different values of a.*

Solution: The given equation is

$$4x^2 + 4xy + y^2 - 4x - 2y + a = 0 \qquad (2.39)$$

Here $a = 4$, $b = 1$, $c = a$, $h = 2$, $g = -2$, $f = -1$.

$$\begin{aligned}
\therefore \ \Delta &= abc + 2fgh - af^2 - bg^2 - ch^2 \\
&= 4.1.a + 2.(-1).(-2).2 - 4.(-1)^2 - 1.(-2)^2 - a(2)^2 \\
&= 4a + 8 - 4 - 4 - 4a = 0.
\end{aligned}$$

So the equation represents a pair of straight lines.
Also $\delta = ab - h^2 = 4.1 - (2)^2 = 0$, i.e., $\delta = 0$ and $\Delta = 0$. Hence the equation represents either a pair of parallel straight lines or a pair of coincidence straight lines.
Now the equation is written as

$$(2x+y)^2 - 2(2x+y) + a = 0$$

or, $(2x + y - 1)^2 + (a - 1) = 0$

or, $\left(\dfrac{2x+y-1}{\sqrt{5}}\right)^2 + \dfrac{a-1}{5} = 0.$

Let us put $\dfrac{2x+y-1}{\sqrt{5}} = X$ [See **Note 2.10.4** and observe that in this case coefficient of $xy > 0$.]

So the given equation reduces to

$$X^2 + \dfrac{a-1}{5} = 0 \qquad (2.40)$$

which is the canonical form of the given equation.

Clearly, if $a < 1$, the given equation represents two parallel straight lines.
 if $a = 1$, the equation (2.40) reduces to $X^2 = 0$ and then equation (2.39) represents two coincident straight lines.
 and if $a > 1$, the given equation represents two imaginary straight lines. □

Example 2.10.7 *Discuss the nature of the conic given by:*
$$6x^2 - 5xy - 6y^2 + 14x + 5y + 4 = 0.$$

Solution: The given equation is
$$6x^2 - 5xy - 6y^2 + 14x + 5y + 4 = 0 \tag{2.41}$$

Here $a = 6$, $b = -6$, $c = 4$, $h = -\dfrac{5}{2}$, $g = 7$, $f = \dfrac{5}{2}$.

$$\begin{aligned}
\therefore \Delta &= abc + 2fgh - af^2 - bg^2 - ch^2 \\
&= 6.(-6).4 + 2.\frac{5}{2}.7.\left(-\frac{5}{2}\right) - 6.\left(\frac{5}{2}\right)^2 - (-6).7^2 - 4.\left(-\frac{5}{2}\right)^2 = 0.
\end{aligned}$$

So the equation represents a pair of straight lines.

Also $\delta = ab - h^2 = 6.(-6) - \left(-\dfrac{5}{2}\right)^2 = -36 - \dfrac{25}{4} = -\dfrac{169}{4} \neq 0$, i.e., $\Delta = 0$ but $\delta \neq 0$. Hence the lines are intersecting lines. So equation (2.41) represents a pair of intersecting lines.

More Discussion:

Let us try to find out the individual straight lines represented by equation (2.41). For this we proceed as follows:

We have
$$\begin{aligned}
6x^2 - 5xy - 6y^2 &= 6x^2 - 9xy + 4xy - 6y^2 \\
&= 3x(2x - 3y) + 2y(2x - 3y) = (2x - 3y)(3x + 2y).
\end{aligned}$$

Now let the individual lines represented by the equation (2.41) be $2x - 3y + \lambda = 0$ and $3x + 2y + \mu = 0$ and then we have

$$6x^2 - 5xy - 6y^2 + 14x + 5y + 4 = (2x - 3y + \lambda)(3x + 2y + \mu).$$

Equating the like powers of x, y and constant terms we get

$$2\mu + 3\lambda = 14 \tag{2.42}$$
$$-3\mu + 2\lambda = 5 \tag{2.43}$$
$$\text{and} \quad \lambda\mu = 4 \tag{2.44}$$

Solving (2.42) and (2.43) we get $\lambda = 4$, $\mu = 1$ which also satisfy the equation (2.44). So we get

$$6x^2 - 5xy - 6y^2 + 14x + 5y + 4 = (2x - 3y + 4)(3x + 2y + 1).$$

General Equation of Second Degree

Therefore L.H.S. of the equation (2.41) has been factorized into two linear factors given by $2x-3y+4$ and $3x+2y+1$. So the individual lines represented by (2.41) are

$$\begin{array}{r} 2x - 3y + 4 = 0 \\ \text{and}\quad 3x + 2y + 1 = 0. \end{array}\Bigg\}$$

Clearly, the lines are intersecting and their point of intersection is given by

$$\left(\frac{-3-8}{4+9}, \frac{12-2}{4+9}\right) = \left(-\frac{11}{13}, \frac{10}{13}\right). \qquad \square$$

Example 2.10.8 *Reduce the equation given in **Example 2.10.7** into its canonical form and state the type of the conic.*

Solution: The given equation is

$$6x^2 - 5xy - 6y^2 + 14x + 5y + 4 = 0 \qquad (2.45)$$

As in **Example 2.10.7**, we see that $\Delta = 0$ and $\delta = -\dfrac{169}{4} \neq 0$. So it represents a central conic (i.e., a pair of intersecting straight lines) and the centre of the conic is the point of intersection of the lines which is given by

$$\left(\frac{hf - bg}{ab - h^2}, \frac{gh - af}{ab - h^2}\right) = \left(\frac{\left(-\frac{5}{2}\right) \times \frac{5}{2} - (-6 \times 7)}{-\frac{169}{4}}, \frac{7.\left(-\frac{5}{2}\right) - 6 \times \frac{5}{2}}{-\frac{169}{4}}\right)$$

$$= \left(\frac{-\frac{25}{4} + 42}{-\frac{169}{4}}, \frac{-\frac{35}{2} - \frac{30}{2}}{-\frac{169}{4}}\right) = \left(\frac{-25 + 168}{-169}, \frac{-70 - 60}{-169}\right) = \left(-\frac{11}{13}, \frac{10}{13}\right).$$

Shifting the origin to this point, i.e., applying the translations

$$\begin{array}{r} x = x' - \frac{11}{13} \\ \text{and}\quad y = y' + \frac{10}{13} \end{array}\Bigg\}$$

the given equation (2.45) becomes $6x'^2 - 5x'y' - 6y'^2 + d = 0$ and $d = \dfrac{\Delta}{\delta} = 0$. Therefore the reduced equation (2.45) is

$$6x'^2 - 5x'y' - 6y'^2 = 0 \qquad (2.46)$$

Now let the axes be rotated through an angle θ, keeping the origin fixed so that the term involving $x'y'$ is removed. In such case θ is obtained as

$$\tan 2\theta = \frac{2h}{a - b} = \frac{-5}{6 - (-6)} = -\frac{5}{12}$$

and the equation (2.46) reduces to $AX^2 + BY^2 = 0$ where

$$A + B = a + b = 6 - 6 = 0$$

and $AB = ab - h^2 = -36 - \left(-\dfrac{5}{12}\right)^2 = -36 - \dfrac{25}{4} = -\dfrac{169}{4}.$

So we get $A + B = 0$ and $AB = -\dfrac{169}{4}$

$\Rightarrow A(-A) = -\dfrac{169}{4} \Rightarrow A^2 = \dfrac{169}{4} \therefore A = \pm\dfrac{13}{2},$ and $B = \mp\dfrac{13}{2}.$

We take positive value of A, i.e., $A = \dfrac{13}{2},$ and $B = -\dfrac{13}{2}$ [§ **Note 2.10.3**] and therefore we get the canonical form of the given equation as

$$\dfrac{13}{2}X^2 - \dfrac{13}{2}Y^2 = 0 \Rightarrow X^2 = Y^2 \Rightarrow X = \pm Y$$

which represents a pair of intersecting straight lines. □

Example 2.10.9 *Show that the conic $ax^2 + 2hxy + by^2 + 2gx\cos^2\alpha + 2fy\sin^2\alpha + c = 0$ where α is a parameter, always passes through two fixed points, provided $g^2 f^2 > c(af^2 + 2fgh + bg^2)$.*

Solution: Let us write the given equation as

$$ax^2 + 2hxy + by^2 + 2gx\cos^2\alpha + 2fy\sin^2\alpha + c = 0$$

or, $ax^2 + 2hxy + by^2 + 2gx + c + 2(fy - gx)\sin^2\alpha = 0.$

The conics of different values of the parameter α will pass through two fixed points provided the straight line $fy - gx = 0$ cuts the conic $ax^2 + 2hxy + by^2 + 2gx + c = 0$ in two distinct points, i.e., the quadratic equation

$$ax^2 + 2hx\dfrac{g}{f}x + b\left(\dfrac{g}{f}x\right)^2 + 2gx + c = 0$$

has two real and distinct values of x. Therefore the discriminant of the equation $(af^2 + 2fgh + bg^2)x^2 + 2gf^2 x + cf^2 = 0$ will be greater than 0, i.e.,

$$4g^2 f^4 - 4cf^2(af^2 + 2fgh + bg^2) > 0$$

or, $g^2 f^2 > c(af^2 + 2fgh + bg^2).$ □

Example 2.10.10 *Reduce the equation $6x^2 + 24xy - y^2 - 60x - 20y + 80 = 0$ to its canonical form and classify the conic represented by it. Find the equation of its axes.*

General Equation of Second Degree

Solution: Here $a = 6$, $b = -1$, $c = 80$, $h = 12$, $g = -30$, $f = -10$.

$$\therefore \Delta = abc + 2fgh - af^2 - bg^2 - ch^2$$
$$= 6.(-1).80 + 2.(-10).(-30).12 - 6.(-10)^2$$
$$-(-1).(-30)^2 - 80.12^2$$
$$= -480 + 7200 - 600 + 900 - 11520$$
$$= 8100 - 12600 = -4500 \neq 0$$

and $\delta = ab - h^2 = 6.(-1) - (12)^2 = -150 \neq 0$.

Since $\delta < 0$ and $\Delta < 0$, so the equation represents a hyperbola and it is a central conic. Its centre is given by

$$\left(\frac{hf - bg}{ab - h^2}, \frac{gh - af}{ab - h^2}\right) = \left(\frac{12.(-10) - (-1).(-30)}{-150}, \frac{12.(-30) - 6.(-10)}{-150}\right)$$
$$= \left(\frac{-120 - 30}{-150}, \frac{-360 + 60}{-150}\right) = (1, 2).$$

So we apply translation $x = x' + 1$, $y = y' + 2$, the given equation then reduces to $6x'^2 + 24x'y' - y'^2 + d = 0$ where $d = \frac{\Delta}{\delta} = \frac{-4500}{-150} = 30$. We get

$$6x'^2 + 24x'y' - y'^2 + 30 = 0 \qquad (2.47)$$

To remove the term containing $x'y'$ we apply rotation of coordinates axes through an angle $\theta = \frac{1}{2}\tan^{-1}\frac{2h}{a-b} = \frac{1}{2}\tan^{-1}\frac{24}{7}$ and the equation (2.47) then becomes

$$AX^2 + BY^2 + 30 = 0 \qquad (2.48)$$

where by the invariants of rotation, $A + B = a + b = 6 - 1 = 5 \Rightarrow B = 5 - A$ and $AB = ab - h^2 = -150$. So we get

$$A(5 - A) = -150 \text{ or, } A^2 - 5A - 150 = 0$$
$$\text{or, } (A + 10)(A - 15) = 0 \Rightarrow A = -10 \text{ or, } 15 \quad \therefore B = 15 \text{ or, } -10$$

We take $A = 15$, $B = -10$ and then equation (2.48) becomes $15X^2 - 10Y^2 + 30 = 0$ or, $\frac{X^2}{-2} + \frac{Y^2}{3} = 1$. Clearly, it represents a hyperbola.

Equation of its axes:

$$\tan 2\theta = \frac{24}{7} \text{ or, } \frac{2\tan\theta}{1-\tan^2\theta} = \frac{24}{7}$$

or, $24 - 24\tan^2\theta - 14\tan\theta = 0$ or, $12\tan^2\theta + 7\tan\theta - 12 = 0$

or, $12\tan^2\theta + 16\tan\theta - 9\tan\theta - 12 = 0$

or, $4\tan\theta(3\tan\theta + 4) - 3(3\tan\theta + 4) = 0$

or, $(3\tan\theta + 4)(4\tan\theta - 3) = 0 \Rightarrow \tan\theta = \frac{3}{4}, -\frac{4}{3}.$

Taking $\tan\theta = \frac{3}{4}$ we get $\sin\theta = \frac{3}{5}$ and $\cos\theta = \frac{4}{5}$ and we have

$$\left.\begin{array}{l} x' = X\cos\theta - Y\sin\theta \\ y' = X\sin\theta + Y\cos\theta \end{array}\right\} \Rightarrow \left.\begin{array}{l} X = x'\cos\theta + y'\sin\theta \\ Y = -x'\sin\theta + y'\cos\theta. \end{array}\right\}$$

Also $x = x' + 1$ and $y = y' + 2$

$$\therefore \quad X = \frac{4}{5}(x-1) + \frac{3}{5}(y-2) = \frac{1}{5}(4x + 3y - 10)$$

and $Y = -\frac{3}{5}(x-1) + \frac{4}{5}(y-2) = \frac{1}{5}(-3x + 4y - 5).$

So the equation of the axes are given by $Y = 0$ and $X = 0$

i.e., $-3x + 4y - 5 = 0$ and $4x + 3y - 10 = 0$

or, $3x - 4y + 5 = 0$ and $4x + 3y - 10 = 0.$

Check:

$$-\frac{1}{2}\frac{(4x+3y-10)^2}{25} + \frac{1}{3}\frac{(-3x+4y-5)^2}{25} = 1$$

$\Rightarrow 6x^2 + 24xy - y^2 - 60x - 20y + 80 = 0.$ □

Example 2.10.11 *Reduce the equation $7x^2 - 2xy + 7y^2 + 22x - 10y + 7 = 0$ to the normal form. Hence find the axes, direction and coordinates of the foci.*

Solution: The given equation is

$$7x^2 - 2xy + 7y^2 + 22x - 10y + 7 = 0 \tag{2.49}$$

$\therefore \ a = 7, \ b = 7, \ c = 7, \ h = -1, \ g = 11, \ f = -5$

$$\begin{aligned}\therefore \ \Delta &= abc + 2fgh - af^2 - bg^2 - ch^2 \\ &= 7.7.7 + 2.(-5).11.(-1) - 7.(-5)^2 - 7.11^2 - 7.(-1)^2 \\ &= 343 + 110 - 175 - 847 - 7 = 453 - 1029 = -576 \neq 0. \\ \text{and } \delta &= ab - h^2 = 7.7 - (-1)^2 = 48 \neq 0.\end{aligned}$$

General Equation of Second Degree

Clearly, it is a central conic. Here $\Delta < 0$ and $\delta > 0$, so it represents an ellipse. Its centre is given by

$$\left(\frac{hf-bg}{ab-h^2}, \frac{gh-af}{ab-h^2}\right) = \left(\frac{(-1).(-5)-7.11}{48}, \frac{11.(-1)-7.(-5)}{48}\right)$$

$$= \left(-\frac{72}{48}, \frac{24}{48}\right) = \left(-\frac{3}{2}, \frac{1}{2}\right).$$

Changing the origin to the point $\left(-\frac{3}{2}, \frac{1}{2}\right)$, without altering the direction of the axes, the equation (2.49) reduces to $7x'^2 - 2x'y' + 7y'^2 + d = 0$, where $d = \frac{\Delta}{\delta} = \frac{-576}{48} = -12$. So we get

$$7x'^2 - 2x'y' + 7y'^2 - 12 = 0 \qquad (2.50)$$

Now we rotate the axes through an angle θ, such that the term $x'y'$ is removed. It gives

$$\tan 2\theta = \frac{2h}{a-b} = \infty, \text{ i.e., } 2\theta = \frac{\pi}{2} \text{ i.e., } \theta = \frac{\pi}{4}$$

and the equation (2.50) transformed into

$$AX^2 + BY^2 - 12 = 0$$

where by the rule of invariant we get

$$A + B = a + b = 7 + 7 = 14 \text{ and } AB = ab - h^2 = 48$$

$$\therefore \quad A(14 - A) = 48 \text{ or, } A^2 - 14A + 48 = 0$$

or $\quad A^2 - 8A - 6A + 48 = 0$

or, $\quad A(A-8) - 6(A-8) = 0$ i.e., $(A-8)(A-6) = 0$

i.e., $\quad A = 8$ or, $A = 6$ and $B = 6$ or, $B = 8$.

We take $A = 6$, $B = 8$ [$\because$ for $A > 0$ $B > 0$, we take smallest value of A].

Hence the final transformed equation, i.e., the canonical form or normal form of the given equation is

$$6X^2 + 8y^2 = 12 \quad \text{or, } \quad \frac{X^2}{2} + \frac{Y^2}{\frac{3}{2}} = 1 \qquad (2.51)$$

The length of semi major axis and semi minor axes are $\sqrt{2}$ and $\sqrt{\dfrac{3}{2}}$ respectively. The eccentricity is given by

$$e^2 = 1 - \dfrac{b^2}{a^2} = 1 - \dfrac{\tfrac{3}{2}}{2} = 1 - \dfrac{3}{4} = \dfrac{1}{4} \quad \text{i.e., } e = \dfrac{1}{2}.$$

Coordinate Axes:

The axes are given by $X = 0$ and $Y = 0$ and we have

$$\left. \begin{array}{l} X = x' \cos\theta + y' \sin\theta \\ Y = -x' \sin\theta + y' \cos\theta \end{array} \right\} \text{ and } \left. \begin{array}{l} x' = x + \tfrac{3}{2} \\ y' = y - \tfrac{1}{2} \end{array} \right\}$$

which gives $X = \left(x + \dfrac{3}{2}\right) \cos \dfrac{\pi}{2} + \left(y - \dfrac{1}{2}\right) \sin \dfrac{\pi}{2} = \dfrac{1}{\sqrt{2}}(x + y + 1)$

and $\quad Y = -\left(x + \dfrac{3}{2}\right) \sin \dfrac{\pi}{2} + \left(y - \dfrac{1}{2}\right) \cos \dfrac{\pi}{2} = \dfrac{1}{\sqrt{2}}(-x + y - 2).$

Hence the axes of the ellipse are given by $x + y + 1 = 0$ and $-x + y - 2 = 0$ or $x - y + 2 = 0$.

Equations of the Directrixes:

The equation of the directrixes are given by $X = \pm \dfrac{a}{e}$, where $a =$ length of semi-major axis $= \sqrt{2}$ and $e = \dfrac{1}{2}$.

So the equations of the directrixes are $X = \pm 2\sqrt{2}$

i.e., $\dfrac{1}{\sqrt{2}}(x + y + 1) = \pm 2\sqrt{2}$ or, $x + y + 1 = \pm 4$.

Coordinates of foci:

Coordinates of foci are given by $(X = \pm ae, Y = 0)$.

Now $X = \pm ae \Rightarrow \dfrac{1}{\sqrt{2}}(x + y + 1) = \pm \sqrt{2}.\dfrac{1}{2}$ or, $x + y + 1 = \pm 1$

and $Y = 0 \Rightarrow \dfrac{1}{\sqrt{2}}(y - x - 2) = 0$ or, $x - y + 2 = 0$.

Solving we get $x = -1, -2$ and $y = 1, 0$.

Hence the foci of the given conic are $(-1, 1)$ and $(-2, 0)$. □

Example 2.10.12 *If for the conic $ax^2 + 2hxy + by^2 + 2gx + 2fy + c = 0$, the characteristic roots of $\begin{pmatrix} a & h \\ h & b \end{pmatrix}$ are equal, then show that the conic is a circle or a point circle or an imaginary circle.*

General Equation of Second Degree

Solution: The characteristic equation of the matrix $\begin{pmatrix} a & h \\ h & b \end{pmatrix}$ is

$$\begin{vmatrix} a-\lambda & h \\ h & b-\lambda \end{vmatrix} = 0 \Rightarrow \lambda^2 - (a+b)\lambda + ab - h^2 = 0.$$

If the roots of this equation are equal, then its discriminant $= 0$

i.e., $(a+b)^2 = 4(ab - h^2)$ or, $(a+b)^2 - 4ab + 4h^2 = 0$
or, $(a-b)^2 + 4h^2 = 0.$

Since a, h, b are real, $a - b = 0$ and $h = 0$.
Hence the equation of the conic takes the form

$$ax^2 + ay^2 + 2gx + 2fy + c = 0$$

or, $x^2 + y^2 + \dfrac{2g}{a}x + \dfrac{2f}{a}y + \dfrac{c}{a} = 0 \quad [\because a \neq 0]$

or, $x^2 + y^2 + 2g'x + 2f'y + c' = 0$, where $g' = \dfrac{g}{a},\ f' = \dfrac{f}{a},\ c' = \dfrac{c}{a}.$

This clearly represents a circle or a point circle or an imaginary circle according as $g'^2 + f'^2 - c >, =, < 0.$ □

Example 2.10.13 *Find the equation of the conic passing through the points $(0,0), (2,3), (0,3), (2,5)$ and $(4,5)$ and determine its nature.*

Solution: Let the equation of the conic be

$$ax^2 + 2hxy + by^2 + 2gx + 2fy + c = 0 \tag{2.52}$$

Since it passes through $(0,0)$ $\therefore\ c = 0$. Hence (2.52) takes the form

$$ax^2 + 2hxy + by^2 + 2gx + 2fy = 0 \tag{2.53}$$

Also it passes through $(2,3), (0,3), (2,5)$ and $(4,5)$. Now for the point $(0,3)$ we get

$$9b + 6f = 0 \Rightarrow b = -\dfrac{2}{3}f \tag{2.54}$$

For the point $(2,3)$, we get

$$4a + 12h + 9b + 4g + 6f = 0$$
or, $4a + 4g + 12h = 0 \quad$ [by (2.54)]
or, $a + g = -3h \tag{2.55}$

For the point $(2, 5)$, we get
$$4a + 20h + 25b + 4g + 10f = 0$$
or, $4(a + g) + 20h + 25 \cdot \left(-\dfrac{2}{3}\right) f + 10f = 0$ [by (2.54)]

or, $-12h + 20h - \dfrac{50}{3} f + 10f = 0$ [by (2.55)]

or, $24h - 20f = 0 \Rightarrow h = \dfrac{5}{6} f$ \hfill (2.56)

For the point $(4, 5)$, we get
$$16a + 40h + 25b + 8g + 10f = 0$$
or, $8(a + g) + 8a + 40h + 25 \cdot \left(-\dfrac{2}{3}\right) f + 10f = 0$ [by (2.54)]

or, $8 \cdot (-3h) + 40h + 8a - \dfrac{50}{3} f + 10f = 0$ [by (2.55)]

or, $40h - 24h + 8a - \dfrac{20}{3} f = 0$

or, $16h + 8a = \dfrac{20}{3} f$ or, $4h + 2a = \dfrac{5}{3} f$

or, $2a = \dfrac{5}{3} f - 4h = \dfrac{5}{3} f - 4 \times \dfrac{5}{6} f$ [using (2.56)]

$\qquad = \dfrac{5}{3} f - \dfrac{10}{3} f = -\dfrac{5}{3} f$

$\therefore \quad a = -\dfrac{5}{6} f$ and so $g = -3h - a$ [using (2.55)]

$\qquad = -3 \times \dfrac{5}{6} f + \dfrac{5}{6} f = -\dfrac{5}{3} f.$

Hence we get $a = -\dfrac{5}{6} f$, $b = -\dfrac{2}{3} f$, $h = \dfrac{5}{6} f$, $g = -\dfrac{5}{3} f$ and $c = 0$. So (2.52) becomes

$$-\dfrac{5}{6} f x^2 + 2 \cdot \dfrac{5}{6} f xy + \left(-\dfrac{2}{3} f\right) y^2 + 2 \cdot \left(-\dfrac{5}{3} f\right) x + 2fy = 0$$

or, $-5x^2 + 10xy - 4y^2 - 20x + 12y = 0$ $\quad [f \neq 0]$

or, $5x^2 - 10xy + 4y^2 + 20x - 12y = 0$

which is the required equation of the conic.

An Alternative Method:

The equation of the line passing through $(2, 3)$ and $(0, 3)$ is $\left. \begin{array}{l} y - 3 = 0 \\ y - 5 = 0. \end{array} \right\}$
The equation of the line passing through $(2, 5)$ and $(4, 5)$ is

The equation of the line passing through the four points $(2,3), (0,3), (2,5)$ and $(4,5)$ is $(y-3)(y-5) = 0$.

The equation of the line passing through $(0,3)$ and $(4,5)$ is given by

$$y - 3 = \frac{5-3}{4-0}(x-0) = \frac{x}{2} \text{ or, } x - 2y + 6 = 0$$

and the equation of the line passing through $(2,3)$ and $(2,5)$ is

$$y - 3 = \frac{5-3}{2-2}(x-2) \Rightarrow x - 2 = 0.$$

Therefore the equation of the pair of lines passing through the same four points $(2,3), (0,3), (2,5)$ and $(4,5)$ is

$$(x-2)(x - 2y + 6) = 0.$$

Let the equation of the conic passing through these four points be

$$(y-3)(y-5) + \lambda(x-2)(x - 2y + 6) = 0.$$

It passes through the point $(0,0)$, then we get

$$15 - 12\lambda = 0 \Rightarrow \lambda = \frac{5}{4}.$$

Therefore the required equation of the conic passing through the given five points is

$$(y-3)(y-5) + \frac{5}{4}(x-2)(x - 2y + 6) = 0$$

or, $4(y^2 - 8y + 15) + 5(x^2 - 2xy + 6x - 2x + 4y - 12) = 0$

or, $5x^2 - 10xy + 4y^2 + 20x - 12y = 0$

which is the required conic.

Second Part: *Nature of the Conic*

For the conic obtained, we have $a = 5$, $b = 4$, $c = 0$, $h = -5$, $g = 10$, $f = -6$.

$$\therefore \Delta = abc + 2fgh - af^2 - bg^2 - ch^2$$
$$= 0 + 2.(-6) \times 10 \times (-5) - 5 \times (-6)^2 - 4.(10)^2$$
$$= 600 - 180 - 400 = 200 - 180 = 20 \neq 0 \text{ and } > 0.$$
and $\delta = ab - h^2 = 5.4 - (-5)^2 = 20 - 25 = -5 \neq 0 \text{ and } < 0.$

Since $\delta \neq 0$ it is a central conic and since $\Delta > 0$ and $\delta < 0$, so it represents a hyperbola. □

Example 2.10.14 *Find the equation of the conic which passes through the point $(-1, -1)$ and also through the intersections of the conic $25x^2 - 14xy + 25y^2 + 64x - 64y - 228 = 0$ with the lines $x + 3y - 2 = 0$ and $3x + y - 4 = 0$. Find also the parabola passing through the same points of intersection.*

Solution: First Part: The equation of the conic will be of the form

$$25x^2 - 14xy + 25y^2 + 64x - 64y - 228 + \lambda(x+3y-2)(3x+y-4) = 0 \quad (2.57)$$

If it passes through the point $(-1, -1)$, then

$$25 - 14 + 25 - 64 + 64 - 228 + \lambda(-1-3-2)(-3-1-4) = 0$$
or, $48\lambda = 192$ i.e., $\lambda = 4$.

So the required equation is

$$37x^2 + 26xy + 37y^2 + 24x - 120y - 196 = 0.$$

Second Part: The equation of the parabola will also be of the form (2.57), i.e.,

$$(25+3\lambda)x^2 + (10\lambda - 14)xy + (25+3\lambda)^2 + (64 - 10\lambda)x - (64+14\lambda)y + 8\lambda - 228 = 0$$

for which the second term must be a perfect square, the condition for which is its Discriminant $= 0$, i.e.,

$$(10\lambda - 14)^2 - 4(25 + 3\lambda)(25 + 3\lambda) = 0$$
or, $(5\lambda - 7)^2 - (25 + 3\lambda)^2 = 0$
or, $(5\lambda - 7 + 25 + 3\lambda)(5\lambda - 7 - 25 - 3\lambda) = 0$
or, $(8\lambda + 18)(2\lambda - 32) = 0 \Rightarrow (4\lambda + 9)(\lambda - 16) = 0 \Rightarrow \lambda = -\dfrac{9}{4}$ or, 16.

For $\lambda = -\dfrac{9}{4}$, equation (2.57) takes the form

$$73x^2 - 146xy + 73y^2 + 346x - 130y - 984 = 0$$

and for $\lambda = 16$, equation (2.57) takes the form

$$73x^2 + 146xy + 73y^2 - 96x - 288y - 100 = 0.$$

These are the required equations of parabola. □

Example 2.10.15 *Determine the number of parabolas passing through the intersections of the conics $3x^2 + 10xy + 3y^2 - 2x - 14y - 13 = 0$ and $25x^2 - 14xy + 25y^2 + 64x - 64y - 224 = 0$.*

Solution: Proceeding exactly in a similar manner as in the forgoing **Example 2.10.14** (second part) we get,

The equation of any conic passing through the intersection of the given conics as

$$(3x^2 + 10xy + 3y^2 - 2x - 14y - 13)$$
$$+\lambda(25x^2 - 14xy + 25y^2 + 64x - 64y - 224) = 0 \qquad (2.58)$$

where λ is a variable parameter.

$$(3+25\lambda)x^2 + 2(5-7\lambda)xy + (3+25\lambda)y^2 + 2(32\lambda - 1)x$$
$$+2(-32\lambda - 7)y - 13 - 224\lambda = 0 \qquad (2.59)$$

Here $a = 3 + 25\lambda$, $h = 5 - 7\lambda$ and $b = 3 + 25\lambda$.

The equation (2.58) represents a parabola if $ab - h^2 = 0$

or, $(3+25\lambda)^2 - (5-7\lambda)^2 = 0$

or, $(3+25\lambda+5-7\lambda)(3+25\lambda-5+7\lambda) = 0$

or, $(8+18\lambda)(32\lambda - 2) = 0$ $\therefore \lambda = -\dfrac{4}{9}, \dfrac{1}{16}.$

Substituting these values of λ we get two parabolas which will be obtained as

$$73x^2 - 146xy + 73y^2 + 274x - 130y - 779 = 0 \qquad (2.60)$$
$$\text{and } 73x^2 + 146xy + 73y^2 + 32x - 288y - 432 = 0 \qquad (2.61)$$

are the required equations. □

Example 2.10.16 *Find the equation of the conic which passes through the point $(-2, 0)$, touches the y-axis at the origin and has its centre at the point $(1, 1)$.*

Solution: Let the equation of the conic be

$$ax^2 + 2hxy + by^2 + 2gx + 2fy + c = 0 \qquad (2.62)$$

Since it passes through the origin $(0,0)$ $\therefore$ $c = 0$.

The tangent at the origin is obtained by equating the terms of the lowest degree to zero, i.e., $gx + fy = 0$.

It is identical to y-axis, i.e., $x = 0 \Rightarrow f = 0$. The centre lies on

$$\left.\begin{aligned} ax + hy + g &= 0 \\ \text{and} \quad hx + by &= 0 \end{aligned}\right\}$$

These equations are satisfied by $(1, 1)$

$$\therefore \quad a + h + g = 0 \qquad (2.63)$$
$$\text{and} \quad h + b = 0. \qquad (2.64)$$

Again the point $(-2, 0)$ lies on the conic

$$\therefore \quad 4a - 4g = 0 \quad \therefore \quad a - g = 0 \qquad (2.65)$$

From (2.63), (2.64) and (2.65) we get $a = g = -\dfrac{h}{2}$ and $b = -h$. Also we have $c = 0, f = 0$. Substituting all these in (2.62) we get

$$-\frac{h}{2}x^2 + 2hxy - hy^2 - hx = 0$$

or, $\quad x^2 - 4xy + 2y^2 + 2x = 0 \qquad (2.66)$

which is the required equation.

Now the discriminant of equation (2.66) is

$$\triangle = abc + 2fgh - af^2 - bg^2 - ch^2 = -2.1^2 = -2 < 0$$

$$[\because \text{ here } a = 1, \ b = 2, \ c = 0, \ h = -2, \ g = 1, \ f = 0]$$

and $ab - h^2 = 1.2 - (-2)^2 = 2 - 4 = -2 < 0$.

Since $\delta < 0$ and $\triangle < 0$ it represents a hyperbola. □

Example 2.10.17 *Show that the conic* $(c^2 + d^2)(x^2 + y^2) = (dx + cy - bc)^2$ *is a parabola of latus rectum* $\dfrac{2bc}{\sqrt{c^2 + d^2}}$.

Solution: The given conic is

$$(c^2 + d^2)(x^2 + y^2) = (dx + cy - bc)^2$$

or, $\quad c^2x^2 + c^2y^2 + d^2x^2 + d^2y^2 = d^2x^2 + c^2y^2 + b^2c^2$
$$+ 2cdxy - 2bcdx - 2bc^2y$$

or, $\quad c^2x^2 + d^2y^2 - 2cdxy = -2bc\left(dx + cy - \dfrac{bc}{2}\right)$

or, $\quad (cx - dy)^2 = -2bc\left(dx + cy - \dfrac{bc}{2}\right)$

or, $\quad \left(\dfrac{cx - dy}{\sqrt{c^2 + d^2}}\right)^2 = -\dfrac{2bc}{\sqrt{c^2 + d^2}} \times \dfrac{dx + cy - \frac{bc}{2}}{\sqrt{c^2 + d^2}} \qquad (2.67)$

Let $\quad Y = \dfrac{cx - dy}{\sqrt{c^2 + d^2}} \quad \text{and} \quad X = \dfrac{dx + cy - \frac{bc}{2}}{\sqrt{c^2 + d^2}}.$

General Equation of Second Degree

So the equation (2.67) takes the form

$$Y^2 = -\frac{2bc}{\sqrt{c^2 + d^2}} X$$

which is a parabola and its latus rectum is given by $\dfrac{2bc}{\sqrt{c^2 + d^2}}$. □

2.11 Exercises

Section A: Objective Type Questions

1. Show that the curve $4x^2 + 4xy + y^2 + 4x + 2y + 20 = 0$ has infinitely many centers.
2. Prove that $x^2 - 2xy + 2y^2 - 4x - 6y + 3 = 0$ represents an ellipse.
3. If the equation $6x^2 + kxy - 3y^2 + 4x + 5y - 2 = 0$ represents a pair of straight lines, then find the value of k.
4. If the expression $7x^2 - 6xy - y^2 - 2 = 0$ reduces to the form $AX^2 + BY^2 - 2 = 0$, where $A > 0$, $B < 0$, then find the values of A and B.
5. Find the nature of the conic $3x^2 + 7xy + 3y^2 - 16x + 20 = 0$.
6. Find the centre of the conic $3x^2 + 2xy + 3y^2 - 16x + 20 = 0$.
7. Show that the transformed equation of the conic $3x^2 + 2xy + 3y^2 - 16x + 20 = 0$ with respect to its centre, taken as the new origin is $3X^2 + 2XY + 3Y^2 = 4$.
8. Show that the nature of the conic $6x^2 - 5xy - 6y^2 + 14x + 5y + 4 = 0$ is a pair of intersecting straight lines.
9. Show that the centre of the conic $6x^2 - 5xy - 6y^2 + 14x + 5y + 4 = 0$ is $\left(-\dfrac{11}{13}, \dfrac{10}{13}\right)$.
10. Find the transformed equation of the equation $6x^2 - 5xy - 6y^2 + 14x + 5y + 4 = 0$ with respect to its centre.
11. Show that the nature of the conic $9x^2 + 24xy + 16y^2 - 126x + 82y - 59 = 0$ is a parabola.
12. Find the constant term of the canonical form of the equation of the conic $x^2 - 2xy + 2y^2 - 4x - 6y + 3 = 0$.
13. Find the value of a so that the equation $ax^2 + 4xy + y^2 - 6x - 2y + 2 = 0$ may represent a point ellipse.
14. Show that the conic $2x^2 - 6xy + 5y^2 + 12x - 4y + 9 = 0$ has a unique centre.
15. Show that the conic $4x^2 + 4xy + y^2 + 4x + 2y + 1 = 0$ has infinitely many centers.
16. Show that the conic $x^2 - 2xy + y^2 + 2x - 4y + 3 = 0$ has no centers.
17. Show that the conic $4x^2 + 4xy + y^2 - 4x - 2y + a = 0$ $(a > 1)$ has no geometrical meaning.
18. Show that the equation $2x^2 + 3xy + y^2 = 0$ represents a pair of straight lines. Find them.
19. Find the value of a so that the equation $ax^2 + xy - 12y^2 + 2x - 31y - 20 = 0$ represents a pair of straight lines.

20. What type of conic does $x^2 + 2y^2 = 3$ represent?
21. Find the coordinates of the focus of the parabola $x^2 + y + 1 = 0$.
22. Find the coordinates of the foci of the ellipse $4x^2 + 3y^2 = 24$.
23. Find the equation of the directrix and the coordinates of the focus of the parabola $x^2 + 4x + y = 0$.
24. Find the equation of one of the directrix of the ellipse $4x^2 + 3y^2 = 24$.

Section B: Broad Answer Type Questions

1. Determine whether each of the following conics has a single centre, no centers or infinitely many centers:

(i) $7x^2 - 2xy + 7y^2 - 16x + 16y - 8 = 0$.
(ii) $9x^2 + 6xy + y^2 + 6x + 2y + 20 = 0$.
(iii) $x^2 - 2xy + y^2 + 2x - 4y + 3 = 0$.

2. Correct or justify the following statements:

(i) $(x - x_1)(x - x_2) + (y - y_1)(y - y_2) = 0$, where (x_1, y_1) and (x_2, y_2) are two fixed points in two dimensional plane, is a circle.
(ii) $(x - 1)(x - 2) + (y - 3)(y - 4) = 0$ is a circle.
(iii) $11x^2 - 4xy + 14y^2 - 58x - 44y + 71 = 0$ is a pair of straight lines.
(iv) $9x^2 - 24xy + 16y^2 - 18x - 101y + 19 = 0$ represents a hyperbola.

3. Find the values of b and f for which the equation $x^2 + 6xy + by^2 + 3x + 2fy - 4 = 0$ represents *(i)* a central conic, *(ii)* a parabola, *(iii)* a curve with infinitely many centers.

4. Reduce each of the following equations to its canonical form and determine the type of the conics represented by it.

(i) $3x^2 - 8xy - 3y^2 + 10x - 13y + 8 = 0$.
(ii) $x^2 + 4xy + 4y^2 + 4x + y - 15 = 0$.
(iii) $x^2 - 2xy + 2y^2 - 4x - 6y + 3 = 0$.
(iv) $6x^2 - 5xy - 6y^2 + 14x + 5y + 4 = 0$.
(v) $4x^2 + 6xy + y^2 - 4x - 2y + a = 0$. for different values of a.

5. Find the equation of the conic which passes through the points $(0,0), (0,2), (-1,0), (-2,1)$ and $(-1,3)$ and determine its nature.

6. Find the equation of the conic which has its centre at the point $(1,-3)$, passes through the point $(0,2)$ and touches the x-axis at the origin.

7. Reduce the equation $4x^2 - 4xy + y^2 - 8x - 6y + 5 = 0$ to its canonical form and show that it represents a parabola. Find the latus rectum and the equation of the axis of the parabola.

8. Reduce the equation $3x^2 + 10xy + 3y^2 - 2x - 14y - 13 = 0$ to its canonical form and determine the type of the conic represented by it.

9. Find the equation of the hyperbola whose directrix is $2x + y = 1$, focus $(1, 1)$ and eccentricity is $\sqrt{3}$.

10. Reduce the equation $x^2 + 4xy + 4y^2 + 3x + 4y - 2 = 0$ to its canonical form and determine the nature of the conic.

11. Reduce the equation $x^2 + 4xy + y^2 - 2x + 2y + 6 = 0$ to its canonical form. Name the conic and find its eccentricity.

12. Reduce the equation $16x^2 - 24xy + 9y^2 - 104x - 172y + 44 = 0$ to its canonical form and determine the nature of the conic. Find the equation of its axes, directrix and latus rectum.

13. Show that the conic $ax^2 + 2hxy + by^2 + 2f\cos^2\theta x + 2g\sin^2\theta y + c = 0$ where θ is a parameter, always passes through two fixed points provided $g^2f^2 > c(af^2 + 2fgh + bg^2)$.

14. Reduce the equation $3(x^2 + y^2) + 2xy = 4\sqrt{2}(x+y)$ to its canonical form and determine the type of the conic represented by it. Name the conic, find the eccentricity of it and also find the equations of its axes.

15. Find the equation of the conic which passes through the intersection of the conics $x^2 - 2xy + 2y^2 - 4x - 6y + 3 = 0$ and $3x^2 + 10xy + 3y^2 - 2x - 14y - 13 = 0$ and also through the point $(1, 1)$.

16. Find the equation of the conic which passes through the point $(1, 1)$ and also through the intersection of the conic $x^2 + 2xy + 5y^2 - 7x - 8y + 6 = 0$ with the straight lines $2x - y - 5 = 0$ and $3x + y - 11 = 0$.

17. Show that the equation $(a^2 + b^2)(x^2 + y^2) = (ax + by - ab)^2$ represents a parabola of latus rectum $\dfrac{2ab}{\sqrt{a^2+b^2}}$.

18. Find the equation of the conic passing through the point of intersection of the straight lines $x - 3y - 4 = 0$ and $x + y = 0$ and the point of intersection of the conics $x^2 - 3xy + y^2 - 6x - 4y + 5 = 0$ and $3x^2 + 7xy - 3y^2 - 14x - 2y + 23 = 0$.

19. Find the condition that the general equation of second degree in x and y given by $ax^2 + 2hxy + by^2 + 2gx + 2fy + c = 0$ may represents (i) a pair of straight lines, (ii) a hyperbola.

20. Prove that in general two parabolas can be drawn to pass through

General Equation of Second Degree

the intersection of the conics

$$S \equiv ax^2 + 2hxy + by^2 + 2gx + 2fy + c = 0$$
$$\text{and} \quad S' \equiv a'x^2 + 2h'xy + b'y^2 + 2g'x + 2f'y + c' = 0$$

[*Hints:* The equation of any conic through the intersection of the given conics will be

$$S + \lambda S' = 0 \tag{2.68}$$

where λ is a constant.

$$(a + a'\lambda)x^2 + 2(h + h'\lambda)xy + (b + b'\lambda)y^2 + 2(g + g'\lambda)x$$
$$+ 2(f + f'\lambda)y + (c + c'\lambda) = 0 \tag{2.69}$$

It will represent a parabola, if

$$(a + a'\lambda)(b + b'\lambda) - (h + h'\lambda)^2 = 0$$
$$\Rightarrow (a'b' - h'^2)\lambda^2 + (ab' + a'b - 2hh')\lambda + (ab - h^2) = 0 \tag{2.70}$$

This is a quadratic equation in λ, showing that λ has two values, indicating that in general two parabolas can be drawn through the points of intersections of these conic.]

21. Show that the conic represented by

$$(a^2 + 1)x^2 + 2(a + b)xy + (b^2 + 1)y^2 = c, \; c > 0$$

is an ellipse of area $\dfrac{\pi c}{|ab - 1|}$, provided $ab \neq 1$.

[*Hints:* The equation is $(ax + y)^2 + (by + x)^2 = c$.

Let us apply the rotation $\left.\begin{array}{l} x = x' \cos\theta - y' \sin\theta \\ y = x' \sin\theta + y' \cos\theta. \end{array}\right\}$ where

$$\tan 2\theta = \frac{2(a + b)}{a^2 - b^2} = \frac{2}{a - b}.$$

Then the equation becomes

$$\{(ax' + y')\cos\theta - (ay' - x')\sin\theta\}^2$$
$$+ \{(bx' - y')\sin\theta + (by + x')\cos\theta\}^2 = c$$

or, $\dfrac{1}{2}\{(a^2 + 1)x'^2 + (b^2 + 1)y'^2\}\left(1 + \dfrac{a - b}{k}\right)$

$$+ \dfrac{1}{2}\{(b^2 + 1)x'^2 + (a^2 + 1)y'^2\}\left(1 - \dfrac{a - b}{k}\right)$$

$$+\{(a+b)x'^2 - (a-b)y^2\}\frac{2}{k} = c, \text{ where } k = \sqrt{(a-b)^2+4}$$

[considering the coefficient of $x'y' = 0$ for above θ]

or, $\left\{\frac{1}{2}(a^2+b^2+2) + \frac{a-b}{2k}(a^2-b^2) + \frac{2}{k}(a-b)\right\}x'^2$

$+ \left\{\frac{1}{2}(a^2+b^2+2) - \frac{a-b}{2k}(a^2-b^2) - \frac{2}{k}(a-b)\right\}y'^2 = c$

or, $\{k(a^2+b^2+2) + (a+b)(a-b)^2 + 4(a+b)\}x'^2$
$+\{k(a^2+b^2+2) - (a+b)(a-b)^2 + 4(a+b)\}y'^2 = 2kc$

or, $[(a^2+b^2+2)\sqrt{(a-b)^2+4} + (a+b)\{(a-b)^2+4\}]x'^2$
$+[(a^2+b^2+2)\sqrt{(a-b)^2+4} - (a+b)\{(a-b)^2+4\}]y'^2 = 2kc.$

We see that coefficient of x'^2 is positive. So the coefficient of y'^2 will also be positive, i.e.,

$$a^2 + b^2 + 2 > (a+b)\sqrt{(a-b)^2+4}$$

or, $(a^2+b^2+2)^2 > (a^2+b^2+2ab)(a^2-2ab+b^2+4)$

or, $(a^2+b^2+2)^2 - (a^2+b^2+2ab)(a^2-2ab+b^2+4) > 0$

$\Rightarrow (ab-1)^2 > 0$ which is true for $ab \neq 1$. So the equation represents an ellipse.

Area of the Ellipse: Area of the ellipse =

$$\pi \cdot \frac{\sqrt{2kc}}{\sqrt{(a^2+b^2+2)\sqrt{(a-b)^2+4} + (a+b)\{(a-b)^2+4\}}}$$

$$\times \frac{\sqrt{2kc}}{\sqrt{(a^2+b^2+2)\sqrt{(a-b)^2+4} - (a+b)\{(a-b)^2+4\}}}$$

$$= \frac{2k\pi c}{\sqrt{(a^2+b^2+2)^2\{(a-b)^2+4\} - (a-b)^2\{(a-b)^2+4\}^2}}$$

$$= \frac{2kc\pi}{k\sqrt{(a^2+b^2+2)^2 - (a+b)^2\{(a-b)^2+4\}}} \quad [\because k = \sqrt{(a-b)^2+4}]$$

$$= \frac{2\pi c}{\sqrt{4(ab-1)^2}} = \frac{\pi c}{|ab-1|}.$$

22. Determine the number of parabolas passing through the points $(-4, 0)$, $\left(-\frac{1}{4}, \frac{5}{4}\right)$, $\left(\frac{4}{9}, -\frac{10}{9}\right)$ and $(1, 0)$ and find their equations.

ANSWERS

Section A: 3. $k = -3, -7$; **4.** $A = 8, B = -2$; **5.** Ellipse; **6.** $(3, -1)$; **10.** $6x'^2 - 5x'y' - 6y'^2 = 0$; **12.** 26; **13.** 5; **18.** $x + y = 0, 2x + y = 0$; **19.** $a = 6$; **20.** Ellipse; **21.** $\left(0, -\frac{5}{4}\right)$; **22.** $(0, \pm\sqrt{2})$; **23.** $y = \frac{17}{4}, \left(-2, \frac{15}{4}\right)$; **24.** $4\sqrt{2}$.

Section B: 1. (i) Single centre; (ii) Infinitely many centers; (iii) No centers; **2.** (i) Correct; (ii) Incorrect, it is an ellipse; (iii) Incorrect, it is a hyperbola; (iv) Incorrect, it is a parabola; **3.** (i) $b \neq 9$; (ii) $b = 9, f \neq \frac{9}{2}$; (iii) $b = 9, f = \frac{9}{2}$; **4.** (i) $x'^2 - y'^2 = \frac{33}{500}$ or $x'^2 - y'^2 = -\frac{33}{500}$, a hyperbola; (ii) $x'^2 = \frac{7}{5\sqrt{5}}y'$ or $y'^2 = \frac{7}{5\sqrt{5}}x'$, a parabola; (iii) $\frac{x'^2}{52} + \frac{y'^2}{\frac{52}{5}} = 1$ or $\frac{3+\sqrt{5}}{2}x'^2 + \frac{3-\sqrt{5}}{2}y'^2 = 26$, an ellipse; (iv) $x'^2 - y'^2 = 0$, a pair of straight lines; (v) $x'^2 = \lambda$ where $\lambda = \frac{1-a}{5}$, when $a = 1 \Rightarrow$ it represents two coincident lines, when $a > 1 \Rightarrow$ it has no geometrical object, when $a < 1 \Rightarrow$ it represents a pair of parallel straight lines; **5.** $3x^2 + 2xy + 2y^2 + 3x - 4y = 0$, Canonical form: $\frac{10+3\sqrt{2}}{8}x'^2 + \frac{10-3\sqrt{2}}{8}y'^2 = \frac{9}{2}$, an ellipse; **6.** $6x'^2 + 4x'y' + y'^2 + 2y' = 0$; **7.** Equation of the parabola is $y'^2 = -\frac{4}{5\sqrt{5}}x'$, axis is $2x + y - \frac{11}{5} = 0$; **8.** $x'^2 - \frac{y'^2}{4} = 1$; **9.** $7x'^2 - 2y'^2 + 12x'y' - 2x' + 4y' - 7 = 0$; **10.** $y'^2 = \frac{2}{3\sqrt{5}}x'$, a parabola; **11.** $\frac{x'^2}{8} - \frac{y'^2}{\frac{8}{3}} = 1$, a hyperbola with $e = \frac{2}{\sqrt{3}}$; **12.** $y'^2 = 8x'$, a parabola with axis $4x - 3y + 2 = 0$, the equation of the directrix is $3x + 4y + 9 = 0$, equation of the latus rectum $3x + 4y - 11 = 0$; **13.** The point of intersection of $ax^2 + 2hxy + by^2 + 2gx + 2fy + c = 0$ and $gx = fy$. Also see Worked Out Example 2.10.9; **14.** $\frac{x'^2}{1} + \frac{y'^2}{2} = 1$, an ellipse with $e = \frac{1}{\sqrt{2}}$ and equation of the axes are $x + y = \sqrt{2}$ and $x - y = 0$; **15.** $5x'^2 + 86x'y' - 8y'^2 + 40x' - 6y' - 117 = 0$; **16.** $34x'^2 + 55x'y' + 139y'^2 - 233x' - 218y' + 223 = 0$; **18.** $5x'^2 + 17x'y' - 7y'^2 - 22x' + 41 = 0$; **19.** (i) $\Delta = 0, \delta \geq 0$, where

$$\Delta = \begin{vmatrix} a & h & g \\ h & b & f \\ g & f & c \end{vmatrix} \text{ and } \delta = ab - h^2 = \begin{vmatrix} a & h \\ h & b \end{vmatrix};$$

(ii) $\delta < 0$; **22.** Two, $x^2 + 2xy + y^2 + 3x - 2y - 4 = 0$ and $169x^2 + 26xy + y^2 + 27x - 146y - 196 = 0$.

Chapter 3

Polar Coordinates and Equations

3.1 Polar Coordinates

Polar coordinates express the position of a point P as (r, θ), where r is the distance of the point, from the origin O, known as the *pole* and θ is the angle between the positive x-axis and the line OP (See **Figure 3.1 (i)**). This angle θ is called the *vectorial angle*, positive x-axis is called the *polar axis* or the *initial line*. The distance $r = OP$, which is the distance from the origin O (pole) to the point P in consideration, is called the *radius vector*.

3.2 Discussion on Polar Coordinates

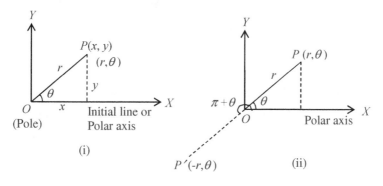

Figure 3.1

(1) We extend the meaning of polar coordinates (r, θ) to the case in which r is negative agreeing that the points $(-r, \theta)$ and (r, θ) lie in the same line through O and at the same distance $|r|$ form O, but on the opposite sides of O (See **Figure 3.1 (ii)**). If $r > 0$ the point (r, θ) lines in the same quadrant at θ. If $r < 0$, it lies in the quadrant on the opposite side of the pole. Thus $(-r, \theta)$ is the same as $(r, \pi + \theta)$.

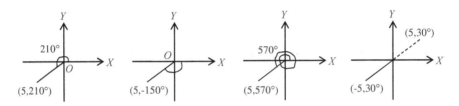

Figure 3.2

(2) In the Cartesian coordinate system, every point has only one representation, but in the polar coordinate system, each point has many representation. For example, let us consider a point $(5, 210°)$. This point can also be written as $(5, -150°), (5, 570°), (-5, 30°)$ as shown in the **Figure 3.2**.

Note 3.2.1 *It may seen unnecessary to introduce the negative concept of the radius vector, since it is always possible to denote the point by using a positive radius vector, after making suitable change in the vectorial angle. That it is necessary will shown in the study of polar equation of conics, to be discussed in this present chapter.*

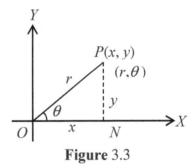

Figure 3.3

3.3 Relation between Polar and Cartesian Coordinates of a Point

We consider the pole and the polar axis (or the initial line) of the polar system be the same as the origin and the positive direction of the x-axis of the Cartesian system respectively. Also, let us suppose that the polar coordinates of a point be (r, θ) and its Cartesian coordinates be (x, y). Then

Polar Coordinates

it is evident from the **Figure 3.3** that

$$\left.\begin{array}{l} x = r\cos\theta \\ y = r\sin\theta. \end{array}\right\} \text{ Also, } \left.\begin{array}{l} x^2 + y^2 = r^2 \\ \dfrac{y}{x} = \tan\theta. \end{array}\right\}$$

These formulae are used to convert from one system to the other.

3.4 Distance between Two Points

Let $P_1(r_1, \theta_1)$ and $P_2(r_2, \theta_2)$ be the given two points, O, the pole and OX is the initial line.

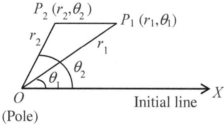

Figure 3.4

From the triangle OP_1P_2 we have

$$\begin{aligned} P_1P_2^2 &= OP_1^2 + OP_2^2 - 2OP_1.OP_2\cos\angle P_1OP_2 \\ &= r_1^2 + r_2^2 - 2r_1r_2\cos(\theta_2 - \theta_1). \end{aligned}$$

$\therefore$ Distance $P_1P_2 = \sqrt{r_1^2 + r_2^2 - 2r_1r_2\cos(\theta_2 - \theta_1)}.$

Alternatively,

If (x_1, y_1) and (x_2, y_2) be the Cartesian coordinates of the points P_1 and P_2 respectively, then we have

$$\begin{aligned} P_1P_2 &= \sqrt{(x_2 - x_1)^2 + (y_2 - y_1)^2} \\ &= \sqrt{(r_2\cos\theta_2 - r_1\cos\theta_1)^2 + (r_2\sin\theta_2 - r_1\sin\theta_1)^2} \\ &= \sqrt{r_1^2 + r_2^2 - 2r_1r_2\cos(\theta_2 - \theta_1)}. \end{aligned}$$

3.5 Area of a triangle

Let $A(r_1, \theta_1), B(r_2, \theta_2)$ and $C(r_3, \theta_3)$ be the vertices of the triangle ABC. Then we have from the **Figure 3.5**,

$$\begin{aligned}\triangle ABC &= \triangle OAB + \triangle OAC - \triangle OBC \\ &= \frac{1}{2}r_1 r_2 \sin(\theta_2 - \theta_1) + \frac{1}{2}r_1 r_3 \sin(\theta_1 - \theta_3) - \frac{1}{2}r_2 r_3 \sin(\theta_2 - \theta_3) \\ &= -\frac{1}{2}[r_1 r_2 \sin(\theta_1 - \theta_2) + r_3 r_1 \sin(\theta_3 - \theta_1) + r_2 r_3 \sin(\theta_2 - \theta_3)]. \end{aligned}$$

Since area can not be negative we get, area of the $\triangle ABC$

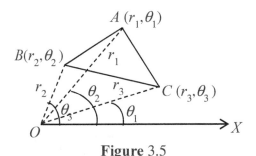

Figure 3.5

$$= \frac{1}{2}[r_1 r_2 \sin(\theta_1 - \theta_2) + r_2 r_3 \sin(\theta_2 - \theta_3) + r_3 r_1 \sin(\theta_3 - \theta_1)].$$

Note 3.5.1 *If the three points be collinear, then the area of the triangle ABC vanishes, so we get*

$$r_1 r_2 \sin(\theta_1 - \theta_2) + r_2 r_3 \sin(\theta_2 - \theta_3) + r_3 r_1 \sin(\theta_3 - \theta_1) = 0$$

which gives the condition for collinearity of the given points.

3.6 Straight Lines

3.6.1 Equation of a straight line in polar coordinates in general form

The general equation of a straight line in Cartesian form is given by

$$ax + by + c = 0.$$

Polar Coordinates

Converting this in polar form, i.e., by putting $x = r\cos\theta$ and $y = r\sin\theta$ we get

$$ar\cos\theta + br\sin\theta + c = 0 \Rightarrow \frac{A}{r} = a\cos\theta + b\sin\theta, \text{ where } A = -c.$$

This is the equation of the straight line in polar coordinates in general form.

Example 3.6.1 *Show that the polar equation of a straight line parallel to the line* $\frac{A}{r} = a\cos\theta + b\sin\theta$ *can be written as* $\frac{B}{r} = a\cos\theta + b\sin\theta$.

Solution: Transferring in Cartesian coordinates the given equation of the line is written as $A = ax + by$. So its parallel line is given by $B = ax + by$ i.e., $\frac{B}{r} = a\cos\theta + b\sin\theta$ is the required equation. □

Note 3.6.1 *The perpendicular line to*

$$\frac{A}{r} = a\cos\theta + b\sin\theta \text{ is } \frac{C}{r} = -b\cos\theta + a\sin\theta.$$

3.6.2 Equation of a Line Passing through the Two Given Points

1st Method: The general form of the equation of a line in polar coordinates is given by

$$\frac{A}{r} = a\cos\theta + b\sin\theta \qquad (3.1)$$

Let it passes through the given points (r_1, θ_1) and (r_2, θ_2). So we get

$$\frac{A}{r_1} = a\cos\theta_1 + b\sin\theta_1 \qquad (3.2)$$

$$\frac{A}{r_2} = a\cos\theta_2 + b\sin\theta_2 \qquad (3.3)$$

Eliminating A, a and b from (3.1), (3.2) and (3.3) we get

$$\begin{vmatrix} \cos\theta & \sin\theta & \dfrac{1}{r} \\ \cos\theta_1 & \sin\theta_1 & \dfrac{1}{r_1} \\ \cos\theta_2 & \sin\theta_2 & \dfrac{1}{r_2} \end{vmatrix} = 0$$

$$\Rightarrow \frac{1}{r}(\sin\theta_2\cos\theta_1 - \cos\theta_2\sin\theta_1) + \frac{1}{r_1}(\sin\theta\cos\theta_2 - \cos\theta\sin\theta_2)$$
$$+ \frac{1}{r_2}(\sin\theta_1\cos\theta - \cos\theta_1\sin\theta) = 0$$

$$\Rightarrow \frac{\sin(\theta_2 - \theta_1)}{r} + \frac{\sin(\theta - \theta_2)}{r_1} - \frac{\sin(\theta - \theta_1)}{r_2} = 0$$

$$\Rightarrow \frac{\sin(\theta_1 - \theta_2)}{r} = \frac{\sin(\theta - \theta_2)}{r_1} + \frac{\sin(\theta_1 - \theta)}{r_2}$$

which is the required equation of the line in polar coordinates.

Note 3.6.2 *To remember take the variable point (r, θ) in the middle as follows:*

then follow the scheme:
$$\begin{array}{ccc} \theta_1 & \theta & \theta_2 \\ r_1 & r & r_2 \end{array}$$

and
$$\left[\frac{\theta_1 - \theta_2}{r}\right] = \left[\frac{\theta - \theta_2}{r_1}\right] + \left[\frac{\theta_1 - \theta}{r_2}\right]$$

with sines of the difference of the angles.

2nd Method: The polar equation can also be established form the corresponding equation in Cartesian form. Let the straight line passes through the two given points (x_1, y_1) and (x_2, y_2). Then its equation is given by

$$\frac{x - x_1}{x_2 - x_1} = \frac{y - y_1}{y_2 - y_1}.$$

Transforming into polar coordinates when $(x_1, y_1) \equiv (r_1, \theta_1)$ and $(x_2, y_2) \equiv (r_2, \theta_2)$ we get

$$\frac{r\cos\theta - r_1\cos\theta_1}{r_2\cos\theta_2 - r_1\cos\theta_1} = \frac{r\sin\theta - r_1\sin\theta_1}{r_2\sin\theta_2 - r_1\sin\theta_1}.$$

After simplification, this gives

$$\frac{\sin(\theta_1 - \theta_2)}{r} = \frac{\sin(\theta - \theta_2)}{r_1} + \frac{\sin(\theta_1 - \theta)}{r_2}.$$

3rd Method: By using properties of triangle.

Let AB be the line passing through the two points (r_1, θ_1) and (r_2, θ_2). Let $P(r, \theta)$ be any point on the line AB.

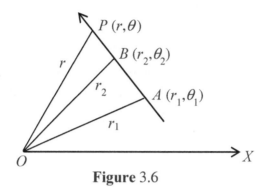

Figure 3.6

Then form the **Figure 3.6**, we have

$$\triangle OAP = \triangle OAB + \triangle OBP$$
$$\Rightarrow \frac{1}{2} r_1 r \sin(\theta - \theta_1) = \frac{1}{2} r_1 r_2 \sin(\theta_2 - \theta_1) + \frac{1}{2} r_2 r \sin(\theta - \theta_2)$$
$$\Rightarrow \frac{\sin(\theta_1 - \theta_2)}{r} = \frac{\sin(\theta - \theta_2)}{r_1} + \frac{\sin(\theta_1 - \theta)}{r_2} \quad \text{(dividing by } \frac{1}{2} r r_1 r_2\text{)}.$$

This is the required equation.

3.6.3 Polar equation of a line in normal form

The equation of a line in terms of p, the perpendicular distance of the line form the pole and the angle α, made by the normal to the line through the pole with the initial line is known as the normal form.

Let $P(r, \theta)$ be any point on the line and ON is the normal to the line through O, then $\triangle OPN$ is right angled.

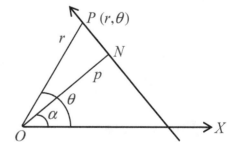

Figure 3.7

$$\therefore \quad \frac{ON}{OP} = \cos \angle NOP \Rightarrow \frac{p}{r} = \cos(\theta - \alpha) \Rightarrow r \cos(\theta - \alpha) = p$$

which is the required equation of the straight line in normal form.

Particular cases:

(i) If the line is perpendicular to the initial line then $\alpha = 0$ and the equation of the line becomes $r\cos\theta = p$.

(ii) If the line is parallel to the initial line then $\alpha = \dfrac{\pi}{2}$ and the equation of the line becomes $r\sin\theta = p$.

(iii) The polar equations of parallel lines are of the forms $r\cos(\theta - \alpha) = p$ and $r\cos(\theta - \alpha) = p'$ and the polar equations of two mutually perpendicular lines are of the form $r\cos(\theta - \alpha) = p$ and $r\sin(\theta - \alpha) = p$.

Example 3.6.2 *Show that the equation of the line inclined at an angle β with the polar axis and passes through the point (r_1, θ_1) is given by*

$$r\sin(\beta - \theta) = r_1 \sin(\beta - \theta_1).$$

Solution:

Let the normal to the line be inclined at an angle α with the polar axis and its distance form the pole is p, then its equation is

$$r\cos(\theta - \alpha) = p \qquad (3.4)$$

If it passes through (r_1, θ_1) then

$$r_1 \cos(\theta_1 - \alpha) = p \qquad (3.5)$$

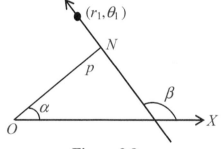

Figure 3.8

From (3.4) and (3.5) we get

$$r\cos(\theta - \alpha) = r_1 \cos(\theta_1 - \alpha).$$

Now $\beta = \dfrac{\pi}{2} + \alpha$, i.e., $\alpha = -\left(\dfrac{\pi}{2} - \beta\right)$.

So we get $r\cos\left[\theta + \left(\dfrac{\pi}{2} - \beta\right)\right] = r_1 \cos\left(\theta_1 - \beta + \dfrac{\pi}{2}\right)$.

Therefore $r\sin(\beta - \theta) = r_1 \sin(\beta - \theta_1)$. □

3.7 The Circle

3.7.1 Polar equation of a circle

Let $C(\rho, \alpha)$ be the centre of the circle with radius d in **Figure 3.9**. Let $P(r, \theta)$ be any point on the circle, O being the pole and OX is the polar axis. From the $\triangle OPC$, we have

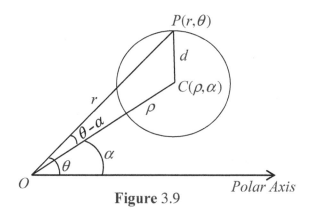

Figure 3.9

$$PC^2 = OP^2 + OC^2 - 2.OP.OC.\cos \angle POC$$
$$\therefore \quad d^2 = r^2 + \rho^2 - 2\rho r \cos(\theta - \alpha)$$
$$\text{or,} \quad r^2 - 2\rho r \cos(\theta - \alpha) + \rho^2 - d^2 = 0.$$

The is the required polar equation of the circle in general form in polar coordinates.

Particular cases:

(i) Let the polar axis touches the circle then we get $d = \rho \sin \alpha$ and the equation of the circle becomes

$$r^2 - 2\rho r \cos(\theta - \alpha) + \rho^2 - \rho^2 \sin^2 \alpha = 0$$
$$\Rightarrow r^2 - 2\rho r \cos(\theta - \alpha) + \rho^2 \cos^2 \alpha = 0.$$

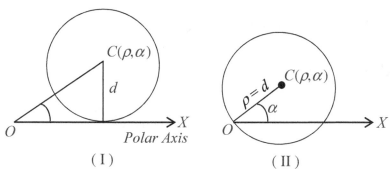

Figure 3.10

(ii) Let the pole O is on the circumference of the circle (*vide.* **Figure 3.10 (II)**), then $\rho = d$ and the equation of the circle becomes

$$r^2 - 2dr\cos(\theta - \alpha) + d^2 - d^2 = 0 \Rightarrow r = 2d\cos(\theta - \alpha).$$

(iii) Let the pole is on the circumference of the circle and the polar axis passes through the centre, then $\alpha = 0$ also and we get the equation of the circle as $r = 2d\cos\theta$.

(iv) When the pole is at the centre of the circle, then $\rho = 0$ and the equation of the circle becomes $r^2 = d^2 \Rightarrow r = d$.

Example 3.7.1 *Show that the polar equation of the circle passing through the pole can be expressed in the form $r = A\cos\theta + B\sin\theta$, where A and B are constants.*

Solution: The equation of a circle passing through the pole is given by

$$r = 2d\cos(\theta - \alpha) = 2d\cos\theta\cos\alpha + 2d\sin\theta\sin\alpha = A\cos\theta + B\sin\theta$$

where $A = 2d\cos\alpha$ and $B = 2d\sin\alpha$. □

Example 3.7.2 *Find the polar coordinates of the centre of the circle*

$$r = 4\cos\theta + 3\sin\theta.$$

Solution: Let us put $4 = a\cos\alpha$, $3 = a\sin\alpha$ which gives $a = \sqrt{4^2 + 3^2} = 5$ and $\tan\alpha = \dfrac{3}{4}$, i.e., $\alpha = \tan^{-1}\dfrac{3}{4}$. Now the equation of the circle is written as

$$r = a\cos\theta\cos\alpha + a\sin\theta\sin\alpha = a\cos(\theta - \alpha).$$

So its centre is given by $\left(\dfrac{a}{2}, \alpha\right) = \left(\dfrac{5}{2}, \tan^{-1}\dfrac{3}{4}\right)$ and radius $= \dfrac{5}{2}$. □

Example 3.7.3 *Find the polar equation of the circle which passes through the pole and the two points where polar coordinates are $(d, 0)$ and $\left(2d, \dfrac{\pi}{3}\right)$. Find also the radius of the circle.*

Solution: Let the equation of the circle be $r = a\cos\theta + b\sin\theta$. It passes through $(d, 0)$ ∴ $d = a$. Also it passes through $\left(2d, \dfrac{\pi}{3}\right)$.

$$\therefore\ 2d = d\cos\dfrac{\pi}{3} + b\sin\dfrac{\pi}{3} = \dfrac{d}{2} + b.\dfrac{\sqrt{3}}{2}$$

$$\therefore\ \sqrt{3}b = 4d - d = 3d \Rightarrow b = \sqrt{3}d.$$

Polar Coordinates

Hence the equation of the circle becomes

$$r = d\cos\theta + \sqrt{3}d\sin\theta = 2d\left(\frac{1}{2}\cos\theta + \frac{\sqrt{3}}{2}\sin\theta\right) = 2d\cos\left(\theta - \frac{\pi}{3}\right).$$

Its radius is d and centre is $\left(d, \dfrac{\pi}{3}\right)$. □

3.8 Polar Equation of a Conic

3.8.1 Definition of a conic

The locus of a point which moves in a plane such that the ratio of its distance form a fixed point to its perpendicular distance from a fixed straight line (not passing through the given fixed point) is always constant is known a *conic section* or a *conic*.

The fixed point is called the *focus* and the fixed line is called the *directrix* of the conic.

Also the constant ratio is called the *eccentricity* of the conic which is denoted by e and we have the followings:

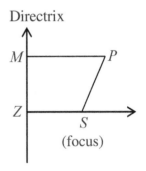

Figure 3.11

(i) If $e = 1$, the conic is a *parabola*.
(ii) If $0 < e < 1$, the conic is an *ellipse*.
(ii) If $e > 1$, the conic is a *hyperbola*.
(iv) If $e = 0$, the conic is a *circle*.
(v) If $e = \infty$, the conic is a *pair of straight lines*.

In the adjacent **Figure 3.11**, we get $\dfrac{SP}{PM} = e$, i.e., $SP = e.PM$.

Note 3.8.1 *Though a circle [case (iv)] and a pair of straight lines [case (v)] also represent special cases of the conic sections, they are generally treated separately. Thus by the term conic section, we would mean only the three curves mentioned in cases (i), (ii) and (iii).*

3.8.2 Polar equation of a conic whose semi latus rectum is l and focus is the pole

Let $P(r, \theta)$ be any point on the conic, A, the vertex, S, the focus, which is considered as pole, SX, the polar axis, MZ, the directrix and SL, the semi-latus rectum of length l.

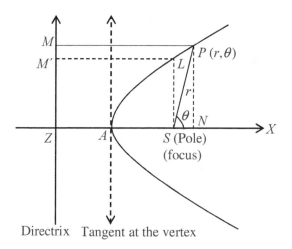

Figure 3.12

From the **Figure 3.12**, we see that $\dfrac{SP}{PM} = e$, the eccentricity of the conic (the definition of a conic).

$$\Rightarrow r = e.PM = e(SN + SZ) = e(r\cos\theta + LM') \tag{3.6}$$

Now, form the same definition of conic, we get

$$\dfrac{SL}{LM'} = e \quad \therefore \quad LM' = \dfrac{SL}{e} = \dfrac{l}{e} \tag{3.7}$$

Using (3.6) and (3.7) we get

$$r = e\left(r\cos\theta + \dfrac{l}{e}\right) = er\cos\theta + l \Rightarrow 1 = e\cos\theta + \dfrac{l}{r}$$

$$\therefore \dfrac{l}{r} = 1 - e\cos\theta,$$

which is the required equation of the conic.

Note 3.8.2 *If the polar axis is taken form S to Z as shown in the **Figure** 3.13, then θ is replaced by $\pi - \theta$ and then the equation of the conic becomes*

$$\dfrac{l}{r} = 1 - e\cos(\pi - \theta) = 1 + e\cos\theta.$$

So, in this case, the equation of the conic will be

$$\dfrac{l}{r} = 1 + e\cos\theta.$$

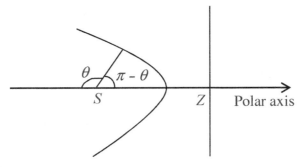

Figure 3.13

Particular cases:

(i) Let the conic is a parabola, then its equation will be $\frac{l}{r} = 1 - \cos\theta$ or $\frac{l}{r} = 1 + \cos\theta$ according as the focus is on the right hand side or left hand side of the directrix respectively, i.e., according as the polar axis is taken along ZS or SZ respectively, i.e., according as the polar axis is taken from directrix to focus or focus to directrix respectively.

(ii) Let the conic is an ellipse, then its equation will be $\frac{l}{r} = 1 - e\cos\theta$ or $\frac{l}{r} = 1 + e\cos\theta$, according as the left hand focus or the right hand focus is the pole respectively.

(iii) Let the conic be a hyperbola, then the equation of its right branch is given by $\frac{l}{r} = 1 - e\cos\theta$ and the equation of its left branch will be given by $\frac{l}{r} = 1 + e\cos\theta$. However, if the equation of its right branch is given by $\frac{l}{r} = 1 - e\cos\theta$, then the equation of the left branch with respect to the right branch is given by $\frac{l}{r} = -(1 + e\cos\theta)$.

(iv) Let the initial line makes an angle γ with the axis of the conic, then the equation of the conic will be obtained by changing θ into $\theta \pm \gamma$ (taking both clockwise or anticlockwise sense of γ) as $\frac{l}{r} = 1 - e\cos(\theta \pm \gamma)$ or $\frac{l}{r} = 1 + e\cos(\theta \pm \gamma)$ according as the positive direction of the polar axis is chosen.

3.9 Polar Equation of the Chord of a Given Conic

Let the equation of the conic be

$$\frac{l}{r} = 1 - e\cos\theta \qquad (3.8)$$

Also let P and Q be the two extremities of the chord PQ of the conic with coordinates $P(r_1, \alpha - \beta)$ and $Q(r_2, \alpha + \beta)$. The equation of the line PQ will be of the form

$$\frac{l}{r} = a\cos\theta + b\sin\theta \qquad (3.9)$$

The coordinates of P and Q will satisfy both (3.8) and (3.9), i.e., we get

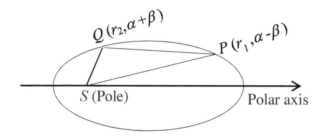

Figure 3.14

$$\frac{l}{r_1} = 1 - e\cos(\alpha - \beta) = a\cos(\alpha - \beta) + b\sin(\alpha - \beta)$$

or, $(a+e)\cos(\alpha - \beta) + b\sin(\alpha - \beta) - 1 = 0$ for the point P

and similarly for point Q, $(a+e)\cos(\alpha + \beta) + b\sin(\alpha + \beta) - 1 = 0$.

By cross-multiplication, we get

$$\frac{a+e}{\sin(\alpha+\beta) - \sin(\alpha-\beta)} = \frac{b}{\cos(\alpha-\beta) - \cos(\alpha+\beta)} = \frac{1}{\sin(\alpha+\beta)\cos(\alpha-\beta) - \cos(\alpha-\beta)\sin(\alpha+\beta)}$$

$$\Rightarrow \frac{a+e}{2\cos\alpha\sin\beta} = \frac{b}{2\sin\alpha\sin\beta} = \frac{1}{\sin 2\beta} = \frac{1}{2\sin\beta\cos\beta}$$

$$\therefore\ a+e = \frac{2\cos\alpha\sin\beta}{2\cos\beta\sin\beta} = \cos\alpha\sec\beta \quad (\because \beta \neq 0 \Rightarrow \sin\beta \neq 0)$$

$$\therefore\ a = \cos\alpha\sec\beta - e$$

and $$b = \frac{2\sin\alpha\sin\beta}{2\cos\beta\sin\beta} = \sin\alpha\sec\beta \quad (\because \sin\beta \neq 0).$$

Polar Coordinates

So the equation of the chord is obtained as

$$\frac{l}{r} = (\cos\alpha \sec\beta - e)\cos\theta + \sin\alpha \sec\beta \sin\theta$$
$$= \sec\beta(\cos\alpha\cos\theta + \sin\alpha\sin\theta) - e\cos\theta$$
$$= \sec\beta\cos(\alpha - \theta) - e\cos\theta.$$

Thus the polar equation of the chord joining the points whose vectorial angles are given as $\alpha - \beta$ and $\alpha + \beta$ on the conic $\frac{l}{r} = 1 - e\cos\theta$ is

$$\frac{l}{r} = \sec\beta\cos(\theta - \alpha) - e\cos\theta.$$

Corollary 3.9.1 *The polar equation of the chord of the conic $\frac{l}{r} = 1 - e\cos\theta$ with θ_1 and θ_2 as the vectorial angles of its extremities is*

$$\frac{l}{r} = \sec\frac{\theta_1 - \theta_2}{2}\cos\left(\theta - \frac{\theta_1 + \theta_2}{2}\right) - e\cos\theta.$$

Proof: Let us put $\theta_1 = \alpha + \beta$ and $\theta_2 = \alpha - \beta$. Then we get $\alpha = \frac{\theta_1 + \theta_2}{2}$ and $\beta = \frac{\theta_1 - \theta_2}{2}$ and so the equation of the chord becomes

$$\frac{l}{r} = \sec\frac{\theta_1 - \theta_2}{2}\cos\left(\theta - \frac{\theta_1 + \theta_2}{2}\right) - e\cos\theta. \qquad \square$$

Note 3.9.1 If the equation of the given conic be $\frac{l}{r} = 1 + e\cos\theta$, then the equation of the chord joining $\alpha - \beta$ and $\alpha + \beta$ as the vectorial angles will be

$$\frac{l}{r} = \sec\beta\cos(\theta - \alpha) + e\cos\theta.$$

Also if the vectorial angles be θ_1 and θ_2, then the equation of the chord will be

$$\frac{l}{r} = \sec\frac{\theta_1 - \theta_2}{2}\cos\left(\theta - \frac{\theta_1 + \theta_2}{2}\right) + e\cos\theta.$$

3.10 Polar Equation of the Tangent to a Given Conic

Polar equation of the chord of the conic $\frac{l}{r} = 1 - e\cos\theta$ with $\alpha - \beta$ and $\alpha + \beta$ as the vectorial angles of the extremities is given by

$$\frac{l}{r} = \sec\beta\cos(\theta - \alpha) - e\cos\theta \qquad (3.10)$$

If we put $\beta = 0$, then the line corresponding to the chord becomes a tangent line to the conic at the point whose vectorial angle is α. The equation (3.10) then takes the form
$$\frac{l}{r} = \cos(\theta - \alpha) - e\cos\theta$$
which is the required equation of the tangent.

Note 3.10.1 *The polar equation of the tangent to the conic* $\frac{l}{r} = 1 + e\cos\theta$ *at the point 'α' will be obtained as*
$$\frac{l}{r} = \cos(\theta - \alpha) + e\cos\theta.$$

Note 3.10.2 *If* $\frac{l}{r} = 1 - e\cos\theta$ *be the equation of the right branch of a hyperbola, then the equation of the tangent at any point α on its left branch is gievn by*
$$\frac{l}{r} = -\cos(\theta - \alpha) - e\cos\theta.$$

Note 3.10.3 *If the equation of the conic be* $\frac{l}{r} = 1 - e\cos(\theta - \gamma)$, *the equation of the tangent at α will be*
$$\frac{l}{r} = \cos(\theta - \alpha) - e\cos(\theta - \gamma).$$

3.11 Polar Equation of the Normal to a Given Conic

The polar equation of the tangent to the conic (3.8) at the point $\theta = \alpha$ is given by
$$\begin{aligned}\frac{l}{r} &= \cos(\theta - \alpha) - e\cos\theta = \cos\theta\cos\alpha + \sin\theta\sin\alpha - e\cos\theta \\ &= (\cos\alpha - e)\cos\theta + \sin\alpha\sin\theta \end{aligned} \qquad (3.11)$$

Equation of any line perpendicular to (3.11) is given by
$$\frac{l'}{r} = -\sin\alpha\cos\theta + (\cos\alpha - e)\sin\theta = \sin(\theta - \alpha) - e\sin\theta.$$

It passes through (r_1, α), so we get
$$\frac{l'}{r_1} = -e\sin\alpha \qquad (3.12)$$

Polar Coordinates

Again the point (r_1, α) is a point on the conic

$$\therefore \frac{l}{r_1} = 1 - e\cos\alpha \qquad (3.13)$$

From (3.12) and (3.13) we get

$$l' = -er_1 \sin\alpha = -e\frac{l\sin\alpha}{1 - e\cos\alpha}.$$

So, the line perpendicular to (3.11) becomes

$$-\frac{le\sin\alpha}{r(1 - e\cos\alpha)} = \sin(\theta - \alpha) - e\sin\theta$$

which is the required equation of the normal.

Note 3.11.1 *If the equation of the conic be* $\frac{l}{r} = 1 + e\cos\theta$, *then the equation of the normal at any point* $\theta = \alpha$ *on it will be*

$$\frac{le\sin\alpha}{r(1 + e\cos\alpha)} = \sin(\theta - \alpha) + e\sin\theta.$$

3.12 Chord of Contact

Definition 3.12.1 *The Chord of contact is a chord joining the points of contact of tangents to a conic from a given point not lying on the conic.*

3.12.1 Equation of Chord of Contact

Let $\alpha - \beta$ and $\alpha + \beta$ be the vectorial angles of the points of contact of the tangents to the conic (3.8) from the point (r_1, θ_1). Then the equation of the chord of contact, i.e., the equation of the line PQ is given by

$$\frac{l}{r} = \sec\beta \cos(\theta - \alpha) - e\cos\theta \qquad (3.14)$$

The equation of the tangent at the point $\alpha - \beta$ is given by

$$\frac{l}{r} = \cos\{\theta - (\alpha - \beta)\} - e\cos\theta.$$

Since it passes through (r_1, θ_1) so we get

$$\frac{l}{r_1} = \cos(\theta_1 - \alpha + \beta) - e\cos\theta_1 \qquad (3.15)$$

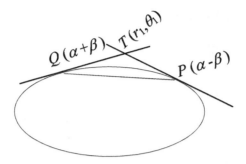

Figure 3.15

Similarly for the point $Q(\alpha + \beta)$ we get

$$\frac{l}{r_1} = \cos(\theta_1 - \alpha - \beta) - e\cos\theta_1 \qquad (3.16)$$

From (3.15) and (3.16) we get

$$\cos(\theta_1 - \alpha + \beta) = \cos(\theta_1 - \alpha - \beta) \Rightarrow \theta_1 - \alpha + \beta = \pm(\theta_1 - \alpha - \beta).$$

Since $\theta_1 - \alpha + \beta = \theta_1 - \alpha - \beta \Rightarrow \beta = 0$ which is not admissible, so we take $\theta_1 - \alpha + \beta = -(\theta_1 - \alpha - \beta) \Rightarrow \theta_1 = \alpha$. So from (3.15)

$$\frac{l}{r_1} = \cos(\theta_1 - \theta_1 + \beta) - e\cos\theta_1 = \cos\beta - e\cos\theta_1$$

$$\therefore \cos\beta = \frac{l}{r_1} + e\cos\theta$$

and hence form (3.14) we get

$$\left(\frac{l}{r} + e\cos\theta\right)\left(\frac{l}{r_1} + e\cos\theta_1\right) = \cos(\theta - \theta_1)$$

which is the required equation of the chord of contact.

Note 3.12.1 *If the equation of the conic be taken as $\frac{l}{r} = 1 + e\cos\theta$, then the equation of the chord of contact of the tangents to the conic form the point (r_1, θ_1) is*

$$\left(\frac{l}{r} - e\cos\theta\right)\left(\frac{l}{r_1} - e\cos\theta_1\right) = \cos(\theta - \theta_1).$$

Note 3.12.2 *Equating (3.15) and (3.16) from above, if we consider the general values,*

$$\cos(\theta_1 - \alpha + \beta) = \cos(\theta_1 - \alpha - \beta) \Rightarrow \theta_1 - \alpha + \beta = 2n\pi \pm (\theta_1 - \alpha - \beta).$$

We get, $\theta_1 - \alpha + \beta = 2n\pi - (\theta_1 - \alpha - \beta) \Rightarrow 2\theta_1 = 2n\pi + 2\alpha \Rightarrow \theta_1 = n\pi + \alpha.$
$\therefore \alpha = \theta_1 - n\pi.$

$\therefore$ form the equation (3.15), $\dfrac{l}{r_1} + e\cos\theta = \cos(n\pi + \beta) = (-1)^n \cos\beta.$

Hence equation (3.14) becomes

$$\left(\dfrac{l}{r} + e\cos\theta\right)\left(\dfrac{l}{r_1} - e\cos\theta_1\right) = (-1)^n \cos(\theta - \theta_1 + n\pi) = \cos(\theta - \theta_1)$$

which is same as before.

Note 3.12.3 *If TP and TQ be the two tangents, then we have $\theta_1 - \alpha = n\pi$ [Note 3.12.2 above].*

$$\left.\begin{array}{rl}\therefore & \theta_1 - (\alpha - \beta) = n\pi + \beta \\ \text{and} & \alpha + \beta - \theta_1 = \beta - n\pi.\end{array}\right\}$$
$\therefore \{\theta_1 - (\alpha - \beta)\} - \{\alpha + \beta - \theta_1\} = 2n\pi.$

Thus $\angle PST = \angle TSQ$ where S is the focus, unless the curve is a hyperbola and the tangents be drawn to different branches, in which case $\angle PST = \angle TSQ'$ where Q' is a point on QS produced.

Note 3.12.4 *We have another approach of Note 3.12.3 by considering the tangents at α and β.*

The tangents at α and β are given by

$$\dfrac{l}{r} = \cos(\theta - \alpha) - e\cos\theta \text{ and } \dfrac{l}{r} = \cos(\theta - \beta) - e\cos\theta.$$

Where these meet, we get $\cos(\theta - \alpha) = \cos(\theta - \beta)$

$$\therefore \theta - \alpha = \pm(\theta - \beta) \Rightarrow 2\theta = \alpha + \beta,$$

taking only the lower sign, for the upper sign is inadmissible. So we get $\theta = \dfrac{1}{2}(\alpha + \beta).$

Hence if T be the point of intersection of the tangents at the two points P and Q of a conic, ST will bisect the angle $\angle PSQ$ (exterior angle $\angle PSQ$ in case of a hyperbola).

3.13 Asymptotes

Definition 3.13.1 *A straight line which meets a curve (conic) at two points at infinity but which does not be wholly at infinity is called an asymptote to this curve, or*

Alternatively, if P be a point on a branch of a curve extending to infinity and if a straight line at a finite distance from the origin exists towards which the tangent line to the curve at P approaches as a limit when $P \to \infty$, then the straight line is an asymptote of the curve.

3.13.1 Equation of the asymptote

These can be obtained from the equation of the tangent. The tangent at α to a conic $\dfrac{l}{r} = 1 - e\cos\theta$ is given by

$$\frac{l}{r} = \cos(\theta - \alpha) - e\cos\theta \qquad (3.17)$$

The point α is a point at ∞ on the conic, if

$$0 = 1 - e\cos\alpha \Rightarrow \cos\alpha = \frac{1}{e}$$

$$\therefore \sin\alpha = \sqrt{1 - \frac{1}{e^2}} = \frac{\sqrt{e^2 - 1}}{e}.$$

So from (3.17) we get

$$\frac{l}{r} = \cos\theta\cos\alpha + \sin\theta\sin\alpha - e\cos\theta$$

$$= \frac{1}{e}\cos\theta + \frac{\sqrt{e^2 - 1}}{e}\sin\theta - e\cos\theta$$

$$= \left(\frac{1}{e} - e\right)\cos\theta + \frac{\sqrt{e^2 - 1}}{e}\sin\theta$$

or, $\dfrac{le}{r} = (1 - e^2)\cos\theta + \sqrt{e^2 - 1}\sin\theta$

or, $\left[\dfrac{el}{r} - (1 - e^2)\cos\theta\right]^2 = (e^2 - 1)\sin^2\theta.$

This is the required equation of the asymptote.

3.14 Polar Equation of the Directrices of an Ellipse

Let ZP and $Z'P'$ be the directrixes of the given ellipse $\dfrac{l}{r} = 1 - e\cos\theta$ corresponding to the foci S and S' respectively. Let Q be one extremity of a latus rectum. The perpendicular form Q on the directrix ZP intersect it at M. Also let the polar axis intersect the directrix ZP at Z. Let $P(r,\theta)$ be any point on the directrix. Then from $\triangle SPZ$, we get

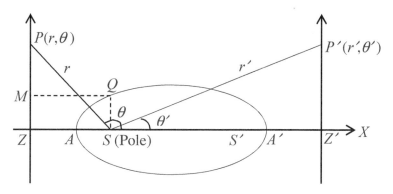

Figure 3.16

$$\dfrac{|SZ|}{SP} = \cos(\pi-\theta) = -\cos\theta \Rightarrow \dfrac{r}{|SZ|} = -\sec\theta, \text{ i.e., } r = -|SZ|\sec\theta \quad (3.18)$$

But $|SZ| = |QM| = \dfrac{|SQ|}{e}$ $\left[\because \text{ from the definition of conic } \dfrac{|SQ|}{|QM|} = e\right]$

$$= \dfrac{l}{e}, \text{ where } l \text{ is the semi-latus rectum.}$$

Hence from (3.18) we get $r = -\dfrac{l}{e}\sec\theta.$ \quad (3.19)

This is the polar equation of the directrix ZP.

Let the polar axis intersect the other directrix $Z'P'$ at Z'. Let A, A' be the vertices of the ellipse and let $P'(r',\theta')$ be a point on the directrix. Then we have from the right angled $\triangle SZ'P'$,

$$\dfrac{r'}{SZ'} = \sec\theta' \text{ or, } r' = SZ'\sec\theta' \quad (3.20)$$

From the equation of the conic we have

$$\frac{l}{SA} = 1 - e\cos\pi \quad \text{and} \quad \frac{l}{SA'} = 1 - e\cos 0.$$

[∵ The coordinates of A and A' are (SA, π) and $(SA', 0)$ respectively.]

$$SA = \frac{l}{1+e} \tag{3.21}$$

$$\text{and} \quad SA' = \frac{l}{1-e} \tag{3.22}$$

Now $\dfrac{|SA'|}{|A'Z|} = e \quad \therefore \quad |A'Z| = \dfrac{|SA'|}{e} = \dfrac{l}{e(1-e)}$ [using (3.22)]

So $|A'Z'| = |AZ| = \dfrac{|SA|}{e} = \dfrac{l}{e(1+e)}$ [using (3.21)]

$$\therefore |ZZ'| = |ZA'| + |A'Z'| = \frac{l}{e(1-e)} + \frac{l}{e(1+e)} = \frac{2l}{e(1-e^2)} \tag{3.23}$$

Hence $|SZ'| = |ZZ'| - |ZS| = |ZZ'| - |MQ|$

$$= \frac{2l}{e(1-e^2)} - \frac{l}{e} \quad \text{[using (3.23)]}$$

$$= \frac{l}{e}\frac{1+e^2}{1-e^2}.$$

Hence from (3.20) we get $r' = \dfrac{l}{e}\dfrac{1+e^2}{1-e^2}\sec\theta'$.

Hence the locus of $P'(r', \theta')$ is $r = \dfrac{l}{e}\dfrac{1+e^2}{1-e^2}\sec\theta$ which is the required equation of this second directrix.

Note 3.14.1 *In case of the conic $\dfrac{l}{r} = 1 + e\cos\theta$, it can be shown that the equations of the directrixes are $r = \dfrac{l}{e}\sec\theta$ and $r = \dfrac{l}{e}\dfrac{1+e^2}{1-e^2}\sec\theta$.*

In case of parabola $\dfrac{l}{r} = 1 - \cos\theta$, the equation of the directrix will be $r = -l\sec\theta$ and for the parabola $\dfrac{l}{r} = 1 + \cos\theta$ its equation will be $r = l\sec\theta$.

3.15 A Few Properties

1. Let PSQ be a focal chord, then if the vectorial angle of P be θ, that of Q will be $\pi + \theta$ and then

$$\frac{l}{SP} = 1 - e\cos\theta \quad \text{and} \quad \frac{l}{SQ} = 1 - e\cos(\pi + \theta) = 1 + e\cos\theta$$

$$\therefore \quad \frac{l}{SP} + \frac{l}{SQ} = 2 \quad \Rightarrow \quad \frac{1}{SP} + \frac{1}{SQ} = \frac{2}{l}.$$

If, however, the curve be a hyperbola and P be on the nearer and Q be on the further branch, then the vectorial angle of P and Q are still θ and $\pi + \theta$, but now the radius vector of Q is negative.

$$\frac{l}{SP} = 1 - e\cos\theta \quad \text{and} \quad -\frac{l}{SQ} = 1 - e\cos(\pi+\theta) = 1 + e\cos\theta.$$

Hence $\quad \dfrac{l}{SP} - \dfrac{l}{SQ} = 2 \Rightarrow \dfrac{1}{SP} - \dfrac{1}{SQ} = \dfrac{2}{l}.$

Thus the proposition may be stated as follows:

The semi-latus rectum of any conic is a harmonic mean between the algebraic focal distances of the extremities of a focal chord.

2. The length of the focal chord $PSQ =$

$$SP + SQ = l\left(\frac{1}{1-e\cos\theta} + \frac{1}{1+e\cos\theta}\right) = \frac{2l}{1 - e^2\cos^2\theta}.$$

3. The sum of the reciprocals of two perpendicular focal chords PSQ and $P'SQ'$

$$= \frac{1 - e^2\cos^2\theta}{2l} + \frac{1 - e^2\sin^2\theta}{2l} = \frac{2 - e^2}{2l} = \text{constant}.$$

4. $\dfrac{1}{PS.SQ} + \dfrac{1}{P'S.SQ'} = \dfrac{1 - e^2\cos^2\theta}{l^2} + \dfrac{1 - e^2\sin^2\theta}{l^2} = \dfrac{2 - e^2}{l^2} = \text{constant}.$

5. If the tangent at any point P of a conic $\dfrac{l}{r} = 1 - e\cos\theta$ meets the directrix at K, then the angle $\angle KSP$ is a right angle.

Proof: Let the vectorial angle of P be α, then the equation of the tangent at P is

$$\frac{l}{r} = \cos(\theta - \alpha) - e\cos\theta.$$

The equation of the directrix is $\dfrac{l}{r} = -e\cos\theta$. They meet at the point K, where $\cos(\theta - \alpha) = 0$

$$\therefore \theta - \alpha = (2n+1)\dfrac{\pi}{2}.$$

Hence $\angle KSP =$ one right angle. □

3.16 Worked-Out Examples

Example 3.16.1 *Transform (i) the equation $(x^2 + y^2)^2 = ax^2$ to its polar equation.*

(ii) The polar equation $\sqrt{r}\cos\dfrac{\theta}{2} = \sqrt{a}$ to a Cartesian equation.

Solution: (i) Putting $x = r\cos\theta$ and $y = r\sin\theta$ we get

$$(r^2)^2 = a(r\cos\theta)^2 \Rightarrow r^2 = a\cos^2\theta \Rightarrow r = \sqrt{a}\cos\theta,$$

which is the required equation in polar form.

(ii) The given equation is

$$r\cos^2\dfrac{\theta}{2} = a \Rightarrow \dfrac{r}{2}(1 + \cos\theta) = a \Rightarrow r = 2a - r\cos\theta.$$

Now putting $x = r\cos\theta$ and $y = r\sin\theta$, i.e., $r^2 = x^2 + y^2$ we get

$$\sqrt{x^2 + y^2} = 2a - x$$

or, $x^2 + y^2 = (2a - x)^2 = 4a^2 - 4ax + x^2$ [squaring both sides]

$\Rightarrow y^2 = -4a(x - a)$ which is the required equation. □

Example 3.16.2 *Convert the following points from polar coordinates to the corresponding Cartesian coordinates (i) $\left(2, \dfrac{\pi}{3}\right)$, (ii) $\left(0, \dfrac{\pi}{2}\right)$.*

Solution: (i) $\left(2, \dfrac{\pi}{3}\right) \equiv (r, \theta)$

$$\therefore \ x = r\cos\theta = 2\cos\dfrac{\pi}{3} = 2 \times \dfrac{1}{2} = 1$$

$$y = r\sin\theta = 2\sin\dfrac{\pi}{3} = 2 \times \dfrac{\sqrt{3}}{2} = \sqrt{3}.$$

Hence the point is $(1, \sqrt{3})$ in the Cartesian system.

(ii) Here $x = 0 \cdot \cos\dfrac{\pi}{2} = 0, y = 0 \cdot \sin\dfrac{\pi}{2} = 0$. So $\left(0, \dfrac{\pi}{2}\right)$ is equivalent to $(0,0)$ in Cartesian coordinate. □

Polar Coordinates

Example 3.16.3 *Convert the following Cartesian coordinates to the corresponding polar coordinates using positive r and negative r, (i) $(-1, 1)$, (ii) $(2, -3)$.*

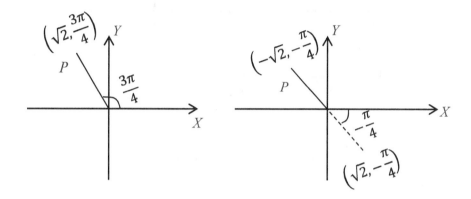

Solution: (i) For the second quadrant, point $(x, y) \equiv (-1, 1)$, we have

$$\tan\theta = \frac{y}{x} = -1 \implies \theta = \frac{3\pi}{4}.$$

Using positive r, we have $r = \sqrt{x^2 + y^2} = \sqrt{(-1)^2 + 1^2} = \sqrt{2}$.
So one set of polar coordinates is $(r, \theta) = \left(\sqrt{2}, \frac{3\pi}{4}\right)$.

Using negative r, we have coordinates $\left(-\sqrt{2}, -\frac{\pi}{4}\right)$.

(ii) For the fourth quadrant, point $(x, y) \equiv (2, -3)$, we have

$$\tan\theta = \frac{-3}{2} \implies \theta = \tan^{-1}\left(-\frac{3}{2}\right) = -\tan^{-1}\frac{3}{2}.$$

Using positive r, we have $r = \sqrt{x^2 + y^2} = \sqrt{(-3)^2 + 2^2} = \sqrt{13}$. So one set of polar coordinates is $(r, \theta) = \left(\sqrt{13}, -\tan^{-1}\frac{3}{2}\right)$.

Using negative r, we have coordinates $\left(-\sqrt{13}, \pi - \tan^{-1}\frac{3}{2}\right)$. □

Example 3.16.4 *Find the maximum distance of any point on the curve $x^2 + 2xy + 2y^2 = 1$ from the origin.*

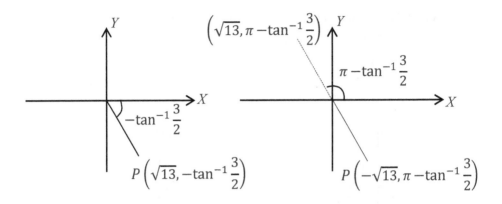

Solution: Let the polar coordinates of any point on the curve be $P(r, \theta)$. Then its Cartesian coordinates are $P(r\cos\theta, r\sin\theta)$. Since P is a point on the curve, so we get

$$r^2\cos^2\theta + 2r^2\sin^2\theta + 2r^2\sin\theta\cos\theta = 1$$

$$\Rightarrow r^2 = \frac{1}{\cos^2\theta + 2\sin^2\theta + \sin 2\theta} = \frac{1}{\sin^2\theta + 1 + \sin 2\theta}$$

$$= \frac{2}{2\sin^2\theta + 2 + 2\sin 2\theta} = \frac{2}{1 - \cos 2\theta + 2 + 2\sin 2\theta}$$

$$= \frac{2}{3 - \cos 2\theta + 2\sin 2\theta}.$$

To find the maximum value of r, i.e., r^2, we consider the expression $3 - \cos 2\theta + 2\sin 2\theta$. Let us put $2 = \rho\cos\alpha$ and $1 = \rho\sin\alpha$, which imply $\rho = \sqrt{1^2 + 2^2} = \sqrt{5}$ and $\tan\alpha = \frac{1}{2}$. Now,

$$2\sin 2\theta - \cos 2\theta = \rho\cos\alpha\sin 2\theta - \rho\sin\alpha\cos 2\theta = \rho\sin(2\theta - \alpha).$$

We have $\qquad -1 \leq \sin(2\theta - \alpha) \leq 1$

$$\Rightarrow -\sqrt{5} \leq \sqrt{5}\sin(2\theta - \alpha) \leq \sqrt{5}$$

$$\Rightarrow -\sqrt{5} + 3 \leq \rho\sin(2\theta - \alpha) + 3 \leq \sqrt{5} + 3$$

$$\Rightarrow 3 - \sqrt{5} \leq 3 - \cos 2\theta + 2\sin 2\theta \leq 3 + \sqrt{5}$$

$$\Rightarrow \frac{2}{3 - \sqrt{5}} \leq \frac{2}{3 - \cos 2\theta + 2\sin 2\theta} \leq \frac{2}{3 + \sqrt{5}}$$

$$\Rightarrow r^2_{max} = \frac{2}{3 - \sqrt{5}} \quad \text{or,} \quad r_{max} = \frac{\sqrt{2}}{\sqrt{3 - \sqrt{5}}}.$$

Hence the required maximum distance is $\dfrac{\sqrt{2}}{\sqrt{3 - \sqrt{5}}}$. □

Polar Coordinates

Example 3.16.5 *Find the polar equation of the straight line joining the points whose polar coordinates are $\left(1, \frac{\pi}{2}\right)$ and $(2, \pi)$.*

Solution: Let the polar equation of the line be $\frac{1}{r} = A\cos\theta + B\sin\theta$, where A and B are constants. It passes through $\left(1, \frac{\pi}{2}\right)$ and $(2, \pi)$, so we get

$$1 = A\cos\frac{\pi}{2} + B\sin\frac{\pi}{2} \;\Rightarrow\; B = 1$$

and $\quad \frac{1}{2} = A\cos\pi + B\sin\pi \;\Rightarrow\; A = -\frac{1}{2}.$

Hence the required line is $\frac{1}{r} = -\frac{1}{2}\cos\theta + \sin\theta.$ $\qquad\square$

Example 3.16.6 *Find the polar equation of the circle when two ends of a diameter are given.*

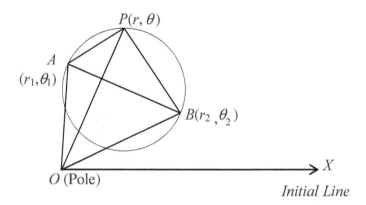

Solution: Let AB be the diameter where the end points A and B have the coordinates (r_1, θ_1) and (r_2, θ_2) respectively.

Let $P(r, \theta)$ be any point on the circle with respect to the pole O and the initial line OX.

Now from the figure we get

$$AP^2 = OA^2 + OP^2 - 2OA.OP\cos\angle POA = r_1^2 + r^2 - 2rr_1\cos(\theta_1 - \theta)$$

$$BP^2 = OB^2 + OP^2 - 2OB.OP\cos\angle POB = r_2^2 + r^2 - 2rr_2\cos(\theta - \theta_2)$$

and $AB^2 = OA^2 + OB^2 - 2OA.OB\cos\angle AOB = r_1^2 + r_2^2 - 2r_2r_1\cos(\theta_1 - \theta_2).$

Since AB is a diameter, $\angle APB$ is a right angle $\therefore AB^2 = AP^2 + BP^2$, i.e.,

$$r_1^2 + r_2^2 - 2r_1 r_2 \cos(\theta_1 - \theta_2) = r_1^2 + r^2 - 2r_1 r \cos(\theta_1 - \theta) + r_2^2 + r^2 - 2rr_2 \cos(\theta - \theta_2)$$

$$\Rightarrow r^2 - [r_1 \cos(\theta - \theta_1) + r_2 \cos(\theta - \theta_2)]r + r_1 r_2 \cos(\theta_1 - \theta_2) = 0$$

which is the required equation of the circle. □

Example 3.16.7 *Show that the equation of the line through the pole and the point of intersection of the lines* $\dfrac{C}{r} = A \cos \theta + B \sin \theta$ *and* $\dfrac{D}{r} = A' \cos \theta + B' \sin \theta$ *is*

$$\left(A - \frac{A'C}{D}\right) \cos \theta + \left(B - \frac{B'C}{D}\right) \sin \theta = 0.$$

Solution: The equation of the given lines when expressed in Cartesian axes are

$$Ax + By - C = 0 \tag{3.24}$$
$$\text{and} \quad A'x + B'y - D = 0 \tag{3.25}$$

So their point of intersection is given by $\left(\dfrac{B'C - BD}{AB' - A'B}, \dfrac{AD - CA'}{AB' - A'B}\right)$.

Hence the equation of the line passing through the pole $(0,0)$ and this point of intersection is given by

$$y - 0 = \frac{\frac{AD - CA'}{AB' - A'B} - 0}{\frac{B'C - BD}{AB' - A'B} - 0}(x - 0) \Rightarrow (AD - A'C)x + (BD - B'C)y = 0$$

$$\Rightarrow \left(A - \frac{A'C}{D}\right) r \cos \theta + \left(B - \frac{B'C}{D}\right) r \sin \theta = 0$$

$$\Rightarrow \left(A - \frac{A'C}{D}\right) \cos \theta + \left(B - \frac{B'C}{D}\right) \sin \theta = 0. \quad \square$$

Example 3.16.8 *If the three lines* $r \cos(\theta - \theta_1) = p_1$, $r \cos(\theta - \theta_2) = p_2$ *and* $r \cos(\theta - \theta_3) = p_3$ *are concurrent, show that*

$$p_1 \sin(\theta_2 - \theta_3) + p_2 \sin(\theta_3 - \theta_1) + p_3 \sin(\theta_1 - \theta_2) = 0.$$

Solution: The three lines are given by

$$r \cos \theta \cos \theta_1 + r \sin \theta \sin \theta_1 - p_1 = 0 \tag{3.26}$$
$$r \cos \theta \cos \theta_2 + r \sin \theta \sin \theta_2 - p_2 = 0 \tag{3.27}$$
$$r \cos \theta \cos \theta_3 + r \sin \theta \sin \theta_3 - p_3 = 0 \tag{3.28}$$

Polar Coordinates

Eliminating $r\cos\theta$ and $r\sin\theta$ from (3.26), (3.27) and (3.28) we get

$$\begin{vmatrix} \cos\theta_1 & \sin\theta_1 & -p_1 \\ \cos\theta_2 & \sin\theta_2 & -p_2 \\ \cos\theta_3 & \sin\theta_3 & -p_3 \end{vmatrix} = 0$$

$\Rightarrow p_1\sin(\theta_2 - \theta_3) + p_2\sin(\theta_3 - \theta_1) + p_3\sin(\theta_1 - \theta_2) = 0.$

Alternatively,

Putting $r\cos\theta = x$ and $r\sin\theta = y$ and expressing (3.26), (3.27) and (3.28) in Cartesian form of the equation of the lines are given as

$$x\cos\theta_1 + y\sin\theta_1 - p_1 = 0 \tag{3.29}$$
$$x\cos\theta_2 + y\sin\theta_2 - p_2 = 0 \tag{3.30}$$
$$x\cos\theta_3 + y\sin\theta_3 - p_3 = 0 \tag{3.31}$$

Solving (3.29) and (3.30) we get

$$\frac{x}{p_1\sin\theta_2 - p_2\sin\theta_1} = \frac{y}{p_2\cos\theta_1 - p_1\cos\theta_2} = \frac{1}{\sin(\theta_2 - \theta_1)}.$$

So the point of intersection of (3.29) and (3.30) is given by

$$\left(x = \frac{p_1\sin\theta_2 - p_2\sin\theta_1}{\sin(\theta_2 - \theta_1)}, y = \frac{p_2\cos\theta_1 - p_1\cos\theta_2}{\sin(\theta_2 - \theta_1)} \right) \tag{3.32}$$

If the three lines be concurrent then (3.32) will satisfy (3.31), i.e.,

$$\frac{p_1\sin\theta_2 - p_2\sin\theta_1}{\sin(\theta_2 - \theta_1)}\cos\theta_3 + \frac{p_2\cos\theta_1 - p_1\cos\theta_2}{\sin(\theta_2 - \theta_1)}\sin\theta_3 - p_3 = 0$$

$p_1[\sin\theta_2\cos\theta_3 - \cos\theta_2\sin\theta_3] + p_2[\sin\theta_3\cos\theta_1 - \cos\theta_3\sin\theta_1]$
$\qquad\qquad\qquad\qquad\qquad +p_3[\sin\theta_1\cos\theta_2 - \cos\theta_1\sin\theta_2] = 0$

$\Rightarrow p_1\sin(\theta_2 - \theta_3) + p_2\sin(\theta_3 - \theta_1) + p_3\sin(\theta_1 - \theta_2) = 0.$ □

Example 3.16.9 *Prove that the polar equation of the locus of the feet of perpendiculars drawn from the pole on the lines passing through a given point (r', θ') is $r = r'\cos(\theta - \theta')$.*

Solution: Let (p, α) be the coordinates of foot of the perpendicular upon one of the lines, then the equation of the line is $r\cos(\theta - \alpha) = p$. If it passes through the point (r', θ'), then $r'\cos(\theta' - \alpha) = p$. Making (p, α) the current coordinates (r, θ), the locus of (p, α) becomes $r = r'\cos(\theta' - \theta)$, i.e., $r = r'\cos(\theta - \theta')$. □

Example 3.16.10 *Show that the locus of a point whose distance from the pole is equal to its distance form the line $r\cos\theta + k = 0$ is $2r\sin^2\dfrac{\theta}{2} = k$.*

Solution: The line is $r\cos\theta + k = 0$. Putting $x = r\cos\theta$ we get $x + k = 0$.

Let the point be (x_1, y_1) where $x_1 = r_1\cos\theta_1$ and $y_1 = r_1\sin\theta_1$. According to the problem

$$\dfrac{x_1+k}{\sqrt{1^2+0^2}} = r_1 \Rightarrow r_1\cos\theta_1 + k = r_1 \Rightarrow k = r_1(1-\cos\theta_1) = 2r_1\sin^2\dfrac{\theta}{2}.$$

Therefore the locus of (x_1, y_1), i.e., of (r_1, θ_1) is $2r\sin^2\dfrac{\theta}{2} = k$. □

Example 3.16.11 *Find the equation of the chord joining the points on the circle $r = 2d\cos\theta$, whose vectorial angles are θ_1 and θ_2 and deduce the equation of the tangent at the point θ_1.*

Solution: Let P and Q be the two points on the circle

$$r = 2d\cos\theta \tag{3.33}$$

with (r_1, θ_1) and (r_2, θ_2) as their respective coordinates.

Then the equation of the line PQ can be written as

$$\dfrac{\sin(\theta_1-\theta_2)}{r} = \dfrac{\sin(\theta-\theta_2)}{r_1} + \dfrac{\sin(\theta_1-\theta)}{r_2}$$

$$\Rightarrow r_1 r_2 \sin(\theta_1 - \theta_2) = rr_2 \sin(\theta - \theta_2) + rr_1 \sin(\theta - \theta_1) \tag{3.34}$$

Since P and Q are points on (3.33), we have $r_1 = 2d\cos\theta_1$ and $r_2 = 2d\cos\theta_2$. Putting these values of r_1 and r_2 in (3.34) we get

$$4d^2\cos\theta_1\cos\theta_2\sin(\theta_1-\theta_2) = 2dr[\cos\theta_2\sin(\theta-\theta_2)$$
$$+\cos\theta_1\sin(\theta_1-\theta)]$$

or, $\quad 4d\cos\theta_1\cos\theta_2\sin(\theta_1-\theta_2) = r[2\cos\theta_2\sin(\theta-\theta_2)$
$$+2\cos\theta_1\sin(\theta_1-\theta)]$$
$$= r[\sin\theta + \sin(\theta-2\theta_2) + \sin(2\theta_1-\theta) - \sin\theta]$$
$$= r[\sin(\theta-2\theta_2) + \sin(2\theta_1-\theta)]$$
$$= 2r\sin(\theta_1-\theta_2)\cos(\theta-\theta_1-\theta_2),$$

i.e., we get $r\cos(\theta-\theta_1-\theta_2) = 2d\cos\theta_1\cos\theta_2 \tag{3.35}$
$$[\because \sin(\theta_1-\theta_2) \neq 0].$$

which is the required equation of the chord PQ.

Polar Coordinates

Second Part: (Equation of the tangent)

Putting $\theta_2 = \theta_1$ in (3.35) we shall get the equation of the tangent at the point whose vectorial angle is θ_1. Hence the required equation of the tangent is $r\cos(\theta - 2\theta_1) = 2d\cos^2\theta_1$. □

Example 3.16.12 *Find the condition that the line $r\cos(\theta - \alpha) = p$ touches the circle $r^2 + 2r(g\cos\theta + f\sin\theta) + c = 0$.*

Solution: If the centre and radius of the given circle be (ρ_1, θ_1) and d respectively, then the equation of it is

$$r^2 - 2\rho_1 r\cos(\theta - \theta_1) + \rho_1^2 - d^2 = 0$$

or, $\quad r^2 - 2r[\rho_1\cos\theta_1\cos\theta + \rho_1\sin\theta_1\sin\theta] + \rho_1^2 - d^2 = 0.$

Comparing with the given form of the equation of the circle we get $\rho_1\cos\theta_1 = -g$, $\rho_1\sin\theta_1 = -f$ and $\rho_1^2 - d^2 = c$ which imply $\rho_1^2 = g^2 + f^2$, $\tan\theta_1 = \dfrac{f}{g}$ and $d^2 = \rho_1^2 - c = g^2 + f^2 - c$.

The equation of a line through the centre parallel to the given line is

$$r\cos(\theta - \alpha) = \rho_1\cos(\theta_1 - \alpha).$$

So, the line $r\cos(\theta - \alpha) = p$ will touch the circle if

$$\rho_1\cos(\theta_1 - \alpha) - p = \text{the radius of the circle} = d$$

or, $(g\cos\alpha + f\sin\alpha - p)^2 = g^2 + f^2 - c.$ □

Remark: The equation of the circle given in the previous **Example 3.16.12** is

$$r^2 + 2r(g\cos\theta + f\sin\theta) + c = 0.$$

Putting $r\cos\theta = x$ and $r\sin\theta = y$ we get the general form of a circle with centre $(-g, -f)$ and radius $\sqrt{g^2 + f^2 - c}$ as

$$x^2 + y^2 + 2gx + 2fy + c = 0.$$

Example 3.16.13 *Chords of the circle $r^2 + 2r(g\cos\theta + f\sin\theta) + c = 0$ are drawn through the pole. Prove that the locus of their middle points is the circle $r + g\cos\theta + f\sin\theta = 0$.*

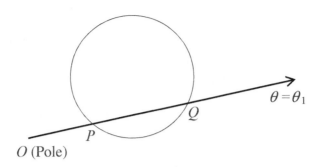

Solution: Let $\theta = \theta_1$ be the equation of any chord passing through the pole and let it cut the circle in P and Q. Then OP and OQ are the roots of the equation

$$r^2 + 2r(g\cos\theta + f\sin\theta) + c = 0$$

If (r_1, θ_1) be the middle point of the chord, then

$$r_1 = \frac{1}{2}(OP + OQ) = -(g\cos\theta_1 + f\sin\theta_1).$$

Hence the locus of (r_1, θ_1) is $r + g\cos\theta + f\sin\theta = 0$, which is obviously a circle passing through the pole. □

Example 3.16.14 *Determine the nature of the following conics and find the length of latus rectum.*

(i) $\dfrac{3}{r} = 2 + 4\cos\theta$ 　　　　(ii) $\dfrac{2}{r} = 3 - 3\cos\theta$

(iii) $\dfrac{8}{r} = 4 - 5\cos\theta$ 　　　　(iv) $\dfrac{12}{r} = 4 - \sqrt{2}\cos\theta.$

Solution: (i) The given equation can be written as $\dfrac{3/2}{r} = 1 + 2\cos\theta$.

Comparing with the standard equation of the conic $\dfrac{l}{r} = 1 + e\cos\theta$ we see that, here $e = 2 > 1$ and $l = \dfrac{3}{2}$.

Hence the conic is a hyperbola with its latus rectum $= 2l = 3$ units.

(ii) The equation is $\dfrac{2/3}{r} = 1 - \cos\theta$ which represents a parabola with its latus rectum $= 2 \times \dfrac{2}{3} = \dfrac{4}{3}$.

(iii) The equation of the conic is $\dfrac{8}{r} = 4 - 5\cos\theta$ which is written as $\dfrac{2}{r} = 1 - \dfrac{5}{4}\cos\theta$ represents a branch of hyperbola with its latus rectum $= 2 \times 2 = 4$.

(iv) The equation is $\dfrac{3}{r} = 1 - \dfrac{1}{2\sqrt{2}}\cos\theta$ and it is an ellipse with latus rectum $= 2 \times 3 = 6$. □

Example 3.16.15 *Find the points on the conic $\dfrac{15}{r} = 1 - 4\cos\theta$ whose radius vector is 5.*

Solution: We are to find the value of the vectorial angle θ for which r is 5.

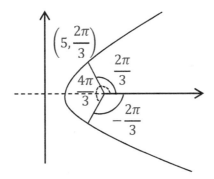

Putting $r = 5$ in the equation of the conic, we get

$$\dfrac{15}{5} = 1 - 4\cos\theta \Rightarrow \cos\theta = -\dfrac{1}{2} \Rightarrow \theta = \dfrac{2\pi}{3}, \dfrac{4\pi}{3} \quad (0 \le \theta < 2\pi).$$

Hence the required points are $\left(5, \dfrac{2\pi}{3}\right)$ and $\left(5, \dfrac{4\pi}{3}\right)$. □

Note 3.16.1 *The points $\left(5, \dfrac{4\pi}{3}\right)$ and $\left(5, -\dfrac{2\pi}{3}\right)$ are equivalent as is seen from the figure given.*

Example 3.16.16 *On the conic $r = \dfrac{6}{1 - \cos\theta}$, find the point with the smallest radius vector.*

Solution: The given conic is $r = \dfrac{6}{1 - \cos\theta}$. Now, r will be smallest when $1 - \cos\theta$ is greatest and $1 - \cos\theta$ is greatest when $\cos\theta = -1$, i.e., when $\theta = \pi$.

Now from the equation of the conic we get $r = \dfrac{6}{1-(-1)} = 3$ at $\theta = \pi$. Hence the required point is $(3, \pi)$. □

Example 3.16.17 *Identify the conics and find its vertex (vertices) and the length of the latus rectum, as the case appears*

(i) $\dfrac{3}{r} = 2 + 4\cos\theta$ (ii) $\dfrac{12}{r} = 2 - \cos\theta$

(iii) $\dfrac{21}{r} = 5 - 2\cos\theta$ (iv) $\dfrac{12}{r} = 2 - 2\cos\theta$.

Solution: (i) $\dfrac{3}{r} = 2 + 4\cos\theta \Rightarrow \dfrac{\frac{3}{2}}{r} = 1 + 2\cos\theta$. Here $e = 2 > 1$, so the

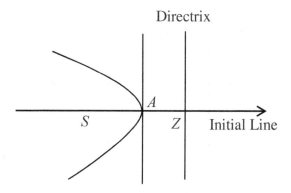

For the vertex A, $\theta = 0$.

conic is a hyperbola on the left branch whose latus rectum is $2l = 3$. Put $\theta = 0$, $\dfrac{3}{r} = 2 + 4 \Rightarrow r = \dfrac{1}{2}$. So its vertex is $\left(\dfrac{1}{2}, 0\right)$.

(ii) The equation is $\dfrac{12}{r} = 2 - \cos\theta$, i.e., $\dfrac{6}{r} = 1 - \dfrac{1}{2}\cos\theta$. Here $e < \dfrac{1}{2} < 1$. So the conic is an ellipse and its left hand focus is considered as the pole. Its latus rectum $= 2 \times 6 = 12$ units. For the vertices A and A', $\theta = \pi$ and 0 respectively.

At $\theta = \pi$, we get $\dfrac{6}{r} = 1 - \dfrac{1}{2}\cos\pi = 1 + \dfrac{1}{2} = \dfrac{3}{2} \Rightarrow r = 4$

and at $\theta = 0$, we get $\dfrac{6}{r} = 1 - \dfrac{1}{2}\cos 0 = 1 - \dfrac{1}{2} = \dfrac{1}{2} \Rightarrow r = 12$.

Hence the coordinates of the vertices A and A' are $(4, \pi)$ and $(12, 0)$.

(iii) **Do Yourself.**

Polar Coordinates

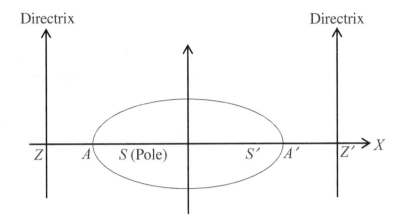

(iv) The conic is $\dfrac{6}{r} = 1 - \cos\theta$, which is a parabola of latus rectum = 12 and coordinates of the vertex is obtained by putting $\theta = \pi$ and we get $\dfrac{6}{r} = 1 + 1 = 2 \Rightarrow r = 3$. So vertex is $(3, \pi)$. □

Example 3.16.18 *Find the points of intersection of the parabolas $\dfrac{1}{r} = 1 - \cos\theta$ and $\dfrac{3}{r} = 1 + \cos\theta$.*

Solution: Eliminating r from the above two equations we get

$$3 - 3\cos\theta = 1 + \cos\theta \;\Rightarrow\; 4\cos\theta = 2 \;\Rightarrow\; \cos\theta = \dfrac{1}{2} \;\Rightarrow\; \theta = \pm\dfrac{\pi}{3}$$

$$\therefore\; \dfrac{1}{r} = 1 - \dfrac{1}{2} = \dfrac{1}{2} \;\Rightarrow\; r = 2.$$

So the points of intersection are $\left(2, \dfrac{\pi}{3}\right), \left(2, -\dfrac{\pi}{3}\right)$. □

Example 3.16.19 *Find the polar equation of the ellipse $\dfrac{x^2}{100} + \dfrac{y^2}{36} = 1$ if*

(i) *its right hand focus is considered as the pole and the positive direction of the x-axis is taken as the direction of the polar axis.*

(ii) *its pole is at the left hand focus and the positive direction of the x-axis is the polar axis.*

(iii) *the pole is at the centre of the curve and the positive direction of the x-axis is taken as the polar axis.*

Solution: (i) Since the pole is at the right hand focus, the equation of the ellipse will be of the form $\dfrac{l}{r} = 1 + e\cos\theta$.

Now for the given ellipse $\dfrac{x^2}{100} + \dfrac{y^2}{36} = 1$, we get $a^2 = 100, b^2 = 36$ so the eccentricity e is given by

$$b^2 = a^2(1-e^2) \Rightarrow 36 = 100(1-e^2) \Rightarrow e = \dfrac{4}{5}$$

and l = semi latus-rectum = $\dfrac{b^2}{a} = \dfrac{36}{10} = \dfrac{18}{5}$.

Hence its polar equation is $\dfrac{18}{5r} = 1 + \dfrac{4}{5}\cos\theta$.

(ii) In this case, since the pole is considered at the left hand focus of the ellipse, so its equation will be

$$\dfrac{l}{r} = 1 - e\cos\theta, \text{ i.e., } \dfrac{18}{5r} = 1 - \dfrac{4}{5}\cos\theta.$$

(iii) Since the pole is here considered at the centre of the ellipse, so we put $x = r\cos\theta, y = r\sin\theta$ and the polar equation of the given ellipse is obtained as

$$\dfrac{r^2\cos^2\theta}{100} + \dfrac{r^2\sin^2\theta}{36} = 1 \text{ or, } r^2(36\cos^2\theta + 100\sin^2\theta) = 3600$$

or, $r^2(9\cos^2\theta + 25\sin^2\theta) = 900$ or, $r^2(9\cos^2\theta + 25 - 25\cos^2\theta) = 900$

or, $r^2(25 - 16\cos^2\theta) = 900$ or, $r^2 = \dfrac{900}{25 - 16\cos^2\theta}$. □

Example 3.16.20 *If PSP' be the focal chord of a conic, show that*

$$\dfrac{1}{SP} + \dfrac{1}{SP'} = \dfrac{2}{l}, \text{ where } l \text{ is the semi latus rectum.}$$

The above problem may be stated alternatively as follows:

Show that the semi latus rectum of a conic is a harmonic mean between the segments of any focal chord.

Solution: § 3.15 Property - 1. □

Example 3.16.21 *Prove that the length of the focal chord of the conic $\dfrac{l}{r} = 1 - e\cos\theta$ which is inclined at an angle α is $\dfrac{2l}{1 - e^2\cos^2\alpha}$.*

Polar Coordinates

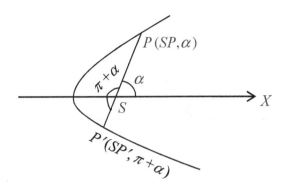

Solution: Let PSP' be the focal chord of the conic

$$\frac{l}{r} = 1 - e\cos\theta \tag{3.36}$$

where $\angle XSP = \alpha$, i.e., the vectorial angle of P is α. Then the vectorial angle of P' is $\pi + \alpha$.

From (3.36) we get

$$\frac{l}{SP} = 1 - e\cos\alpha \quad \therefore SP = \frac{l}{1 - e\cos\alpha}$$

and $\quad \dfrac{l}{SP'} = 1 - e\cos(\pi + \alpha) = 1 + e\cos\alpha \quad \therefore SP' = \dfrac{l}{1 + e\cos\alpha}$

$$\therefore PSP' = SP + SP' = \frac{l}{1 - e\cos\alpha} + \frac{l}{1 + e\cos\alpha} = \frac{2l}{1 - e^2\cos^2\alpha}. \qquad \square$$

Example 3.16.22 *For any conic, prove that* $\dfrac{1}{PP'} + \dfrac{1}{QQ'} = $ *constant, where PP' and QQ' are focal chords such that $PP' \perp QQ'$.*

Alternatvely,

Show that the sum of the reciprocals of two mutually perpendicular focal chords is constant.

Solution: Let α be the vectorial angle of the point P on the conic. Then the vectorial angle of Q, Q' and P' are respectively given by

$$\frac{\pi}{2} + \alpha, \; -\left(\pi - \frac{\pi}{2} + \alpha\right) = -\left(\frac{\pi}{2} - \alpha\right) \text{ and } -(\pi - \alpha).$$

Now since all these points lie on the conic, so we get

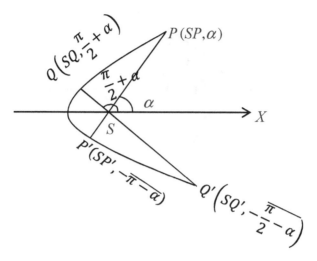

$$\frac{l}{SP} = 1 - e\cos\alpha \quad \therefore \ SP = \frac{l}{1 - e\cos\alpha},$$

$$\frac{l}{SP'} = 1 - e\cos(-\pi - \alpha) = 1 - e\cos(\pi - \alpha) = 1 + e\cos\alpha$$

$$\therefore \quad SP' = \frac{l}{1 + e\cos\alpha}.$$

$$\frac{l}{SQ} = 1 - e\cos\left(\frac{\pi}{2} + \alpha\right) = 1 + e\sin\alpha \ \Rightarrow \ SQ = \frac{l}{1 + e\sin\alpha},$$

$$\frac{l}{SQ'} = 1 - e\cos\left\{-\left(\frac{\pi}{2} - \alpha\right)\right\} = 1 - e\cos\left(\frac{\pi}{2} - \alpha\right) = 1 - e\sin\alpha$$

$$\Rightarrow \quad SQ' = \frac{l}{1 - e\sin\alpha}.$$

$$\therefore \quad PP' = SP + SP' = \frac{l}{1 - e\cos\alpha} + \frac{l}{1 + e\cos\alpha} = \frac{2l}{1 - e^2\cos^2\alpha}$$

$$\therefore \quad \frac{1}{PP'} = \frac{1 - e^2\cos^2\alpha}{2l}.$$

Similarly, $\quad QQ' = SQ + SQ' = \dfrac{l}{1 + e\sin\alpha} + \dfrac{l}{1 - e\sin\alpha} = \dfrac{2l}{1 - e^2\sin^2\alpha}$

$$\therefore \quad \frac{1}{QQ'} = \frac{1 - e^2\sin^2\alpha}{2l}.$$

Therefore, $\quad \dfrac{1}{PP'} + \dfrac{1}{QQ'} = \dfrac{1}{2l}[1 - e^2\cos^2\alpha + 1 - e^2\sin^2\alpha]$

$$= \frac{1}{2l}[2 - e^2(\cos^2\alpha + \sin^2\alpha)] = \frac{2 - e^2}{2l}. \qquad \square$$

Polar Coordinates

Example 3.16.23 *For any conic, prove that*

$$\frac{1}{SP.SP'} + \frac{1}{SQ.SQ'} = \frac{2-e^2}{l^2},$$

where PSP' and QSQ' are two mutually perpendicular focal chords of the conic.

Solution: [Put figure form the **Example 3.16.22**]

From the **Example 3.16.22** we get,

$$SP = \frac{l}{1-e\cos\alpha}, \quad SP' = \frac{l}{1+e\cos\alpha}, \quad SQ = \frac{l}{1+e\sin\alpha}, \quad SQ' = \frac{l}{1+e\sin\alpha}.$$

Now, $\quad \dfrac{1}{SP.SP'} + \dfrac{1}{SQ.SQ'}$

$$= \frac{(1-e\cos\alpha)(1+e\cos\alpha)}{l^2} + \frac{(1-e\sin\alpha)(1+e\sin\alpha)}{l^2}$$

$$= \frac{1-e^2\cos^2\alpha + 1 - e^2\sin^2\alpha}{l^2} = \frac{2-e^2}{l^2}. \qquad \square$$

Example 3.16.24 *If PSQ and $PS'R$ be respectively the two chords of an ellipse through the foci S and S', then prove that $\left(\dfrac{SP}{SQ} + \dfrac{S'P}{S'R}\right)$ is independent of the position of P.*

Solution: Let PSQ be the focal chord, such that vectorial angles of P and

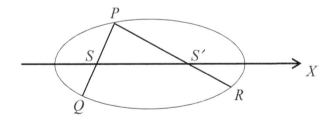

Q are respectively α and $\pi + \alpha$ respectively.

$$\frac{l}{SP} = 1 - e\cos\alpha \quad \text{and} \quad \frac{l}{SQ} = 1 - e\cos(\pi+\alpha) = 1 + e\cos\alpha$$

$$\therefore \quad \frac{1}{SP} + \frac{1}{SQ} = \frac{2}{l}, \quad \therefore \quad \frac{SP}{SP} + \frac{SP}{SQ} = \frac{2}{l}SP$$

(Multiplying both sides by SP.)

$$\therefore \quad \frac{SP}{SQ} = \frac{2}{l}SP - 1 \qquad (3.37)$$

Similarly we shall get $\dfrac{S'P}{S'R} = \dfrac{2}{l}S'P - 1$ \hfill (3.38)

Adding (3.37) and (3.38) we get

$$\dfrac{SP}{SQ} + \dfrac{S'P}{S'R} = \dfrac{2}{l}(SP + S'P) - 2 = \dfrac{2}{l}\cdot 2a - 2 = \text{constant}.$$

as we know that the sum of the focal distances from any point P on the ellipse is equal to the length of the major axis of the ellipse, i.e., $SP + S'P = 2a$, where $2a$ is the major axis of the ellipse.

Since right hand side is independent of the position of P, so we say that $\left(\dfrac{SP}{SQ} + \dfrac{S'P}{S'R}\right)$ is independent of the position of P. □

Example 3.16.25 *Prove that the locus of the middle points of focal chords of a conic is a conic.*

Solution: See § **3.15** Property - 1.

We have $SP = \dfrac{l}{1 - e\cos\theta}$ and $SQ = \dfrac{l}{1 + e\cos\theta}$.

If (r, θ) be the coordinates of the middle points of the focal chord PSQ, then

$$r = \dfrac{1}{2}(SP - SQ) = \dfrac{1}{2}\left(\dfrac{l}{1 - e\cos\theta} - \dfrac{l}{1 - e\cos\theta}\right) = \dfrac{le\cos\theta}{1 - e^2\cos^2\theta}$$

or, $r^2(1 - e^2\cos^2\theta) = ler\cos\theta$.

Transforming into Cartesian coordinates, this equation becomes

$$x^2 + y^2 - e^2 x^2 = lex \Rightarrow (1 - e^2)x^2 + y^2 - lex = 0$$

which clearly represents an ellipse. □

Example 3.16.26 *If r_1 and r_2 be two mutually perpendicular radius vectors of the ellipse $r^2 = \dfrac{b^2}{1 - e^2\cos^2\theta}$, show that $\dfrac{1}{r_1^2} + \dfrac{1}{r_2^2} = \dfrac{1}{a^2} + \dfrac{1}{b^2}$.*

Solution: Let SP and SQ be two mutually perpendicular radius vectors, where P and Q have the coordinates (r_1, α) and $\left(r_2, \dfrac{\pi}{2} + \alpha\right)$. So from the equation of the given ellipse $r^2 = \dfrac{b^2}{1 - e^2\cos^2\theta}$ we get

$$r_1^2 = \dfrac{b^2}{1 - e^2\cos^2\alpha} \quad \text{and} \quad r_2^2 = \dfrac{b^2}{1 - e^2\cos^2\left(\frac{\pi}{2} + \alpha\right)} = \dfrac{b^2}{1 - e^2\sin^2\alpha}$$

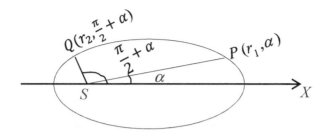

$$\therefore \frac{1}{r_1^2} + \frac{1}{r_2^2} = \frac{1}{b^2}[1 - e^2 \sin^2 \alpha + 1 - e^2 \cos^2 \alpha] = \frac{2 - e^2}{b^2} = \frac{2 - \left(1 - \frac{a^2}{b^2}\right)}{b^2}$$

$$= \frac{1 + \frac{a^2}{b^2}}{b^2} = \frac{a^2 + b^2}{a^2 b^2} = \frac{1}{a^2} + \frac{1}{b^2}. \qquad \square$$

Example 3.16.27 Show that the line $r \cos \theta = \rho + a$ touches the circle $r^2 - 2\rho r \cos \theta + \rho^2 - a^2 = 0$ and find the coordinates of its point of contact.

Solution: The equation of the line and the circle are respectively

$$r \cos \theta = \rho + a \qquad (3.39)$$
$$\text{and} \quad r^2 - 2\rho r \cos \theta + \rho^2 - a^2 = 0 \qquad (3.40)$$

Solving (3.39) and (3.40) for r we get

$$r^2 - 2\rho(\rho + a) + \rho^2 - a^2 = 0$$
$$\Rightarrow \quad r^2 = \rho^2 + 2\rho a + a^2 = (\rho + a)^2 \quad \therefore \quad r = \pm(\rho + a).$$
Form (3.40), $\quad (\rho + a)^2 - 2\rho(\rho + a) \cos \theta + \rho^2 - a^2 = 0$
$$[\text{taking } r = +(\rho + a)]$$
$$\Rightarrow \quad \rho + a - 2\rho \cos \theta + \rho - a = 0 \ [\because \rho + a \neq 0]$$
$$\Rightarrow \quad 2\rho = 2\rho \cos \theta \Rightarrow \cos \theta = 1 = \cos(\pm 0°)$$

which gives the two coincident points $\theta = \pm 0°$ which means that the given lines touches the circle, the coordinates of the point of contact being $(\rho + a, 0°)$.

Alternative Method: (By transferring into Cartesian form)

The equation of the line is

$$x = \rho + a \qquad (3.41)$$

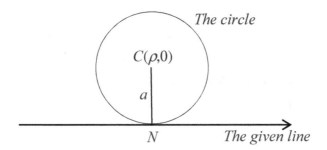

The equation of the circle is

$$r^2 - 2\rho r \cos\theta + \rho^2 - a^2 = 0 \tag{3.42}$$

The radius of the circle is $\sqrt{\rho^2 + 0^2 - (\rho^2 - a^2)} = a$ and its centre is $C(\rho, 0)$. [vide § **3.7.1**]

Now distance of the line from the centre $C(\rho, 0)$ is

$$= CN = \frac{\rho - \rho - a}{\pm\sqrt{1^2 + 0^2}} = a = \text{ radius of the circle.}$$

Hence the line touches the circle.

Solving (3.41) and (3.42) we get the point of contact as $(\rho + a, 0°)$. □

Example 3.16.28 *Show that the straight line $r\cos(\theta - \alpha) = p$ touches the conic $\frac{l}{r} = 1 + e\cos\theta$ if $(l\cos\alpha - ep)^2 + l^2\sin^2\alpha = p^2$.*

Solution: The equation of the given line can be written as

$$r\cos\theta\cos\alpha + r\sin\theta\sin\alpha = p \tag{3.43}$$

Equation of tangent at β to the conic $\frac{l}{r} = 1 + e\cos\theta$ is

$$\frac{l}{r} = e\cos\theta + \cos(\theta - \beta) = e\cos\theta + (\cos\theta\cos\beta + \sin\theta\sin\beta)$$

$$\Rightarrow r\cos\theta(e + \cos\beta) + r\sin\theta\sin\beta = l \tag{3.44}$$

Equation (3.43) will be a tangent to the conic if (3.43) and (3.44) are identical. Comparing the coefficients of (3.43) and (3.44) we get

$$\frac{l}{p} = \frac{e + \cos\beta}{\cos\alpha} = \frac{\sin\beta}{\sin\alpha}$$

$$\therefore \quad \cos\beta = \frac{l}{p}\cos\alpha - e \tag{3.45}$$

$$\text{and} \quad \sin\beta = \frac{l}{p}\sin\alpha \tag{3.46}$$

Polar Coordinates

Squaring and adding (3.45) and (3.46) we get

$$\left(\frac{l}{p}\cos\alpha - e\right)^2 + \left(\frac{l}{p}\sin\alpha\right)^2 = 1$$

or, $\dfrac{l^2}{p^2}(\cos^2\alpha + \sin^2\alpha) - 2e\dfrac{l}{p}\cos\alpha + e^2 = 1$

or, $\dfrac{l^2}{p^2} - \dfrac{2el}{p}\cos\alpha + e^2 = 1$ or, $l^2 - 2lep\cos\alpha + e^2p^2 = p^2$

or, $l^2(\sin^2\alpha + \cos^2\alpha) - 2lep\cos\alpha + e^2p^2 = p^2$

or, $(l\cos\alpha - pe)^2 + l^2\sin^2\alpha = p^2$. □

Example 3.16.29 *Let the normal is drawn at one extremity of the latus rectum PSP' of the conic $\dfrac{l}{r} = 1 + e\cos\theta$, where S is the pole. Show that the distance from the focus S to the other point in which the normal meets the curve is $\dfrac{l(1 + 3e^2 + e^4)}{1 + e^2 - e^4}$.*

Solution: The equation of the normal at $\theta = \alpha$ to the conic

$$\frac{l}{r} = 1 + e\cos\theta \tag{3.47}$$

is $\dfrac{le\sin\alpha}{r(1 + e\cos\alpha)} = \sin(\theta - \alpha) + e\sin\theta.$ [§ **3.11**, Note 3.11.1]

Let the normal be drawn at the point P whose polar coordinates are taken as $\left(l, \dfrac{\pi}{2}\right)$.

Therefore the equation of the normal at P is

$$\frac{el}{r} = e\sin\theta + \sin\left(\theta - \frac{\pi}{2}\right)$$

or, $\dfrac{l}{r} = \dfrac{e\sin\theta - \cos\theta}{e}$ \hfill (3.48)

For the common point of intersection of (3.47) and (3.48) we get

$$\frac{e\sin\theta - \cos\theta}{e} = 1 + e\cos\theta \Rightarrow e(1 + e\cos\theta) = e\sin\theta - \cos\theta$$

or, $(e^2 + 1)\cos\theta = -e(1 - \sin\theta)$ or, $\dfrac{\cos\theta}{1 - \sin\theta} = -\dfrac{e}{1 + e^2}$

or, $\dfrac{\cos^2\theta}{(1 - \sin\theta)^2} = \dfrac{e^2}{(1 + e^2)^2} \Rightarrow \dfrac{1 + \sin\theta}{1 - \sin\theta} = \dfrac{e^2}{(1 + e^2)^2}$

or, $\sin\theta = \dfrac{e^2 - (1 + e^2)^2}{e^2 + (1 + e^2)^2}$ (By Componendo-dividendo method).

$$\therefore \cos\theta = -\frac{e}{1+e^2}(1-\sin\theta) = \frac{e}{1+e^2}(\sin\theta - 1)$$

$$= \frac{e}{1+e^2}\left[\frac{e^2-(1+e^2)^2}{e^2+(1+e^2)^2} - 1\right] = \frac{-2e(1+e^2)}{1+3e^2+e^4}.$$

So from (3.47) required distance $= r = \dfrac{l}{1+e\cos\theta} = \dfrac{l}{1 - \frac{2e^2(1+e^2)}{1+3e^2+e^4}}$

$$= \frac{l(e^4+3e^2+1)}{e^4+3e^2+1-2e^4-2e^2} = \frac{l(1+3e^2+e^4)}{1+e^2-e^4}. \qquad \square$$

Example 3.16.30 *A conic touches the ellipse $\dfrac{l}{r} = 1 + e\cos\theta$ at the point $\theta = \alpha$. They have a common focus at the pole and have the same eccentricity. Show that the length of the latus rectum of the conic is $\dfrac{2l(1-e^2)}{e^2+2e\cos\alpha+1}$.*

Solution: Let the equation of the conic in our consideration be $\dfrac{l'}{r} = 1 + e\cos(\theta-\beta)$ as it has the same focus and same eccentricity of the given ellipse $\dfrac{l}{r} = 1 + e\cos\theta$.

Equations of the tangents at the point $\theta = \alpha$ to the two conics are

$$\frac{l}{r} = e\cos\theta + \cos(\theta-\alpha) \qquad (3.49)$$

and $\qquad \dfrac{l'}{r} = e\cos(\theta-\beta) + \cos(\theta-\alpha) \qquad (3.50)$

From (3.49) and (3.50) we have

$$l = r\cos\theta(e+\cos\alpha) + r\sin\theta\sin\alpha \qquad (3.51)$$
and $\quad l' = r\cos\theta(e\cos\beta+\cos\alpha) + r\sin\theta(\sin\alpha+e\sin\beta) \qquad (3.52)$

Since (3.51) and (3.52) are identical, so comparing the coefficients, we get

$$\frac{l}{l'} = \frac{e+\cos\alpha}{\cos\alpha+e\cos\beta} = \frac{\sin\alpha}{\sin\alpha+e\sin\beta}$$

$\therefore \quad l(\cos\alpha+e\cos\beta) = l'(e+\cos\alpha)$

$\Rightarrow \quad el\cos\beta = l'e + (l'-l)\cos\alpha \qquad (3.53)$

and $\quad l(\sin\alpha + e\sin\beta) = l'\sin\alpha \;\Rightarrow\; el\sin\beta = (l'-l)\sin\alpha \qquad (3.54)$

Polar Coordinates

Squaring (3.53) and (3.54) and adding we get

$$l^2 e^2 = (l'-l)^2 + l'^2 e^2 + 2el'(l'-l)\cos\alpha$$
$$\Rightarrow e^2 l^2 - e^2 l'^2 = (l-l')^2 - 2el'(l-l')\cos\alpha$$
$$\Rightarrow e^2(l+l') = (l-l') - 2el'\cos\alpha \quad [\because l-l' \neq 0]$$
$$\Rightarrow l'(e^2 + 2e\cos\alpha + 1) = l(1-e^2)$$
$$\Rightarrow l' = \frac{l(1-e^2)}{e^2 + 2e\cos\alpha + 1} = \text{length of the semi-latus rectum.}$$

So the length of latus rectum of the conic in our consideration

$$= 2l' = \frac{2l(1-e^2)}{e^2 + 2e\cos\alpha + 1}. \qquad \square$$

Example 3.16.31 *The tangents at two points P and Q of a conic meet in T and S is the focus. Prove that*

(i) If the conic is a parabola, then $ST^2 = SP.SQ$.

(ii) If the conic be central, then $\dfrac{1}{SP.SQ} - \dfrac{1}{ST^2} = \dfrac{1}{b^2}\sin^2\dfrac{1}{2}\angle PSQ$ where b is the semi-minor axis.

Solution: (i) Let the equation of the parabola be $\dfrac{l}{r} = 1 + \cos\theta$ and let

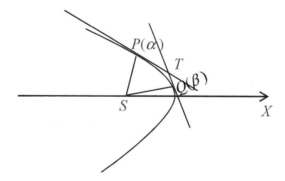

the vectorial angles at P and Q be α and β respectively. At T, where the tangents meet we have

$$\frac{l}{r} = \cos\theta + \cos(\theta - \alpha) = \cos\theta + \cos(\theta - \beta) \Rightarrow \theta - \alpha = \pm(\theta - \beta).$$

Since, $\alpha \neq \beta$, so $\theta - \alpha = -\theta + \beta \Rightarrow \theta = \dfrac{1}{2}(\alpha + \beta).$

So the vectorial angle of T is $\dfrac{\alpha+\beta}{2}$.

Now for the point P, $\quad \dfrac{l}{SP} = 1+\cos\alpha \Rightarrow SP = \dfrac{l}{1+\cos\alpha}$

and for the point Q, $\quad \dfrac{l}{SQ} = 1+\cos\beta \Rightarrow SQ = \dfrac{l}{1+\cos\beta}$.

Therefore $\quad SP.SQ = \dfrac{l^2}{(1+\cos\alpha)(1+\cos\beta)} = \dfrac{l^2}{4\cos^2\frac{\alpha}{2}\cos^2\frac{\beta}{2}}$.

Again the equation of the tangent at $P(\alpha)$ is $\dfrac{l}{r} = \cos\theta + \cos(\theta-\alpha)$.

It meets $T\left(\dfrac{\alpha+\beta}{2}\right)$. So,

$$\dfrac{l}{ST} = \cos\dfrac{\alpha+\beta}{2} + \cos\left(\dfrac{\alpha+\beta}{2}-\alpha\right)$$

$$= \cos\dfrac{\alpha+\beta}{2} + \cos\dfrac{\alpha-\beta}{2} = 2\cos\dfrac{\alpha}{2}\cos\dfrac{\beta}{2}$$

$$\therefore \quad ST^2 = \dfrac{l^2}{4\cos^2\frac{\alpha}{2}\cos^2\frac{\beta}{2}} = SP.SQ.$$

(ii) Let the equation of the central conic be $\dfrac{l}{r} = 1 + e\cos\theta$. Let α and β be the vectorial angles of the two points P and Q. At T, the two tangents meet. So we get

$$\cos(\theta-\alpha) + e\cos\theta = \cos(\theta-\beta) + e\cos\theta \Rightarrow \theta = \dfrac{1}{2}(\alpha+\beta) \text{ [as in part (i)]}$$

So the vectorial angle of T is $\dfrac{\alpha+\beta}{2}$. Now for the point P,

$$\dfrac{l}{SP} = 1 + e\cos\alpha \Rightarrow SP = \dfrac{l}{1+e\cos\alpha}; \quad \text{and similarly } SQ = \dfrac{l}{1+e\cos\beta}.$$

$$\therefore \quad \dfrac{1}{SP.SQ} = \dfrac{(1+e\cos\alpha)(1+e\cos\beta)}{l^2} \qquad (3.55)$$

The equation of the tangent at $P(\alpha)$ is $\dfrac{l}{r} = e\cos\theta + \cos(\theta-\alpha)$.

It meets $T\left(\dfrac{\alpha+\beta}{2}\right)$. So,

$$\dfrac{l}{ST} = e\cos\dfrac{\alpha+\beta}{2} + \cos\left(\dfrac{\alpha+\beta}{2}-\alpha\right) = e\cos\dfrac{\alpha+\beta}{2} + \cos\dfrac{\alpha-\beta}{2}$$

Polar Coordinates

$$\therefore \quad \frac{1}{ST} = \frac{1}{l}\left[e\cos\frac{\alpha+\beta}{2} + \cos\frac{\alpha-\beta}{2}\right]$$

$$\therefore \quad \frac{1}{SP.SQ} - \frac{1}{ST^2}$$

$$= \frac{1}{l^2}\left[(1+e\cos\alpha)(1+e\cos\beta) - \left(e\cos\frac{\alpha+\beta}{2} + \cos\frac{\alpha-\beta}{2}\right)^2\right]$$

$$= \frac{1}{l^2}\Big[1 + e(\cos\alpha + \cos\beta) + e^2\cos\alpha\cos\beta - \cos^2\frac{\alpha-\beta}{2}$$
$$\qquad - e^2\cos^2\frac{\alpha+\beta}{2} - 2e\cos\frac{\alpha-\beta}{2}\cos\frac{\alpha-\beta}{2}\Big]$$

$$= \frac{1}{l^2}\Big[1 + 2e\cos\frac{\alpha+\beta}{2}\cos\frac{\alpha-\beta}{2} + e^2\cos\alpha\cos\beta - \cos^2\frac{\alpha-\beta}{2}$$
$$\qquad -e^2\cos^2\frac{\alpha+\beta}{2} - 2e\cos\frac{\alpha+\beta}{2}\cos\frac{\alpha-\beta}{2}\Big]$$

$$= \frac{1}{l^2}\left[1 + e^2\cos\alpha\cos\beta - \cos^2\frac{\alpha-\beta}{2} - e^2\cos^2\frac{\alpha+\beta}{2}\right]$$

$$= \frac{1}{l^2}\left[1 - \cos^2\frac{\alpha-\beta}{2} + e^2\left(\cos\alpha\cos\beta - \cos^2\frac{\alpha+\beta}{2}\right)\right]$$

$$= \frac{1}{l^2}\left[1 - \cos^2\frac{\alpha-\beta}{2} + e^2\left\{\cos\alpha\cos\beta - \frac{1}{2}(1+\cos\overline{\alpha+\beta})\right\}\right]$$

$$= \frac{1}{l^2}\left[1 - \cos^2\frac{\alpha-\beta}{2} + \frac{e^2}{2}\{2\cos\alpha\cos\beta - 1 - \cos\alpha\cos\beta + \sin\alpha\sin\beta\}\right]$$

$$= \frac{1}{l^2}\left[1 - \cos^2\frac{\alpha-\beta}{2} + \frac{e^2}{2}\{\cos\alpha\cos\beta + \sin\alpha\sin\beta - 1\}\right]$$

$$= \frac{1}{l^2}\left[1 - \cos^2\frac{\alpha-\beta}{2} - \frac{e^2}{2}\{1 - \cos(\alpha-\beta)\}\right]$$

$$= \frac{1}{l^2}\left[\sin^2\frac{\alpha-\beta}{2} - e^2\sin^2\frac{\alpha-\beta}{2}\right] = \frac{1-e^2}{l^2}\sin^2\frac{\alpha-\beta}{2}.$$

Now $b^2 = a^2(1-e^2) \Rightarrow 1 - e^2 = \dfrac{b^2}{a^2}$ and $l^2 = \left(\dfrac{b^2}{a}\right)^2 = \dfrac{b^4}{a^2}$

$$\therefore \quad \frac{1}{SP.SQ} - \frac{1}{ST^2} = \frac{1}{b^2}\sin^2\frac{\alpha-\beta}{2}$$

$$= \frac{1}{b^2}\sin^2\frac{1}{2}\angle PSQ \quad (\because \angle PSQ = \alpha - \beta). \qquad \square$$

Example 3.16.32 S is the focus and P and Q are two points on a conic such that the $\angle PSQ$ is constant and equal to 2δ. Prove that

(i) locus of intersection of the tangents at P and Q is a conic section whose focus is S.

(ii) the line PQ always touches a conic whose focus is S.

Solution: (i) It is evident that the vectorial angles of P and Q can be expressed as $\gamma + \delta$ and $\gamma - \delta$ respectively where γ is a variable [See § **3.12 Note 3.12.4**]. Then the tangents at P and Q are respectively given by

$$\frac{l}{r} = e\cos\theta + \cos(\theta - \gamma - \delta) \text{ and } \frac{l}{r} = e\cos\theta + \cos(\theta - \gamma + \delta).$$

Let (r_1, θ_1) be their point of intersection. Then

$$\theta_1 = \frac{1}{2}(\gamma + \delta + \gamma - \delta) = \gamma \text{ and } \frac{l}{r_1} = e\cos\theta_1 + \cos\delta$$

$$\therefore \frac{l\sec\delta}{r_1} = 1 + e\sec\delta\cos\theta_1.$$

Hence the locus of (r_1, θ_1) is

$$\frac{l'}{r} = 1 + e'\cos\theta \quad \text{where } l' = l\sec\delta \text{ and } e' = e\sec\delta.$$

(ii) The equation of the chord PQ is

$$\frac{l}{r} = e\cos\theta + \sec\delta\cos(\theta - \gamma) \quad [\S \text{ 3.9 Note 3.9.1}]$$

$$\therefore \frac{l\cos\delta}{r} = e\cos\delta\cos\theta + \cos(\theta - \gamma) \text{ or, } \frac{l_1}{r} = e_1\cos\theta + \cos(\theta - \gamma).$$

which is the form of a line touching the conic $\dfrac{l_1}{r} = 1 + e_1\cos\theta$ at γ_1, where $l_1 = l\cos\delta$ and $e_1 = e\cos\delta$. □

Example 3.16.33 If the tangents at any two points P and Q of a conic meet in a point T and the straight line PQ meets the directrix corresponding to S in a point K, the angle $\angle KST$ is a right angle.

Solution: Let the equation of the conic be $\dfrac{l}{r} = 1 + e\cos\theta$. If the vectorial angles of the points P and Q respectively be α and β, then the vectorial angle of T is $\dfrac{1}{2}(\alpha + \beta)$. The equation of the chord PQ is

$$\frac{l}{r} = e\cos\theta + \sec\frac{\beta - \alpha}{2}\cos\left(\theta - \frac{\alpha + \beta}{2}\right). \quad [\text{See § 3.9 Corollary and Note}]$$

Polar Coordinates

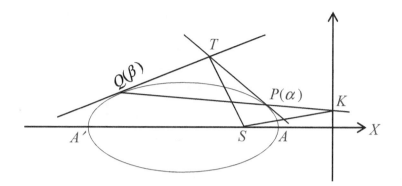

and the equation of the directrix is $\dfrac{l}{r} = e\cos\theta$ [§ **3.14** Note **3.14.1**].

The vectorial angle of the point K is obtained from the relation

$$\sec\left(\dfrac{\beta-\alpha}{2}\right)\cos\left(\theta - \dfrac{\alpha+\beta}{2}\right) = 0 \;\Rightarrow\; \theta - \dfrac{\alpha+\beta}{2} = \pm\dfrac{\pi}{2}$$

$$\therefore\; \theta = \angle KSA = \pm\dfrac{\pi}{2} + \dfrac{\alpha+\beta}{2}.$$

Also we have $\angle TSA = \dfrac{\alpha+\beta}{2}$ $\;\therefore\; \angle KST = \angle KSA - \angle TSA = \pm\dfrac{\pi}{2}.$ □

Example 3.16.34 P, Q *are two points on the conic* $\dfrac{l}{r} = 1 - e\cos\theta$ *with* $\alpha-\gamma$ *and* $\alpha+\gamma$ *as vectorial angles, where* γ *is a constant and* α *is a variable. Show that the line PQ always touches another conic* $\dfrac{l\cos\gamma}{r} = 1 - e\cos\gamma\cos\theta$, *which has in common with the given conic, a focus, namely the pole and the corresponding directrix.*

Solution: The equation of the chord PQ is

$$\dfrac{l}{r} = \sec\gamma\cos(\theta-\alpha) - e\cos\theta \;[vide. \;\S\; \mathbf{3.9}]$$

$$\therefore\; \dfrac{l\cos\gamma}{r} = \cos(\theta-\alpha) - e\cos\theta\cos\gamma \;\Rightarrow\; \dfrac{l'}{r} = \cos(\theta-\alpha) - e'\cos\theta$$

where $l' = l\cos\gamma$ and $e' = e\cos\gamma$. Clearly this is a tangent to the conic $\dfrac{l'}{r} = 1 - e'\cos\theta$, i.e., $\dfrac{l\cos\gamma}{r} = 1 - e\cos\gamma\cos\theta$ at the point whose vectorial angle is $\dfrac{\alpha+\beta}{2}$.

Equation of the directrix of the given conic with the corresponding focus is

$$\frac{l}{r} = -e\cos\theta \quad \Rightarrow \quad r = -\frac{l}{e}\sec\theta.$$

Equation of the directrix of the second conic with the corresponding focus is

$$r = -\frac{l'}{e'}\sec\theta \quad \Rightarrow \quad r = -\frac{l\cos\gamma}{e\cos\gamma}\sec\theta \quad \Rightarrow \quad r = -\frac{l}{e}\sec\theta.$$

So the directrixes of the two conics are the same. □

Example 3.16.35 *Prove that three normals can be drawn to a parabola from a given point. If the normals at θ_1, θ_2, θ_3 on the parabola $\frac{l}{r} = 1 + \cos\theta$ meet in the point (ρ, ϕ), prove that $\theta_1 + \theta_2 + \theta_3 = 2\phi$.*

Solution: The equation of the normal at $\theta = \alpha$ to the given parabola is

$$\frac{l}{r}\frac{\sin\alpha}{1+\cos\alpha} = \sin\theta + \sin(\theta - \alpha) \tag{3.56}$$

This can be written as $\quad \dfrac{l}{r}\tan\dfrac{\alpha}{2} = 2\sin\left(\theta - \dfrac{\alpha}{2}\right)\cos\dfrac{\alpha}{2}.$

If it is passes through (ρ, ϕ) then

$$\frac{l}{\rho}\tan\frac{\alpha}{2} = 2\sin\left(\phi - \frac{\alpha}{2}\right)\cos\frac{\alpha}{2} = 2\left(\sin\phi\cos^2\frac{\alpha}{2} - \cos\phi\sin\frac{\alpha}{2}\cos\frac{\alpha}{2}\right).$$

Multiplying both sides by $\sec^2\dfrac{\alpha}{2}$, we get

$$\frac{l}{\rho}\tan\frac{\alpha}{2}\left(1 + \tan^2\frac{\alpha}{2}\right) = 2\left(\sin\phi - \cos\phi\tan\frac{\alpha}{2}\right)$$

or, $\quad l\tan^3\dfrac{\alpha}{2} + (l + 2\rho\cos\phi)\tan\dfrac{\alpha}{2} - 2\rho\sin\phi = 0 \tag{3.57}$

Let us now put $\tan\dfrac{\alpha}{2} = t$, then equation (3.57) becomes

$$lt^3 + (l + 2\rho\cos\phi)t - 2\rho\sin\phi = 0 \tag{3.58}$$

which is a cubic equation in t, having three roots. Corresponding to each of these three roots we get a normal passing through the point (ρ, ϕ), showing that in general three normals can be drawn to a parabola through a given point.

Polar Coordinates

Now if t_1, t_2, t_3 be the roots of the equation (3.58), i.e., $t_1 = \tan\frac{\theta_1}{2}, t_2 = \tan\frac{\theta_2}{2}$ and $t_3 = \tan\frac{\theta_3}{2}$, then we get

$$\sum t_1 = 0, \sum t_1 t_2 = \frac{l + 2\rho\cos\phi}{l} = 1 + 2\rho\frac{\cos\phi}{l} \text{ and } t_1 t_2 t_3 = \frac{2\rho\sin\phi}{l}.$$

and then $\tan\left(\frac{\theta_1 + \theta_2 + \theta_3}{2}\right) = \frac{t_1 + t_2 + t_3 - t_1 t_2 t_3}{1 - (t_2 t_3 + t_3 t_1 + t_1 t_2)}$

$$= \frac{\frac{-2\rho\sin\phi}{l}}{1 - \left(1 + \frac{2\rho\cos\phi}{l}\right)} = \tan\phi.$$

$$\therefore \quad \frac{\theta_1 + \theta_2 + \theta_3}{2} = n\pi + \phi, \text{ where } n \text{ is an integer.}$$

Putting $n = 0$ we get $\theta_1 + \theta_2 + \theta_3 = 2\phi$. □

Example 3.16.36 *Show that, if the normals at the points whose vectorial angles are $\theta_1, \theta_2, \theta_3, \theta_4$ on the conic $\frac{l}{r} = 1 + e\cos\theta$ meet in the point (ρ, ϕ) then $\theta_1 + \theta_2 + \theta_3 + \theta_4 - 2\phi = (2n+1)\pi$.*

Solution: The normal at α is

$$\frac{e\sin\alpha}{1 + e\cos\alpha} \cdot \frac{l}{r} = e\sin\theta + \sin(\theta - \alpha).$$

If it passes through (ρ, ϕ), then

$$\frac{e\sin\alpha}{1 + e\cos\alpha} \cdot \frac{l}{\rho} = e\sin\phi + \sin(\phi - \alpha)$$

$$= e\sin\phi + \sin\phi\cos\alpha - \cos\phi\sin\alpha.$$

Putting t for $\tan\frac{\alpha}{2}$, we get

$$e\sin\phi + \sin\phi\frac{1-t^2}{1+t^2} - \cos\phi\frac{2t}{1+t^2} = \frac{e\frac{2t}{1+t^2}}{1 + e\frac{1-t^2}{1+t^2}} \times \frac{l}{\rho}$$

$$= \frac{2et}{(1+e) + (1-e)t^2} \times \frac{l}{\rho}$$

$$\left[\because \sin\alpha = \frac{2\tan\frac{\alpha}{2}}{1 + \tan^2\frac{\alpha}{2}} = \frac{2t}{1+t^2} \text{ and } \cos\alpha = \frac{1 - \tan^2\frac{\alpha}{2}}{1 + \tan^2\frac{\alpha}{2}} = \frac{1-t^2}{1+t^2}\right]$$

or, $\rho\{e\sin\phi(1+t^2) + \sin\phi(1-t^2) - \cos\phi 2t\}\{(1+e) + (1-e)t^2\} = 2elt(1+t^2)$

or, $\rho(1-e)^2 \sin\phi\, t^4 + 2[el + (1-e)\rho\cos\phi]t^3 + 2[el + (1+e)\rho\cos\phi]t$
$$-\rho(1+e^2)\sin\phi = 0.$$

If t_1, t_2, t_3, t_4 be the four roots, then $\sum t_1 = -\dfrac{2[el + \rho(1-e)^2 \cos\phi]}{\rho(1-e)\sin\phi}$,

$\sum t_1 t_2 = 0$, $\sum t_1 t_2 t_3 = -\dfrac{2[el + \rho(1+e)\cos\phi]}{\rho(1-e)^2 \sin\phi}$, $t_1 t_2 t_3 t_4 = -\dfrac{(1+e)^2}{(1-e)^2}.$

$\therefore\ \tan\dfrac{1}{2}(\theta_1+\theta_2+\theta_3+\theta_4) = \dfrac{\sum t_1 - \sum t_1 t_2 t_3}{1 - \sum t_1 t_2 + t_1 t_2 t_3 t_4}$

$= \dfrac{-\dfrac{2[el+\rho(1-e)\cos\phi]}{\rho(1-e)^2 \sin\phi} + \dfrac{2[el+\rho(1+e)\cos\phi]}{\rho(1-e)^2 \sin\phi}}{1 - 0 - \dfrac{(1+e)^2}{(1-e)^2}} = \dfrac{\dfrac{1}{\rho\sin\phi(1-e)^2}}{\dfrac{1}{(1-e)^2}} \times \dfrac{2e.2\rho\cos\phi}{(1-e)^2 - (1+e)^2}$

$= \dfrac{4e\rho\cos\phi}{-4e\rho\sin\phi} = -\cot\phi = \tan\left[\phi + (2n+1)\dfrac{\pi}{2}\right].$

Hence, $\dfrac{\theta_1+\theta_2+\theta_3+\theta_4}{2} = \phi + (2n+1)\dfrac{\pi}{2}$

$\Rightarrow \theta_1+\theta_2+\theta_3+\theta_4 - 2\phi = (2n+1)\pi.$ $\square$

Example 3.16.37 *Show that the equation of the circle which passes through the focus of the parabola* $\dfrac{2a}{r} = 1 + \cos\theta$ *and touches it at the point* $\theta = \alpha$ *is given by*

$$r^2 \cos^3\dfrac{\alpha}{2} = a\cos\left(\theta - \dfrac{3\alpha}{2}\right).$$

Solution: Let the equation of the circle be $r = 2d\cos(\theta - \beta)$.

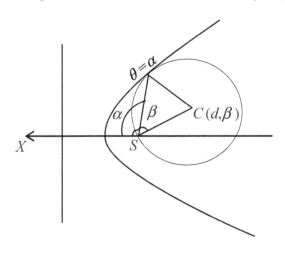

Polar Coordinates

At the point $\theta = \alpha$, the parabola and the circle have the common tangent and hence the common normal.

The normal to the parabola

$$\frac{2a}{r} = 1 + \cos\theta \text{ at } \theta = \alpha \text{ is } \frac{2a\sin\alpha}{(1+\cos\alpha)r} = \sin(\theta - \alpha) + \sin\theta.$$

It is also the normal to the circle at $\theta = \alpha$ and hence passes through its centre (d, β)

$$\therefore \quad \frac{2a\sin\alpha}{(1+\cos\alpha)d} = \sin(\beta - \alpha) + \sin\beta \tag{3.59}$$

At the common point $\theta = \alpha$ both the radius vectors are the same.

$$\therefore \quad \frac{2a}{1+\cos\alpha} = 2d\cos(\alpha - \beta) \tag{3.60}$$

From (3.59) and (3.60) we get

$$d(1+\cos\alpha) = \frac{2a\sin\alpha}{\sin(\beta - \alpha) + \sin\beta} = \frac{a}{\cos(\alpha - \beta)}.$$

The last two gives

$$2a\sin\alpha\cos(\alpha - \beta) = a\{\sin(\beta - \alpha) + \sin\beta\}$$

or, $\sin(2\alpha - \beta) + \sin\beta = \sin(\beta - \alpha) + \sin\beta$

or, $\sin(2\alpha - \beta) = \sin(\beta - \alpha) \Rightarrow 2\alpha - \beta = \beta - \alpha \Rightarrow \beta = \dfrac{3\alpha}{2}.$

Therefore from (3.60),

$$2d = \frac{2a}{(1+\cos\alpha)\cos\left(\alpha - \frac{3\alpha}{2}\right)} = \frac{2a}{2\cos^2\frac{\alpha}{2}\cos\frac{\alpha}{2}} = \frac{a}{\cos^3\frac{\alpha}{2}}$$

Hence the required circle is

$$r = \frac{a}{\cos^3\frac{\alpha}{2}}\cos\left(\theta - \frac{3\alpha}{2}\right)$$

or, $\quad r^2\cos^3\dfrac{\alpha}{2} = a\cos\left(\theta - \dfrac{3\alpha}{2}\right).$

$\square$

3.17 Exercises

Section A: Objective Type Questions

1. Write down the polar equation of a straight line perpendicular to the initial line and at a distance of 5 units form the pole.

2. Write down the polar equation of the circle of radius 5 units with centre on the initial line at a distance of 5 units from the pole on the positive side.

3. Write down the equation of a parabola $y^2 = 8x$ in polar form, if the pole is at the vertex.

4. Find the polar equation of the straight line parallel to the initial line and at a distance of 2 units form the pole.

5. Find the polar coordinates of the point whose Cartesian coordinates are $(-2, 0)$.

6. Show that the distance between the points $\left(1, \dfrac{\pi}{6}\right)$ and $\left(3, -\dfrac{\pi}{2}\right)$ is $\sqrt{7}$ units,

7. Write down the polar equation of the circle of radius 2 units with centre on the initial line at a distance of 2 units from the pole on the positive side.

8. Show that the nature of the conic $\dfrac{8}{r} = 4 - 5\cos\theta$ is a hyperbola.

9. Show that the length of the latus rectum of the conic $\dfrac{6}{r} = 4 - 3\cos\theta$ is 3 units.

10. Find the polar coordinates of the point whose Cartesian coordinates are $(-1, -1)$.

11. Write down the polar equation of the straight line $x = 0$.

12. Find the points on the conic $\dfrac{14}{r} = 3 - 8\cos\theta$ whose radius vector is 2.

13. Find the point with smallest radius vector of the conic $r = \dfrac{6}{1 - \cos\theta}$.

14. Find the polar coordinates of the point symmetric to the point $P\left(2, \dfrac{\pi}{4}\right)$ with respect to the pole.

15. Find the polar equation of the straight line which passes through the pole and makes an angle of 45° with the polar axis.

16. On the parabola $r = \dfrac{1}{3(1 - \cos\theta)}$, find the point with smallest radius vector.

17. What is the polar equation of a circle when the pole is at the centre of the circle.

18. Find the equation of the conic with focus as the pole, eccentricity is 2, and directrix is $r\cos\theta + 5 = 0$.

[*Hints:* Let the equation of the conic be $\dfrac{l}{r} = 1 - 2\cos\theta$. The directrix is $5 = -r\cos\theta \Rightarrow \dfrac{10}{r} = -2\cos\theta \Rightarrow l = 10$. Thus the required equation of the conic is $\dfrac{10}{r} = -2\cos\theta$.]

19. On the ellipse $r = \dfrac{21}{5 - 2\cos\theta}$, find the point with the greatest radius vector.

20. Find the eccentricity and vertex of $r = \dfrac{6}{1 - \cos\theta}$.

Section B: Broad Answer Type Questions

1. Convert the following equations with their equivalent Cartesian form:

(i) $r = 2\cos\theta$ \qquad (ii) $r = 4\tan\theta \sec\theta$

(iii) $r\sin\theta = r\cos\theta + 4$ \qquad (iv) $r = \operatorname{cosec}\theta\, e^{r\cos\theta}$.

2. Convert the following equations into their equivalent polar equations:

(i) $2x^2 + 3y^2 = 6$ \qquad (ii) $y = 10$ \qquad (iii) $(x^2 + y^2)^2 = ax^2$.

3. Find the distance between the points whose polar coordinates are $(2, 40°)$ and $(4, 100°)$.

4. Find the polar equation of a straight line joining the two points $\left(1, \dfrac{\pi}{2}\right)$ and $(2, \pi)$.

5. Show that the polar equation of the straight line passing through the points (r_1, θ_1) and (r_2, θ_2) is $\dfrac{\sin(\theta_1 - \theta_2)}{r} = \dfrac{\sin(\theta - \theta_2)}{r_1} - \dfrac{\sin(\theta - \theta_1)}{r_2}$.

6. The vectorial angle of a point P on the straight line joining the points (r_1, θ_1) and (r_2, θ_2) is $\dfrac{1}{2}(\theta_1 + \theta_2)$. Find the radius vector OP.

7. Find the condition that the points (r_1, θ_1), (r_2, θ_2) and (r_3, θ_3) may be collinear.

8 Show that the condition for concurrence of the lines $r\cos(\theta - \alpha) =$

a, $r\cos(\theta-\beta)=b$ and $r\cos(\theta-\gamma)=c$ is

$$\begin{vmatrix} \cos\alpha & \sin\alpha & a \\ \cos\beta & \sin\beta & b \\ \cos\gamma & \sin\gamma & c \end{vmatrix} = 0.$$

9. Find the polar equation of the straight lines bisecting the angles between the straight line $\theta = \dfrac{\pi}{6}$ and $\theta = \dfrac{\pi}{3}$.

10. Show that the polar equations of the bisectors of the angles between the lines $\theta = \alpha$ and $\theta = \beta$ are $\theta = \dfrac{1}{2}(\alpha+\beta)$ and $\theta = \dfrac{\pi}{2} + \dfrac{1}{2}(\alpha+\beta)$.

11. Show that the lines $r = \cos(\theta-\alpha) = p$ and $r = \cos(\theta-\beta) = p$ intersect at $\left(p\sec\dfrac{1}{2}(\alpha-\beta), \dfrac{1}{2}(\alpha+\beta)\right)$.

12. Show that the perpendicular distance of the point (r_1, θ_1) from $r = \cos(\theta-\alpha) = p$ is $r_1 \cos(\theta_1 - \alpha) - p$.

13. Show that the straight lines $r\cos(\theta-\alpha) = p$ and $r\cos(\theta-\alpha) = p'$ are parallel and the lines $r\sin(\theta-\alpha) = p'$ and $r\cos(\theta-\alpha) = p$ are perpendicular to each other.

14. Find the equations of the circles under the following conditions:

(i) Centre $(4,0)$, radius 4
(ii) Centre $\left(5, \dfrac{\pi}{3}\right)$, radius 2
(iii) Centre $(4, 60°)$ and passes through $(7, 45°)$
(iv) Centre on the line $\theta = \dfrac{\pi}{3}$, passing through $(0,0)$ and $\left(5, \dfrac{\pi}{2}\right)$
(v) Extremities of the diameter are $\left(2, \dfrac{2\pi}{3}\right)$ and $\left(-2, \dfrac{\pi}{3}\right)$.

15. Show that each of the following equations represents circle. Also find its radius:

(i) $r^2 - 2r\sin\theta - 3 = 0$
(ii) $r = 3\sin\theta + 3\sqrt{3}\cos\theta$
(iii) $r^2 - 2(2\cos\theta + 3\sin\theta)r - 12 = 0$.

16. Find the polar coordinates of the centre of each of the following circles:

(i) $r = 3\sin\theta + 4\cos\theta$
(ii) $r = 8\cos\theta$
(iii) $r(r - 4\cos\theta) = 5$.

17. Show that the polar equation of a parabola may be written in the

form $\sqrt{r} \sin \dfrac{\theta}{2} = \sqrt{a}$, where $4a$ is the length of the latus rectum.

18. Determine the nature of the following conics:

(i) $\dfrac{3}{r} = 2 + 4\cos\theta$ (ii) $\dfrac{17}{r} = \sqrt{5} - 2\cos\theta$ (iii) $\dfrac{2}{r} = 3 - 3\cos\theta$

(iv) $\dfrac{8}{r} = 4 - 5\cos\theta$ (v) $r = \dfrac{1}{4 - 5\cos\theta}$.

19. Find the point/points of the following conics with the following conditions:

(i) $\dfrac{15}{r} = 1 - 4\cos\theta$, whose radius vector is 5

(ii) $\dfrac{6}{r} = 1 + 2\cos\theta$ whose vectorial angle is $\dfrac{\pi}{3}$

(iii) $\dfrac{14}{r} = 3 - 8\cos\theta$, whose radius vector is 2

(iv) $\dfrac{l}{r} = 1 - \cos\theta$ which has the smallest radius vector

(v) $r(1 - \cos\theta) = 6$ having the least radius vector.

20. Find the equation of the conics with focus at the pole and having given the following:

(i) Eccentricity $e = 1$, semi latus rectums $= l = 4$

(ii) Eccentricity $e = 1$, directrix $r = \cos\left(\theta - \dfrac{\pi}{4}\right) = 2\sqrt{2}$.

[*Hints:* If $r\cos\theta = -p$ be the directrix, the conic is $r = \dfrac{ep}{1 - e\cos\theta}$.]

21. Find the polar equation of the ellipse $\dfrac{x^2}{36} + \dfrac{y^2}{20} = 1$, if the pole is at its right hand focus and the positive direction of the x-axis is considered as the polar axis.

22. Find the equation of $\dfrac{x^2}{16} - \dfrac{y^2}{9} = 1$ in the polar form with the left hand focus as the pole and the positive direction of the x-axis is taken as the polar axis.

23. Show that the polar equation of a parabola may be written in the form $\sqrt{r}\cos\dfrac{\theta}{2} = \sqrt{a}$, whose $4a$ is the length of the latus rectum.

24. Find the polar equation of left branch of the hyperbola $\dfrac{x^2}{36} - \dfrac{y^2}{64} = 1$, if the pole is at the right hand focus and the positive direction of the x-axis is taken as the polar axis.

25. For the ellipse $r^2 = \dfrac{b^2}{1 - e^2 \cos^2 \theta}$, is the vectorial angle corresponding to the radius vector ae be θ, then show that $\tan^2 \theta = \dfrac{b^4}{a^2(a^2 e^2 - b^2)}$.

26. For each of the following conics, find the polar coordinates of vertex/vertices and the length of the latus rectum.

(i) $\dfrac{3}{r} = 4 - 2\cos\theta$, (ii) $\dfrac{12}{r} = 2 - 2\cos\theta$, (iii) $\dfrac{3}{r} = 1 + 2\cos\theta$,

(iv) $\dfrac{21}{r} = 5 + 2\cos\theta$.

27. For the conics of the foregoing **Example 26**, find the equations of the directrix corresponding to their foci/focus.

28. Prove that the polar equation of the locus of the foot of the perpendicular form the pole on a line which always passes through a given point (d, β) is $r = d\cos(\theta - \beta)$.

29. The latus rectum of a conic is 6 and its eccentricity is $\dfrac{1}{2}$. Find the length of the focal chord inclined at an angle of $45°$ with the major axis.

30. Find the polar equation of the directrixes of the conic $\dfrac{12}{r} = 2 - \cos\theta$.

[*Hints:* Two directions are $r = -\dfrac{l}{e}\sec\theta$ and $r = \dfrac{l}{r} \times \dfrac{1+e^2}{1-e^2}\sec\theta$, See § 3.14]

31. In a conic, prove the followings:

(i) the semi-latus rectum of a conic is a harmonic mean between the segments of any focal chord.

or, the sum of the reciprocals of the segments of a focal chord is a constant.

(ii) the sum of the reciprocals of two perpendicular focal chords is constant.

(iii) the locus of the middle points of the system of focal chords is a similar conic.

32. If PSP' and QSQ' be two perpendicular focal chords of a conic, then prove that

(i) $\dfrac{1}{SP.SP'} + \dfrac{1}{SQ.SQ'} = $ constant and (ii) $\dfrac{1}{PP'} + \dfrac{1}{QQ'} = $ constant.

33. If PSP' and $PS'R$ be two chords of an ellipse through the foci S and S', then prove that $\dfrac{SP}{SQ} + \dfrac{S'P}{S'R}$ is independent of the position of P.

34. A chord PQ of a conic with eccentricity e and semi-latus rectum l subtends a right angle at a focus S. Show that

$$\left(\frac{1}{SP} - \frac{1}{l}\right)^2 + \left(\frac{1}{SQ} - \frac{1}{l}\right)^2 = \frac{e^2}{l^2}.$$

35. A parabola and an ellipse have a common focus S and they intersect in two real points P and Q of which P is the vertex of the parabola. If e be the eccentricity of the ellipse and α, the angle which SP makes with the major axis of the ellipse, prove that

$$\frac{SQ}{SP} = 1 + \frac{4e^2 \sin^2 \alpha}{(1 - e \cos \alpha)^2}.$$

36. Show that the triangle formed by the pole and the points of intersection of the circle $r = 4\cos\theta$ with the line $r\cos\theta = 3$ is an equilateral triangle.

37. A circle of given diameter d passes through the focus of a given conic and cuts in four points whose distances form the focus are r_1, r_2, r_3 and r_4. Prove that

(i) $\dfrac{1}{r_1} + \dfrac{1}{r_2} + \dfrac{1}{r_3} + \dfrac{1}{r_4} = \dfrac{2}{l}$ and (ii) $r_1 r_2 r_3 r_4 = \dfrac{d^2 l^2}{e^2}$

where l is the semi-latus rectum and e is the eccentricity of the conic.

38. If d is the diameter of the circle passing through the pole and the points (r_1, θ_1) and (r_2, θ_2), show that

$$d^2 \sin^2(\theta_1 - \theta_2) = r_1^2 + r_2^2 - 2r_1 r_2 \cos(\theta_1 - \theta_2).$$

39. Show that the locus of the equation $r^2 - ar\cos 2\theta \sec\theta - 2a^2 = 0$ consists of a straight line and a circle.

[*Hints:* The equation is $(r - 2a\cos\theta)(r\cos\theta + a) = 0$. Here $r = 2a\cos\theta$ is a circle and $r\cos\theta + a = 0$ is a straight line.]

40. Show that the locus of the middle points of a family of focal chords of a parabola is again a parabola.

41. Show that the locus of the middle points of a family of focal chords of an ellipse is again an ellipse.

42. Show that the locus of the middle points of a family of focal chords of a hyperbola is again a hyperbola.

Section C: Problems on Tangents and Normals

1. Find the equation of the chord of the conic $\dfrac{l}{r} = 1 + e\cos\theta$, joining two points whose vectorial angles are $(\alpha - \beta)$ and $(\alpha + \beta)$.

Hence deduce the equation of the tangent to the conic at the point whose vectorial angle is α.

[*Hints:* § **3.9** and **Note 3.9.1** & § **3.10** and **Note 3.10.1**]

2. Find the equation of the chord of the conic $\dfrac{l}{r} = 1 - e\cos\theta$ joining two points whose vectorial angles are α and β.

Hence find the equation of the tangent to the conic at the point whose vectorial angle is α.

[*Hints:* § **3.9** & § **3.10**]

3. Find the polar equation of the chord joining the points on the circle $r = 2d\cos\theta$, whose vectorial angles are θ_1 and θ_2 and deduce the equation of the tangent at the point θ_1.

[*Hints:* See Worked Out **Example 3.16.11**]

4. Find the polar equation of the straight line joining two points on the parabola $\dfrac{2a}{r} = 1 + \cos\theta$ with $(\alpha - \beta)$ and $(\alpha + \beta)$ as their vectorial angles.

[*Hints:* See Worked Out **Example 3.16.11**]

Hence show that the equation of tangent to the parabola at α is $r = a\sec(\theta - \alpha)$

[*Hints:* See Worked Out **Example 3.16.11** 2nd Part.]

5. Find the polar equation of the circle through the point $(c, 0°)$ and $(4c, 0°)$ and touches the line $\theta = \alpha$.

[*Hints:* Take the equation of the circle in general form and use the conditions.]

6. Find the polar equation of the normal to the conic $\dfrac{l}{r} = 1 - e\cos\theta$ at the point $\theta = \alpha$.

[*Hints:* § **3.11**]

7. Show that the polar equation of the chord of contact of the tangents to the conic $\dfrac{l}{r} = 1 - e\cos\theta$ form the point (r_1, θ_1) is given by

$$\left(\dfrac{l}{r} + e\cos\theta\right)\left(\dfrac{l}{r_1} + e\cos\theta_1\right) = \cos(\theta - \theta_1).$$

8. If the straight line $r\cos(\theta-\alpha) = p$ touches the parabola $\dfrac{l}{r} = 1+\cos\theta$, show that $p = \dfrac{1}{2}l\sec\alpha$.

9. P, Q, R are points on the parabola $\dfrac{l}{r} = 1+\cos\theta$ with vectorial angles α, β, γ respectively. Show that the equation of the circum-circle of the triangle formed by the tangents at P, Q, R to the parabola is

$$2r\cos\frac{\alpha}{2}\cos\frac{\beta}{2}\cos\frac{\gamma}{2} = l\cos\left(\theta - \frac{\alpha+\beta+\gamma}{2}\right).$$

10. (a) Find the point of intersection of the two tangents at α and β to the conic $\dfrac{l}{r} = 1+e\cos\theta$.

(b) If the tangents at P and Q of a conic meet at a point T and S be the focus of the conic, then prove that

(i) $ST^2 = SP.SQ$, if the conic is a parabola.

(ii) if the conic is a central conic and b its semi-minor axis then

$$\frac{1}{SP.SQ} - \frac{1}{ST^2} = \frac{1}{b^2}\sin^2\frac{\angle PSQ}{2}.$$

[Hints: See Worked Out **Example 3.16.30**]

11. Show that the conics $\dfrac{l_1}{r} = 1 - e_1\cos\theta$ and $\dfrac{l_2}{r} = 1 - e_2\cos(\theta - \alpha)$ will touch one another if $l_1^2(1-e_2^2) + l_2^2(1-e_1^2) = 2l_1l_2(1 - e_1e_2\cos\alpha)$.

12. Show that the equation to the circle which passes through the focus of the curve $\dfrac{l}{r} = 1 - e\cos\theta$ and touches it at the point $\theta = \alpha$ is

$$r(1-e\cos\alpha)^2 = l\cos(\theta-\alpha) - el\cos(\theta-2\alpha).$$

13. Show that the equation of the tangent to the conic $\dfrac{l}{r} = 1+e\cos\theta$ parallel to the tangent at $\theta = \alpha$ is given by

$$l(e^2 + 2e\cos\alpha + 1) = r(e^2-1)[\cos(\theta-\alpha) + e\cos\theta].$$

14. Find the condition that the straight line $\dfrac{l}{r} = A\cos\theta + B\sin\theta$ may be a tangent to the conic $\dfrac{l}{r} = 1+e\cos(\theta-\alpha)$.

15. If the normals at three points of the parabola $r = a\,\text{cosec}^2\dfrac{\theta}{2}$, whose

vectorial angles are α, β, γ meet in a point whose vectorial angle is ϕ, prove that $2\phi = \alpha + \beta + \gamma - \pi$.

16. If the normals at the points α, β, γ, δ on the conic $\dfrac{l}{r} = 1 + e\cos\theta$, meet at (ρ, ϕ), prove that $\alpha + \beta + \gamma + \delta - 2\pi =$ an odd multiple of π and

$$\tan\frac{\alpha}{2}.\tan\frac{\alpha}{2}.\tan\frac{\alpha}{2}.\tan\frac{\alpha}{2} + \left(\frac{1+e}{1-e}\right)^2 = 0.$$

17. The normal at the extremity $\left(l, \dfrac{\pi}{2}\right)$ of the latus rectum of the conic $\dfrac{l}{r} = 1 - e\cos\theta$ meets the curve at the other point P. Show that the distance of P from the pole is $\dfrac{1+3e^2+e^4}{1+e^2-e^4}$.

18. If a focal chord of an ellipse makes an angle α with the axis, the angle between the tangents at its extremities is $\tan^{-1}\dfrac{2e\sin\theta}{1-e^2}$.

OR, PSP' is a focal chord of the conic. Prove that the angle between the tangents at P and P' is $\tan^{-1}\dfrac{2e\sin\theta}{1-e^2}$, where α is the angle between the chord and the major axis.

19. P, Q are two points on the conic $\dfrac{l}{r} = 1 - e\cos\theta$, having vectorial angles $\alpha - \beta$, $\alpha + \beta$. Find the locus of the foot of perpendicular from the pole on the line PQ.

20. (a) If S is the focus and P, Q are two points on a conic such that the angle $\angle PSQ$ is constant and equal to δ, prove that the locus of the intersection of the tangents at P and Q is a conic whose focus is S.

(b) Prove that the locus of the point of intersection of the tangents at the extremities of perpendicular focal radii of a conic is another conic having the same focus.

[*Hints:* **(a)** Locus will be $\dfrac{l\sec\frac{\delta}{2}}{r} = 1 - e\sec\dfrac{\delta}{2}\cos\theta$. It is a conic with the same focus S.

(b) Put $\delta = \dfrac{\pi}{2}$ and get the locus as $\dfrac{\sqrt{2}l}{r} = 1 - \sqrt{2}e\cos\theta$ which has the same focus.]

21. Show that the auxiliary circle of the conic $\dfrac{l}{r} = 1 - e\cos\theta$ is

$$r^2(e^2 - 1) + 2ler\cos\theta + l^2 = 0.$$

[*Hints*: It is the locus of the foot of the perpendicular from the pole on a variable tangent of the conic.]

22. PQ is a variable chord of the conic $\dfrac{l}{r} = 1 - e\cos\theta$, subtending a constant angle 2β at the focus S, where S is the pole, show that the locus of the foot of the perpendicular from S on PQ is the curve

$$r^2(e^2 - \sec^2\beta) + 2ler\cos\theta + l^2 = 0.$$

OR, P, Q are points on the conic $\dfrac{l}{r} = 1 - e\cos\theta$ with $\alpha - \beta$, $\alpha + \beta$ as vectorial angle. Show that the locus of the foot of perpendicular form the pole on the line PQ is

$$r^2(e^2 - \sec^2\beta) + 2ler\cos\theta + l^2 = 0.$$

ANSWERS

Section A: 1. $r\cos\theta = 5$; **2.** $r = 10\cos\theta$; **3.** $r = 8\operatorname{cosec}\theta\cot\theta$; **4.** $r\sin\theta = 2$; **5.** $(2, 0°)$; **7.** $r = 4\cos\theta$; **10.** $(\sqrt{2}, \frac{\pi}{4})$; **11.** $\theta = \frac{\pi}{2}$; **12.** $(2, \pm\frac{2\pi}{3})$; **13.** $(3, \pi)$; **14.** $(2, \frac{5\pi}{4})$; **15.** $\theta = \frac{\pi}{4}$; **16.** $(\frac{1}{3}, \frac{\pi}{2})$; **17.** $r = a$, where a is the radius; **18.** $\frac{10}{r} = 1 - 2\cos\theta$; **19.** $(\frac{21}{5}, \frac{\pi}{2})$; **20.** $e = 1$, $(3, \pi)$.

Section B: 1. (i) $(x-1)^2 + y^2 = 1$; (ii) $x^2 = 4y$; (iii) $y = x + 4$; (iv) $y = e^x$; **2.** (i) $r^2(2 + \sin^2\theta) = 6$; (ii) $r = 10\operatorname{cosec}\theta$; (iii) $r = \sqrt{a}\cos\theta$; **3.** $2\sqrt{3}$ units; **4.** $r(2\sin\theta - \cos\theta) = 2$; **6.** $\frac{2r_1r_2}{r_1+r_2} \cdot \cos\frac{1}{2}(\theta_1 - \theta_2)$; **7.** $\frac{\sin(\theta_2-\theta_3)}{r_1} + \frac{\sin(\theta_3-\theta_1)}{r_2} + \frac{\sin(\theta_1-\theta_2)}{r_3} = 0$; **8.** $\theta = \frac{\pi}{2}$, $\theta = \frac{3\pi}{2}$; **14.** (i) $r = 8\cos\theta$; (ii) $r^2 - 10r\cos(\theta - \frac{\pi}{3}) + 21 = 0$; (iii) $r^2 - 8r\cos(\theta - \frac{\pi}{3}) - 49 + 14(\sqrt{2} + \sqrt{6}) = 0$; (iv) $r = \frac{10}{\sqrt{3}}\cos(\theta - \frac{\pi}{3})$; (v) $r^2 + 2r\cos\theta - 2 = 0$; **15.** (i) 2; (ii) 3; (iii) 5; **16.** (i) $(\frac{5}{2}, \tan^{-1}\frac{3}{4})$; (ii) $(4, 0°)$; (iii) $(2, 0°)$; **18.** (i) Hyperbola; (ii) Ellipse; (iii) Parabola; (iv) Hyperbola; (v) Hyperbola; **19.** (i) $(5, \frac{2\pi}{3}), (5, \frac{4\pi}{3})$; (ii) $(3, \frac{\pi}{3})$; (iii) $(2, \frac{2\pi}{3}), (2, \frac{4\pi}{3})$; (iv) $(\frac{l}{2}, \pi)$, (v) $(3, \pi)$; **20.** (i) $\frac{4}{r} = 1 \pm \cos\theta$; (ii) $r = \frac{-4}{\sqrt{2}-(\sin\theta+\cos\theta)}$; **21.** $\frac{10}{r} = 3 + 2\cos\theta$; **22.** $\frac{9}{r} = 4 + 5\cos\theta$; **24.** $\frac{32}{5r} = -(1 + \frac{5}{3}\cos\theta)$; **26.** (i) An ellipse, vertices $(\frac{1}{2}, \pi)$ and $(\frac{3}{2}, 0°)$, latus rectum $= \frac{3}{2}$; (ii) A parabola, vertex $(3, \pi)$, latus rectum $= 12$; (iii) A branch of a hyperbola, vertex $(1, \frac{\pi}{2})$, latus rectum $= 6$; (iv) An ellipse, vertices $(3, 0°)$ and $(7, \pi)$, latus rectum $= \frac{42}{5}$; **27.** (i) $r = -\frac{3}{2}\sec\theta$; (ii) $r = -6\sec\theta$; (iii) $r = \frac{3}{2}\sec\theta$; (iv) $r = \frac{21}{2}\sec\theta$; **29.** $\frac{48}{7}$ units; **30.** $r = -12\sec\theta$ and $r = 20\sec\theta$.

Section C: 1. Equation of the chord: $\frac{l}{r} = \sec\beta\cos(\theta - \alpha) + e\cos\theta$, Equation of the tangent: $\frac{l}{r} = \cos(\theta - \alpha) + e\cos\theta$; **2.** Equation of the chord: $\frac{l}{r} = \sec\beta\cos(\theta-\alpha) - e\cos\theta$, Equation of the tangent: $\frac{l}{r} = \cos(\theta-\alpha) - e\cos\theta$; **3.** Equation of the chord: $r\cos(\theta - \theta_1 - \theta_2) = 2d\cos\theta_1\cos\theta_2$, Equation of the tangent: $r\cos(\theta - 2\theta_1) = 2d\cos^2\theta_1$; **5.** $r^2 - 2cr\cos\theta + r\sin\theta\left(\frac{5e\cos\alpha \mp 4c}{\sin\alpha}\right) + 4e^2 = 0$; **6.** $\frac{-le\sin\alpha}{r(1-e\cos\alpha)} = \sin(\theta - \alpha) - e\sin\theta$; **10.** (a) If the point of intersection of the tangents be (r_1, θ_1) then $\theta_1 = \frac{1}{2}(\alpha+\beta)$ and $\frac{l}{r} = \cos\frac{1}{2}(\beta-\alpha) + e\cos\frac{1}{2}(\beta+\alpha)$; **14.** $A^2 + B^2 - 2e(A\cos\gamma + B\sin\gamma) + e^2 - 1 = 0$; **19.** $r^2(e^2 - \sec^2\beta) + 2ler\cos\theta + l^2 = 0$.

Chapter 4

Pair of Straight Lines

4.1 General Equations

Let $lx + my + n = 0$ and $l'x + m'y + n' = 0$ represent two straight lines. Then their product equation, given by $(lx + my + n)(l'x + m'y + n') = 0$ represents the pair, for any value of x and y which satisfy one or other of the first two equations will satisfy this equation.

Now if the above product is done, the form of the equation becomes

$$ll'x^2 + (lm' + l'm)xy + mm'y^2 + (ln' + l'n)x + (mn' + m'n)y + nn' = 0 \quad (4.1)$$

This can be expressed in the following form:

$$ax^2 + 2hxy + by^2 + 2gx + 2fy + c = 0 \quad (4.2)$$

where

$$\left. \begin{array}{ccc} ll' = a & mm' = b & nn' = c \\ lm' + l'm = 2h & ln' + l'n = 2g & mn' + m'n = 2f \end{array} \right\} \quad (4.3)$$

The equation (4.2) is known as the *general equation of second degree in the two variables x and y*, as it contains second degree, first degree terms of x and y as well as the constant terms (discussed in **Chapter 2**). From the above discussions, it is clear that if equation (4.2) is to represent a *pair of straight lines*, it must be possible to express its L.H.S. as a product of two linear factors of the form $(lx + my + n)(l'x + m'y + n')$ and consequently the corresponding coefficients a, b, c, f, g and h must satisfy the relations obtained as in (4.3). If there exists no such relation between the coefficients, then the equation (4.2) does not represent a pair of straight lines.

Corollary 4.1.1 *Homogeneous Equation*

If the above two lines pass through the origin, remaining parallel to the lines $lx + my + n = 0$ and $l'x + m'y + n' = 0$, then we have $n = n' = 0$ and correspondingly we get $f = g = c = 0$ and the equation (4.2) becomes

$$ax^2 + 2hxy + by^2 = 0 \qquad (4.4)$$

This equation contains only the second degree terms in x and y. It is evident that (4.4) represents a pair of straight lines through the origin parallel to the pair represented by (4.2). Also since (4.4) is a homogeneous second degree equation in two variables x and y, so, we can say that the equation of a pair of straight lines passing through the origin must be a **homogeneous second degree** equation in x and y.

Note 4.1.1 *It is to be noted that equation of any pair of straight lines is a general equation of second degree, but any general equation of second degree may not represent a pair of straight lines.*

Illustration 4.1.1 Let $x + y + 1 = 0$ and $2x - 3y + 5 = 0$ be the two straight lines. Then their combined equation is given by

$$(x + y + 1)(2x - 3y + 5) = 0 \Rightarrow 2x^2 - xy - 3y^2 + 7x + 2y + 5 = 0.$$

Now we consider a general equation of second degree, given by

$$16x^2 - 24xy + 9y^2 - 104x - 172y - 44 = 0,$$

the left hand side of which can not be factorised into two linear factors and so this equation, although, is a second degree equation, does not represent a pair of straight lines.

Remark: From the above discussion the following conclusion can be drawn.

The general equation of second degree

$$f(x, y) := ax^2 + 2hxy + by^2 + 2gx + 2fy + c = 0$$

will represent a pair of straight lines if and only if the left hand side expression be facrorised into two linear factors.

Illustration 4.1.2 Let us consider the equation

$$f(x, y) := x^2 - 5xy + 4y^2 + x + 2y - 2 = 0 \qquad (4.5)$$

Pair of Straight Lines

To know whether the equation (4.5) represents a pair of straight lines or not, we try to factorise it into two linear factors. For this we have

$$x^2 - 5xy + 4y^2 = (x - 4y)(x - y).$$

Let $x^2 - 5xy + 4y^2 + x + 2y - 2 = (x - 4y + n_1)(x - y + n_2)$
$$= x^2 - 5xy + 4y^2 + (n_1 + n_2)x - (n_1 + 4n_2)y + n_1 n_2.$$

Comparing coefficients, $n_1 + n_2 = 1$, $n_1 + 4n_2 = -2$ and $n_1 n_2 = -2$.

From the first two equation we get $n_1 = 2$, $n_2 = -1$ which satisfy the third relation $n_1 n_2 = -2$. Hence

$$x^2 - 5xy + 4y^2 + x + 2y - 2 = (x - 4y + 2)(x - y - 1).$$

Hence the equation (4.5) represents a pair of straight lines.

4.2 To Find a Necessary Condition that the General Equation of Second Degree Should Represent a Pair of Straight Lines

First Method

Let the general equation of second degree be given by

$$ax^2 + 2hxy + by^2 + 2gx + 2fy + c = 0 \qquad (4.6)$$

Treating the equation (4.6) as a quadratic in x, we get

$$ax^2 + 2(hy + g)x + (by^2 + 2fy + c) = 0$$

$$\therefore \quad x = \frac{-2(hy + g) \pm \sqrt{4(hy + g)^2 - 4a(by^2 + 2fy + c)}}{2a}$$

$$= \frac{-(hy + g) \pm \sqrt{(hy + g)^2 - a(by^2 + 2fy + c)}}{a}$$

or, $\qquad ax + hy + g = \pm\sqrt{(hy + g)^2 - a(by^2 + 2fy + c)} \qquad (4.7)$

Now equation (4.6) will represent a pair of straight lines, if its L.H.S. be factorised into two linear factors for which the expression under the radical sign of (4.7) must be a perfect square, i.e., if

$$(hy + g)^2 - a(by^2 + 2fy + c) = (h^2 - ab)y^2 + 2(gh - af)y^2 + g^2 - ac$$

is a perfect square.

For this its discriminant $= 0$

i.e., $4(gh - af)^2 - 4(h^2 - ab)(g^2 - ac) = 0$

or, $(gh - af)^2 - (h^2 - ab)(g^2 - ac) = 0$

or, $g^2h^2 - 2afgh + a^2f^2 - h^2g^2 + h^2ac + g^2ab - a^2bc = 0$

or, $-2afgh + a^2f^2 - h^2ac + g^2ab - a^2bc = 0$

or, $abc + 2fgh - af^2 - bg^2 - ch^2 = 0$ (dividing by $-a$)

which is the required necessary condition.

Second Method

Let us assume that the equation (4.6) represents a pair of intersecting straight lines, (α, β) being their point of intersection.

Shifting the origin to (α, β), i.e., taking the translation formulae $x = x' + \alpha, y = y' + \beta$, we get the transformed equation of (4.6) as

$$a(x' + \alpha)^2 + 2h(x' + \alpha)(y' + \beta) + b(y' + \beta)^2$$
$$+ 2g(x' + \alpha) + 2f(y' + \beta) + c = 0$$
$$\Rightarrow ax'^2 + 2hx'y' + by'^2 + 2(a\alpha + h\beta + g)x' + 2(h\alpha + b\beta + f)y'$$
$$+ a\alpha^2 + 2h\alpha\beta + b\beta^2 + 2g\alpha + 2f\beta + c = 0 \quad (4.8)$$

This is the equation referred to the new set of axes with respect to which the two lines represented by the equation pass through the new origin. Therefore its equation must be homogeneous in form and hence its linear terms and constant terms must vanish and we get

$$a\alpha + h\beta + g = 0 \quad (4.9)$$
$$h\alpha + b\beta + f = 0 \quad (4.10)$$
and $a\alpha^2 + 2h\alpha\beta + b\beta^2 + 2g\alpha + 2f\beta + c = 0$
or, $(a\alpha + h\beta + g)\alpha + (h\alpha + b\beta + f)\beta + (g\alpha + f\beta + c) = 0$
or $g\alpha + f\beta + c = 0 \quad (4.11)$

[using the relations (4.9) and (4.10)]

Now eliminating α and β form (4.9), (4.10) and (4.11) we get

$$\Delta = \begin{vmatrix} a & h & g \\ h & b & f \\ g & f & c \end{vmatrix} = 0 \text{ or } abc + 2fgh - af^2 - bg^2 - ch^2 = 0$$

which is the required necessary condition.

Pair of Straight Lines

Third Method

Let the equation (4.6) represents a pair of straight lines and let the individual lines be $lx + my + n = 0$ and $l'x + m'y + n' = 0$. So we get

$$ax^2 + 2hxy + by^2 + 2gx + 2fy + c = (lx + my + n)(l'x + m'y + n').$$

Equating coefficients we get

$$ll' = a \qquad mm' = b \qquad nn' = c$$
$$lm' + l'm = 2h \quad ln' + l'n = 2g \quad mn' + m'n = 2f.$$

Now we consider the determinants

$$\begin{vmatrix} l & l' & 0 \\ m & m' & 0 \\ n & n' & 0 \end{vmatrix} \quad \text{and} \quad \begin{vmatrix} l' & m' & n' \\ l & m & n \\ 0 & 0 & 0 \end{vmatrix}.$$

Since the value of each of these determinants is zero, the value of their product is also zero, i.e., we have

$$\begin{vmatrix} l & l' & 0 \\ m & m' & 0 \\ n & n' & 0 \end{vmatrix} \times \begin{vmatrix} l' & m' & n' \\ l & m & n \\ 0 & 0 & 0 \end{vmatrix} = 0$$

or,
$$\begin{vmatrix} 2ll' & lm' + ml' & ln' + l'n \\ ml' + m'l & 2mm' & mn' + m'n \\ nl' + ln' & mn' + nm' & 2nn' \end{vmatrix} = 0$$

or,
$$\begin{vmatrix} 2a & 2h & 2g \\ 2h & 2b & 2f \\ 2g & 2f & 2c \end{vmatrix} = 0 \Rightarrow \Delta = \begin{vmatrix} a & h & g \\ h & b & f \\ g & f & c \end{vmatrix} = 0$$

which is the necessary condition.

Note 4.2.1 *It can be shown that the above condition is not sufficient, i.e., if the equation (4.6) satisfies the above condition $\Delta = 0$, then it does not mean that it always represents a real pair of straight lines. To prove it let us consider the following* **Illustration 4.2.1**.

Illustration 4.2.1 Let us consider the equation

$$f(x, y) := 41x^2 + 24xy + 34y^2 + 30x - 40y + 25 = 0 \tag{4.12}$$

Comparing the coefficients with the general equation of second degree

$$ax^2 + 2hxy + by^2 + 2gx + 2fy + c = 0$$

we get $a = 41$, $b = 34$, $c = 25$, $h = 12$, $g = 15$, $f = -20$.

$$\therefore \Delta = \begin{vmatrix} a & h & g \\ h & b & f \\ g & f & c \end{vmatrix} = \begin{vmatrix} 41 & 12 & 15 \\ 12 & 34 & -20 \\ 15 & -20 & 25 \end{vmatrix} = 5 \times \begin{vmatrix} 41 & 12 & 3 \\ 12 & 34 & -4 \\ 15 & -20 & 5 \end{vmatrix}$$

$$= 5 \times 5 \times \begin{vmatrix} 41 & 12 & 3 \\ 12 & 34 & -4 \\ 3 & -4 & 1 \end{vmatrix} = 25 \times 2 \times \begin{vmatrix} 41 & 12 & 3 \\ 6 & 17 & -2 \\ 3 & -4 & 1 \end{vmatrix}$$

$$= 50 \times \begin{vmatrix} 41 & 56 & 3 \\ 6 & 21 & -2 \\ 3 & 0 & 1 \end{vmatrix} \qquad [\text{by } C_2' = C_2 + \overline{C_1 + C_3}].$$

$$= 50 \times 7 \times \begin{vmatrix} 41 & 8 & 3 \\ 6 & 3 & -2 \\ 3 & 0 & 1 \end{vmatrix} = 350 \times [3(-16 - 9) + (123 - 48)] = 0.$$

Again, $\delta = ab - h^2 = 41 \times 34 - 12^2 = 1394 - 144 = 1250 > 0$ (& $\neq 0$).
So the conic is a central conic and its centre is given by

$$\left(\frac{hf - bg}{ab - h^2}, \frac{gh - af}{ab - h^2} \right) = \left(\frac{-240 - 510}{1250}, \frac{180 + 820}{1250} \right) = \left(-\frac{3}{5}, \frac{4}{5} \right).$$

[vide § **2.1.3**, **Chapter 2**, Centre of a conic]

We now shift the origin to the point $\left(-\frac{3}{5}, \frac{4}{5} \right)$, i.e., we apply the translation $x = x' - \frac{3}{5}$, $y = y' + \frac{4}{5}$. The given equation then reduces to

$$41x'^2 + 24x'y' + 34y'^2 + d = 0, \quad \text{where } d = \frac{\Delta}{\delta} = \frac{0}{1250} = 0.$$

Hence we get

$$41x'^2 + 24x'y' + 34y'^2 = 0 \tag{4.13}$$

To remove the term containing $x'y'$ we rotate the coordinate axes through an angle

$$\theta = \frac{1}{2} \tan^{-1} \frac{2h}{a - b} = \frac{1}{2} \tan^{-1} \frac{24}{41 - 34} = \frac{1}{2} \tan^{-1} \frac{24}{7}$$

and let the equation (4.13) becomes $AX^2 + BY^2 = 0$, where by the invariants of rotation, we have

$$\left. \begin{aligned} A + B &= a + b = 41 + 34 = 75 \\ AB &= ab - h^2 = 1250. \end{aligned} \right\}$$

Pair of Straight Lines

or, $A(75 - A) = 1250$ or, $A^2 - 75A + 1250 = 0$

or, $A^2 - 50A - 25A + 1250 = 0$ or, $(A - 50)(B - 25) = 0$

∴ $A = 50$ or 25 and hence $B = 25$ or 50.

We take $A = 50$, $B = 25$. Hence the equation is

$$50X^2 + 25Y^2 = 0 \text{ or, } 2X^2 + Y^2 = 0$$

and this is of the form $2x^2 + y^2 = 0$, which evidently does not represent a real pair of straight lines. Actually, this represents a point ellipse. Therefore $\triangle = 0$ is not a sufficient condition that the general equation of second degree in x and y given as (4.12) represents a pair of straight lines. However, we may state below the following necessary and sufficient conditions:

Theorem 4.2.1 *The necessary and sufficient conditions for the general equation of second degree in x and y, given by*

$$ax^2 + 2hxy + by^2 + 2gx + 2fy + c = 0 \qquad (4.14)$$

should represent a pair of straight lines is that $\triangle = 0$ and $\delta \leq 0$, where $\triangle$ and δ are given by

$$\triangle = \begin{vmatrix} a & h & g \\ h & b & f \\ g & f & c \end{vmatrix} = abc + 2fgh - af^2 - bg^2 - ch^2 \text{ and } \delta = \begin{vmatrix} a & h \\ h & b \end{vmatrix} = ab - h^2.$$

Proof: **Condition Necessary:**

$\triangle = 0$ has been proved earlier.

Now $4(h^2 - ab) = (lm' + l'm)^2 - 4ll'mm' = (lm' - ml')^2$

∴ $h^2 - ab \geq 0 \Rightarrow ab - h^2 \leq 0$.

Therefore, we get, the necessary conditions as $\triangle = 0$ and $\delta \leq 0$.

Condition Sufficient:

Let us suppose that the equation (4.14) is such that the condition $\triangle = 0$ and $\delta \leq 0$ are satisfied.

We now consider the following system of equations

$$\left. \begin{array}{l} ax + hy + g = 0 \\ hx + by + f = 0 \\ gx + fy + c = 0 \end{array} \right\} \qquad (4.15)$$

In virtue of the condition $\triangle = 0$, it follows that the above system has a solution. Let (x_1, y_1) be such a solution. Then we get

$$\left.\begin{array}{l} ax_1 + hy_1 + g = 0 \\ hx_1 + by_1 + f = 0 \\ gx_1 + fy_1 + c = 0 \end{array}\right\} \quad (4.16)$$

Now we apply the translation $x = x' + x_1$, $y = y' + y_1$. Then equation (4.14) is transformed into

$$a(x' + x_1)^2 + 2(x' + x_1)(y' + y_1) + b(y' + y_1)^2 + 2g(x' + x_1)$$
$$+ 2f(y' + y_1) + c = 0$$

or, $\quad ax'^2 + 2hx'y' + by'^2 + 2(ax_1 + hy_1 + g)x' + 2(hx_1 + by_1 + f)y'$
$$+ ax_1^2 + 2hx_1y_1 + by_1^2 + 2gx_1 + 2fy_1 + c = 0$$

or, $\quad ax'^2 + 2hx'y' + by'^2 + 2(0)x' + 2(0)y' + (ax_1 + hy_1 + g)x_1$
$$+ (hx_1 + by_1 + f)y_1 + (gx_1 + fy_1 + c) = 0$$

or, $\quad ax'^2 + 2hx'y' + by'^2 = 0 \quad$ [by (4.16)]

or, $\quad b\dfrac{y'^2}{x'^2} + 2h\dfrac{y'}{x'} + a = 0 \quad$ (4.17)

Since $\delta \leq 0$, i.e., $ab - h^2 \leq 0$ $\therefore h^2 - ab \geq 0$, so equation (4.17) represents a pair of real straight lines. Thus we get the required necessary and sufficient conditions for (4.14) to be a pair of straight lines are that

$$\begin{vmatrix} a & h & g \\ h & b & f \\ g & f & c \end{vmatrix} = 0 \text{ and } ab - h^2 \leq 0 \text{ i.e., } \triangle = 0 \text{ and } \delta \leq 0. \qquad \square$$

Note 4.2.2 *The equation (4.14) represents a pair of intersecting straight lines if and only if $\delta < 0$ and $\triangle = 0$ and it represents a pair of parallel straight lines if and only if $\delta = 0$ and $\triangle = 0$.*

Note 4.2.3 *The homogeneous equation of second degree in x and y, given by $ax^2 + 2hxy + by^2 = 0$ represents a pair of straight lines through the origin if and only if $ab - h^2 \leq 0$, i.e., $\delta \leq 0$.*

The truth of the statement follows by taking $f = g = h = 0$.

Note 4.2.4 *In the cited **Illustration 4.2.1**, we have seen that its $\delta = ab - h^2 = 1250 > 0$, $\triangle = 0$ and it reduces finally to an equation of the form $2x^2 + y^2 = 0$, which geometrically represents a **point-ellipse**, but it can also be called a **pair of imaginary lines**. That is why, sometimes $\triangle = 0$ is considered as a **necessary and sufficient condition** for the equation (4.14) to be a pair of straight lines whatever be the value of δ may be.*

Pair of Straight Lines 173

Point of Intersection

Let us suppose that equation (4.14) represents a pair of real straight lines having equations

$$\left. \begin{array}{l} lx + my + n = 0 \\ l'x + m'y + n' = 0 \end{array} \right\} \qquad (4.18)$$

Solving we get
$$\frac{x}{mn' - m'n} = \frac{y}{l'n - ln'} = \frac{1}{lm' - l'm}.$$

Now by **Note 4.2.2**,

$$lm' - l'm = \sqrt{(lm' + l'm)^2 - 4ll'mm'} = \sqrt{4h^2 - 4ab} = 2\sqrt{h^2 - ab} > 0.$$

$$\therefore \quad x = \frac{n'm - m'n}{lm' - l'm}, \quad y = \frac{nl' - ln'}{lm' - l'm}.$$

Now $4bg - 4hf = 2mm'(ln' + l'n) - (lm' + l'm)(mn' + m'n)$
$\qquad\qquad\qquad = mm'ln' + mm'l'n - lnm'^2 - l'n'm^2$
$\qquad\qquad\qquad = mn'(lm' - l'm) - m'm(lm' - l'm)$
$\qquad\qquad\qquad = (lm' - l'm)(mn' - m'n).$

$$\therefore \quad x = \frac{nm' - mn'}{lm' - l'm} = \frac{(nm' - mn')(lm' - l'm)}{(lm' - l'm)^2} = \frac{4(bg - hf)}{4(h^2 - ab)} = \frac{hf - bg}{ab - h^2}.$$

Similarly we shall get $y = \dfrac{gh - af}{ab - h^2}.$

Thus the point of intersection of the pair of lines (4.18) is given by

$$\left(\frac{hf - bg}{ab - h^2}, \frac{gh - af}{ab - h^2} \right).$$

Note 4.2.5 *We note that the point of intersection of a pair of straight lines is the same as that the centre of a conic of (4.14) representing a central conic.*   *[vide § 2.1.3 of Chapter 2]*

Note 4.2.6 *We also note that the above point of intersection may also be obtained by solving the equations for α and β,*

$$\left. \begin{array}{l} a\alpha + h\beta + g = 0 \\ h\alpha + b\beta + f = 0 \end{array} \right\}$$

from which we get

$$\frac{\alpha}{hf-bg} = \frac{\beta}{gh-af} = \frac{1}{ab-h^2} \Rightarrow \alpha = \frac{hf-bg}{ab-h^2}, \beta = \frac{gh-af}{ab-h^2}$$

and hence the point of intersection is

$$\left(\frac{hf-bg}{ab-h^2}, \frac{gh-af}{ab-h^2}\right).$$

Note 4.2.7 *$\delta = 0$ implies that the lines are parallel.*

4.3 Angle Between a Pair of Lines

Let $lx + my + n = 0$ and $l'x + m'y + n' = 0$ be the lines represented by the pair of lines

$$ax^2 + 2hxy + by^2 + 2gx + 2fy + c = 0 \tag{4.19}$$

Then we get

$$ll' = a \qquad mm' = b \qquad nn' = c$$
$$lm' + l'm = 2h \quad ln' + l'n = 2g \quad mn' + m'n = 2f.$$

Let ϕ be the angles between these lines. Then we get

$$\tan\phi = \pm\frac{lm' - l'm}{ll' + mm'} = \pm\frac{\sqrt{(lm' + l'm)^2 - 4ll'mm'}}{ll' + mm'}$$
$$= \pm\frac{\sqrt{(2h)^2 - 4ab}}{a+b} = \frac{\pm 2\sqrt{h^2 - ab}}{a+b}.$$

Hence the angle between the pair of straight lines represented by (4.19) is given by

$$\tan\phi = \frac{\pm 2\sqrt{h^2 - ab}}{a+b}.$$

Note 4.3.1 *It is evident that the angle between the pair of straight lines represented by the homogeneous equation $ax^2 + 2hxy + by^2 = 0$ is the same.*

Note 4.3.2 *The positive value of $\tan\phi$ gives the acute angle between the pair of straight lines and the negative value will give the obtuse angle.*

Note 4.3.3 *If $\phi = 0$, then $\tan\phi = 0$ and we get $h^2 - ab = 0$, i.e., $\delta = 0$ is the condition for **parallelism** of the lines.*

Pair of Straight Lines 175

Note 4.3.4 $\phi = 90°$, $\tan \phi = \infty \Rightarrow a + b = 0$ which is the condition of perpendicularity of the straight lines, i.e., for **perpendicularity** of the two straight lines we must have

$$\text{Coefficient of } x^2 + \text{Coefficient of } y^2 = 0.$$

4.4 Equation of Bisectors of the Angle between a pair of straight lines

A. For Homogeneous Equation $ax^2 + 2hxy + by^2 = 0$.

Let us consider the pair of straight lines

$$ax^2 + 2hxy + by^2 = 0 \qquad (4.20)$$

Also let $y = mx$ and $y = m'x$ be the lines represented by (4.20), then we get

$$y^2 + \frac{2h}{b}xy + \frac{a}{b}x^2 = (y - mx)(y - m'x)$$

from which we get $m + m' = -\frac{2h}{b}$ and $mm' = \frac{a}{b}$.

The equations of the bisectors of the angles between the lines $y - mx = 0$ and $y - m'x = 0$ are given by

$$\frac{y - mx}{\sqrt{1 + m^2}} = \pm \frac{y - m'x}{\sqrt{1 + m'^2}}.$$

On squaring we get

$$(y - mx)^2 (1 + m'^2) = (y - m'x)^2 (1 + m^2)$$

or, $(y^2 - 2mxy + m^2 x^2)(1 + m'^2) = (y^2 - 2m'xy + m'^2 x^2)(1 + m^2)$

or, $(1 + m'^2 - 1 - m^2)y^2 - [2m(1 + m'^2) - 2m'(1 + m^2)]xy$
$\qquad\qquad\qquad\qquad\qquad + (m^2 - m'^2)x^2 = 0$

or, $(m + m')(m - m')x^2 - [2(m - m')(1 - mm')]xy - (m^2 - m'^2)y^2 = 0$

or, $(m + m')x^2 - 2(1 - mm')xy - (m + m')y^2 = 0$

or, $-\frac{2h}{b}x^2 - 2\left(1 - \frac{a}{b}\right)xy + \frac{2h}{b}y^2 = 0$

or, $h(x^2 - y^2) - (a - b)xy = 0$ or, $\dfrac{x^2 - y^2}{a - b} = \dfrac{xy}{h}$

which are the equations of the required bisectors of the angles between the lines.

B. For General Equation $ax^2 + 2hxy + by^2 + 2gx + 2fy + c = 0$.

Let us consider the most general equation of second degree

$$ax^2 + 2hxy + by^2 + 2gx + 2fy + c = 0 \qquad (4.21)$$

Let (α, β) be the point of intersection between the pair of lines represented by equation (4.21). If we shift the origin to this point (α, β), i.e., we apply the translation $x = x' + \alpha$, $y = y' + \beta$ then equation (4.21) reduces to

$$ax'^2 + 2hx'y' + by'^2 = 0 \qquad (4.22)$$

Now the equations to the bisectors of the angles between the pair represented by (4.22) are given by

$$\frac{x'^2 - y'^2}{a - b} = \frac{x'y'}{h}.$$

So the equations to the bisectors of the pair of given lines, represented by equation (4.22) are given by

$$\frac{(x - \alpha)^2 - (y - \beta)^2}{a - b} = \frac{(x - \alpha)(y - \beta)}{h}.$$

4.5 Equation of Two Lines Joining the Origin to the Points in Which a Line Meets a Conic

We are to determine the equation to the pair of lines joining the origin to the points of intersection of loci represented by a second and a first degree equation in the two variables x and y.

Let the equation of the conic, i.e., the second degree equation be

$$ax^2 + 2hxy + by^2 + 2gx + 2fy + c = 0 \qquad (4.23)$$

and the first degree equation be

$$lx + my + n = 0 \qquad (4.24)$$

The required pair of lines will pass through the origin. So its joint equation must be a homogeneous second degree. Also, these lines would pass through the points of intersection of (4.23) and (4.24). Therefore the values of x and y must satisfy (4.23) and (4.24). So the required pair will

Pair of Straight Lines

be obtained by making equation (4.23) homogeneous with the help of (4.24) by which we get

$$ax^2 + 2hxy + by^2 + (2gx + 2fy)\left(-\frac{lx+my}{n}\right) + c\left(-\frac{lx+my}{n}\right)^2 = 0$$

$$\left[\because \text{by (4.24)}, \frac{lx+my}{-n} = 1\right]$$

This is the equation to the required lines.

Corollary 4.5.1 *A homogeneous equation of n-th degree of the form*

$$f(x,y) = y^n + A_1 xy^{n-1} + A_2 x^2 y^{n-2} + \cdots + A_r x^r y^{n-r} + \cdots + A_n x^n = 0 \quad (4.25)$$

represents n straight lines, real or imaginary, passing through the origin, provided that the L.H.S. of (4.25) can be factorized into n linear factors. For, it can be written as

$$\left(\frac{y}{x}\right)^n + A_1 \left(\frac{y}{x}\right)^{n-1} + A_2 \left(\frac{y}{x}\right)^{n-2} + \cdots + A_r \left(\frac{y}{x}\right)^{n-r} + \cdots + A_n = 0 \quad (4.26)$$

which is an equation of n-th degree in $\frac{y}{x}$. *So it has n single roots ($\because$ L.H.S. of (4.25) contains n linear factors). Let the roots be* $m_1, m_2, \cdots m_n$. *Then the equation (4.26) will be written as*

$$\left(\frac{y}{x} - m_1\right)\left(\frac{y}{x} - m_2\right)\cdots\left(\frac{y}{x} - m_r\right)\cdots\left(\frac{y}{x} - m_n\right) = 0 \quad (4.27)$$

giving n straight lines $y = m_1 x, y = m_2 x, \cdots, y = m_n x$. *Hence the proposition.*

4.6 Worked Out Examples

Example 4.6.1 *Show that the equation* $x^2 - 5xy + 4y^2 + x + 2y - 2 = 0$ *represents a pair of straight lines and find the lines.*

Solution: Here $a = 1, b = 4, c = -2, h = -\frac{5}{2}, g = \frac{1}{2}, f = 1$.

$$\begin{aligned}
\therefore \Delta &= abc + 2fgh - af^2 - bg^2 - ch^2 \\
&= 1.4.(-2) + 2.1.\frac{1}{2}\left(-\frac{5}{2}\right) - 1.1^2 - 4.\left(\frac{1}{2}\right)^2 - (-2).\left(-\frac{5}{2}\right)^2 \\
&= -8 - \frac{5}{2} - 1 - 1 + \frac{25}{2} = -10 + 10 = 0
\end{aligned}$$

and $\delta = ab - h^2 = 1.4 - \left(-\frac{5}{2}\right)^2 = -\frac{9}{4} < 0$.

Hence the given equation represents a pair of real and intersecting straight lines.

The individual lines can be obtained by factorizing its L.H.S. we have

$$x^2 - 5xy + 4y^2 = x^2 - 4xy - xy + 4y^2 = x(x - 4y) - y(x - 4y)$$
$$= (x - 4y)(x - y).$$

Let $x^2 - 5xy + 4y^2 + x + 2y - 2 = (x - 4y + \lambda)(x - y + \mu)$.

Then comparing the coefficients, we get

$$\lambda + \mu = 1 \qquad (4.28)$$
$$-\lambda - 4\mu = 2 \qquad (4.29)$$
$$\text{and} \quad \lambda\mu = -2 \qquad (4.30)$$

Solving (4.28) and (4.29) we get $\lambda = 2$ and $\mu = -1$, which also satisfy the equation (4.30). Hence the system of equations (4.28), (4.29) and (4.30) are consistent, showing that

$$x^2 - 5xy + 4y^2 + x + 2y - 2 = (x - 4y + 2)(x - y - 1).$$

So the individual lines are given by $x - 4y + 2 = 0$ and $x - y - 1 = 0$. □

Example 4.6.2 *Show that the equation to the pair of straight lines through the origin perpendicular to the pair of straight lines $ax^2 + 2hxy + by^2 = 0$ is $bx^2 - 2hxy + ay^2 = 0$.*

Solution: Let $y - m_1 x = 0$ and $y - m_2 x = 0$ be the two lines represented by the equation $ax^2 + 2hxy + by^2 = 0$,

$$\therefore \quad m_1 + m_2 = -\frac{2h}{b}, \quad m_1 m_2 = \frac{a}{b} \qquad (4.31)$$

The equation of the lines perpendicular to these lines are $m_1 y + x = 0$ and $m_2 y + x = 0$. Hence the equation to this pair is

$$(m_1 y + x)(m_2 y + x) = 0 \Rightarrow m_1 m_2 y^2 + (m_1 + m_2)xy + x^2 = 0$$

$$\Rightarrow \frac{a}{b} y^2 - \frac{2h}{b} xy + x^2 = 0 \Rightarrow bx^2 - 2hxy + ay^2 = 0. \qquad □$$

Example 4.6.3 *Show that the equation $2x^2 + 3xy + y^2 = 0$ represents a pair of straight lines. Find the angle between them.*

Pair of Straight Lines

Solution: Here $a = 2, b = 1, c = 0, h = \dfrac{3}{2}, g = 0, f = 0$.

$$\therefore \Delta = abc + 2fgh - af^2 - bg^2 - ch^2$$
$$= 2.1.0 + 2.0.0.\dfrac{3}{2} - 2.0^2 - 1.0^2 - 0.\left(\dfrac{3}{2}\right)^2 = 0.$$

Hence the equation represents a pair of straight lines.

Otherwise, We have $2x^2 + 3xy + y^2 = 0 \Rightarrow (x+y)(2x+y) = 0$ which shows that the lines represented by $2x^2 + 3xy + y^2 = 0$ are $x + y = 0$ and $2x + y = 0$.

Second Part If θ be the angle between them, then we get

$$\tan \theta = \pm\dfrac{2\sqrt{h^2 - ab}}{a+b} = \pm\dfrac{2\sqrt{\left(\dfrac{3}{2}\right)^2 - 2.1}}{2+1} = \pm\dfrac{2}{3}\sqrt{\dfrac{9-8}{4}} = \pm\dfrac{1}{3}$$

$$\therefore \theta = \tan^{-1}\left(\pm\dfrac{1}{3}\right).$$

Otherwise, If we consider the individual lines as $x + y = 0$ and $2x + y = 0$, then the angle between them is given by

$$\tan \theta = \left|\dfrac{m_1 - m_2}{1 + m_1 m_2}\right| = \pm\dfrac{1-2}{1+2} \Rightarrow \theta = \tan^{-1}\left(\pm\dfrac{1}{3}\right). \quad \square$$

Example 4.6.4 *If one of the straight lines $a_1x^2 + 2h_1xy + b_1y^2 = 0$ coincides with one of the straight lines $a_2x^2 + 2h_2xy + b_2y^2 = 0$ and the remaining straight lines are at right angles, then show that*

$$h_1\left(\dfrac{1}{b_1} - \dfrac{1}{a_1}\right) = h_2\left(\dfrac{1}{b_2} - \dfrac{1}{a_2}\right).$$

Solution: Let $y - m_1 x = 0$ and $y - m_2 x = 0$ be the two lines represented by the pair $a_1 x^2 + 2h_1 xy + b_1 y^2 = 0$. Then we get

$$\therefore m_1 + m_2 = -\dfrac{2h_1}{b_1}, \quad m_1 m_2 = \dfrac{a_1}{b_1} \qquad (4.32)$$

Let one line of the pair $a_2 x^2 + 2h_2 xy + b_2 y^2 = 0$ coincides with $y = m_1 x$ and other line is perpendicular to $y = m_2 x$. So we have

$$b_2 y^2 + 2h_2 xy + a_2 x^2 = (y - m_1 x)(m_2 y + x) = m_2 y^2 + (1 - m_1 m_2)xy - m_1 x^2.$$

Comparing and using the relations (4.32),

$$\dfrac{m_2}{b_2} = \dfrac{1 - m_1 m_2}{2h_2} = \dfrac{m_1}{-a_2} = \dfrac{m_1 + m_2}{b_2 - a_2} = \dfrac{-2h_1}{b_1(b_2 - a_2)}.$$

$$\therefore \quad m_1 = \frac{2a_2 h_1}{b_1(b_2 - a_2)} \tag{4.33}$$

$$\text{and} \quad m_2 = \frac{-2b_2 h_1}{b_1(b_2 - a_2)} \tag{4.34}$$

Again $\dfrac{1 - m_1 m_2}{2h_2} = \dfrac{m_2}{b_2} = \dfrac{m_1}{-a_2}$

or, $\dfrac{1 - \frac{a_1}{b_1}}{2h_2} = \dfrac{m_2}{b_2} = \dfrac{m_1}{-a_2}$ or, $\dfrac{b_1 - a_1}{2h_2 b_1} = \dfrac{m_2}{b_2} = \dfrac{m_1}{-a_2}$

$$\therefore \quad m_1 = \frac{-a_2(b_1 - a_1)}{2h_2 b_1} \tag{4.35}$$

$$\text{and} \quad m_2 = \frac{b_2(b_1 - a_1)}{2h_2 b_1} \tag{4.36}$$

From (4.33) and (4.36) we get

$$m_1 m_2 = \frac{2a_2 h_1}{b_1(b_2 - a_2)} \times \frac{b_2(b_1 - a_1)}{2h_2 b_1} = \frac{a_1}{b_1} \quad \text{[using (4.32)]}$$

or, $\dfrac{h_1 a_2 b_2}{b_2 - a_2} = \dfrac{h_2 a_1 b_1}{b_1 - a_1}$ or, $\dfrac{h_1(b_1 - a_1)}{b_1 a_1} = \dfrac{h_2(b_2 - a_2)}{b_2 a_2}.$

Hence $h_1 \left(\dfrac{1}{b_1} - \dfrac{1}{a_1} \right) = h_2 \left(\dfrac{1}{b_2} - \dfrac{1}{a_2} \right).$ □

Example 4.6.5 *Show that the equation $ax^2 + 2hxy + by^2 + 2gx + 2fy + c = 0$ represents two parallel straight lines if $a : h = h : b = g : h$.*

Also show that in such case, the distance between them is $2\sqrt{\dfrac{g^2 - ca}{a(a+b)}}.$

Solution: Here $\Delta = 0$ and $\delta = 0$,

i.e., $abc + 2fgh - af^2 - bg^2 - ch^2 = 0$ and $ab - h^2 = 0$

or, $c(ab - h^2) - (af^2 + bg^2 - 2fgh) = 0$ or, $af^2 + bg^2 - 2fgh = 0$

$\therefore \quad (\sqrt{a}f - \sqrt{b}g)^2 = 0 \Rightarrow \sqrt{a}f - \sqrt{b}g = 0 \Rightarrow \sqrt{a}f = \sqrt{b}g$

$\therefore \quad \dfrac{g}{f} = \dfrac{\sqrt{a}}{\sqrt{b}} = \dfrac{\sqrt{ab}}{b} = \dfrac{h}{b} = \dfrac{a}{\sqrt{ab}} = \dfrac{a}{h}$

So we get $\dfrac{a}{h} = \dfrac{h}{b} = \dfrac{g}{f} \Rightarrow a : h = h : b = g : h.$

Second Part: Let the pair of parallel straight lines represented by the given pair of equations be $lx + my + n = 0$ and $lx + my + n' = 0$. So we get

$$ax^2 + 2hxy + by^2 + 2gx + 2fy + c = (lx + my + n)(lx + my + n').$$

Pair of Straight Lines

Comparing coefficients we get

$$l^2 = a \qquad m^2 = b \qquad nn' = c \\ 2lm = 2h \Rightarrow lm = h \quad l(n+n') = 2g \quad m(n+n') = 2f \} \quad (4.37)$$

Now $\left(-\dfrac{n}{l}, 0\right)$ is a point on the line $lx + my + n = 0$. Then if we denote the required distance by d, then

$$\begin{aligned}
d &= \text{distance between the lines} \\
&= \text{distance from any point of any of the lines to the other} \\
&= \text{distance form the point } \left(-\dfrac{n}{l}, 0\right) \text{ to the line } lx + my + n' = 0 \\
&= \left| \dfrac{l\left(-\frac{n}{l}\right) + m \cdot 0 + n'}{\sqrt{l^2 + m^2}} \right| = \left| \dfrac{n - n'}{\sqrt{l^2 + m^2}} \right| \\
&= \left| \dfrac{\sqrt{(n+n')^2 - 4nn'}}{\sqrt{l^2 + m^2}} \right| = \left| \dfrac{\sqrt{(\frac{2g}{l})^2 - 4c}}{\sqrt{a+b}} \right| \qquad \text{[using (4.37)]} \\
&= \sqrt{\dfrac{\frac{4g^2}{a} - 4c}{a + b}} = \sqrt{\dfrac{4g^2 - 4ac}{a(a+b)}} = 2\sqrt{\dfrac{g^2 - ca}{a(a+b)}}. \qquad \square
\end{aligned}$$

Note 4.6.1 In the above calculation, we have taken $n + n' = \dfrac{2g}{l}$. But if we take $n + n' = \dfrac{2f}{l}$, then we get

$$d = \sqrt{\dfrac{\left(\frac{2f}{m}\right)^2 - 4c}{a+b}} = \sqrt{\dfrac{\frac{4f^2}{b} - 4c}{a+b}} = \sqrt{\dfrac{4f^2 - 4bc}{b(a+b)}} = 2\sqrt{\dfrac{f^2 - bc}{b(a+b)}}.$$

Example 4.6.6 Prove that the product of the perpendiculars from the point (x_1, y_1) on the straight lines $ax^2 + 2hxy + by^2 = 0$ is $\dfrac{ax_1^2 + 2hx_1y_1 + by_1^2}{\sqrt{(a-b)^2 + 4h^2}}$.

Solution: Let $y - m_1 x = 0$ and $y - m_2 x = 0$ be the two lines represented by the given pair. Then we get

$$m_1 + m_2 = -\dfrac{2h}{b}, \quad m_1 m_2 = \dfrac{a}{b} \qquad (4.38)$$

Perpendicular distances from the point (x_1, y_1) to the lines are $\left|\dfrac{y_1 - m_1 x_1}{\sqrt{1+m_1^2}}\right|$ and $\left|\dfrac{y_1 - m_2 x_1}{\sqrt{1+m_2^2}}\right|$. Therefore their product is given by

$$\left|\frac{(y_1 - m_1 x_1)(y_1 - m_2 x_1)}{\sqrt{1+m_1^2}\sqrt{1+m_2^2}}\right| = \left|\frac{y_1^2 - (m_1+m_2)x_1 y_1 + m_1 m_2 x_1^2}{\sqrt{1+m_1^2+m_2^2+(m_1 m_2)^2}}\right|$$

$$= \left|\frac{y_1^2 - (m_1+m_2)x_1 y_1 + m_1 m_2 x_1^2}{\sqrt{1+(m_1+m_2)^2 - 2m_1 m_2 + (m_1 m_2)^2}}\right| = \left|\frac{y_1^2 + \tfrac{2h}{b}x_1 y_1 + \tfrac{a}{b}x_1^2}{\sqrt{1 + \tfrac{4h^2}{b^2} - 2\tfrac{a}{b} + \tfrac{a^2}{b^2}}}\right|$$

$$= \left|\frac{ax_1^2 + 2hx_1 y_1 + by_1^2}{\sqrt{a^2 - 2ab + b^2 + 4h^2}}\right| = \left|\frac{ax_1^2 + 2hx_1 y_1 + by_1^2}{\sqrt{(a-b)^2 + 4h^2}}\right|. \qquad \square$$

Example 4.6.7 *Test whether the equation $x^2 + 6xy + 9y^2 + 4x + 12y - 5 = 0$ represents a pair of parallel straight lines. If so, find the distance between them.*

Solution: The equation of the given pair of lines is

$$x^2 + 6xy + 9y^2 + 4x + 12y - 5 = 0$$
$$\Rightarrow (x+3y)^2 + 4(x+3y) - 5 = 0$$
$$\Rightarrow (x+3y)^2 + 5(x+3y) - (x+3y) - 5 = 0$$
$$\Rightarrow (x+3y)(x+3y+5) - (x+3y+5) = 0$$
$$\Rightarrow (x+3y+5)(x+3y-1) = 0.$$

Thus the lines represented by the given pair are given by $x+3y+5 = 0$ and $x+3y-1 = 0$ from which it is evident that the lines are parallel.

Now $(-5, 0)$ is a point on the first line $x+3y+5 = 0$. Therefore, distance between the lines = distance from the point $(-5, 0)$ to the line $x+3y+5 = 0$.

$$\therefore \text{ Required distance } = \left|\frac{-5 + 3.0 - 1}{\sqrt{1^2 + 3^2}}\right| = \frac{6}{\sqrt{10}} = \frac{3}{5}\sqrt{10} \text{ units.} \qquad \square$$

Example 4.6.8 *Show that the lines joining the origin to the common points of $ax^2 + 2hxy + by^2 + 2gx + 2fy + c = 0$ and $lx + my = 1$ will be at right angles if $(a+b) + 2(fm + gl) + c(l^2 + m^2) = 0$.*

Solution: To get the equation of the line joining the origin to the common points of

$$ax^2 + 2hxy + by^2 + 2gx + 2fy + c = 0 \qquad (4.39)$$

Pair of Straight Lines

and the first degree equation

$$lx + my = 1 \qquad (4.40)$$

we are to make equation (4.39) homogeneous with the help of (4.40) by rewriting it as follows:

$$ax^2 + 2hxy + by^2 + (2gx + 2fy)(lx + my) + c(lx + my)^2 = 0 \; [\S \; 4.1.5]$$
$$\Rightarrow (a + 2gl + cl^2)x^2 + 2(h + fl + gm + clm)xy$$
$$+ (b + 2fm + cm^2)y^2 = 0 \qquad (4.41)$$

The lines (4.41) will be at right angles if the sum of the coefficients of x^2 and y^2 in (4.41) is zero. [*vide* § **4.1.3 Note 4.3.4**]

i.e., if $\quad a + 2gl + cl^2 + b + 2fm + cm^2 = 0$

or, $\quad (a+b) + 2(fm + gl) + c(l^2 + m^2) = 0.$ $\qquad \square$

Example 4.6.9 *If the straight lines joining the origin to the points of intersection of the curve $3x^2 - xy + 3y^2 + 2x - 3y + 4 = 0$ and the straight line $2x + 3y + k = 0$ be at right angles, then show that $6k^2 + 5k + 52 = 0$.*

Solution: Proceeding similarly as above **Example 4.6.8** and making homogeneous the second degree equation by the first degree equation we get

$$3x^2 - xy + 3y^2 + (2x - 3y)\left(\frac{2x+3y}{-k}\right) + 4\left(\frac{2x+3y}{-k}\right)^2 = 0$$
$$\Rightarrow \left(3 - \frac{4}{k} + \frac{16}{k^2}\right)x^2 - \left(1 - \frac{48}{k^2}\right)xy + \left(3 + \frac{9}{k} + \frac{36}{k^2}\right)y^2 = 0$$

[After simplification]

These lines are perpendicular to each other if

coefficient of x^2 + coefficient of $y^2 = 0$

i.e., if $\left(3 - \frac{4}{k} + \frac{16}{k^2}\right) + \left(3 + \frac{9}{k} + \frac{36}{k^2}\right) = 0$

$\Rightarrow \quad 6k^2 + 5k + 52 = 0.$ $\qquad \square$

Example 4.6.10 *Show that the triangle formed by the straight lines $ax^2 + 2hxy + by^2 = 0$ and $lx + my = 1$ is right angled, if*

$$(a+b)(al^2 + 2hlm + bm^2) = 0.$$

Solution: Let $y = m_1 x$ and $y = m_2 x$ be the lines represented by the pair $ax^2 + 2hxy + by^2 = 0$. Then, we have

$$m_1 + m_2 = -\frac{2h}{b}, \quad m_1 m_2 = \frac{a}{b} \tag{4.42}$$

The third line is given by $lx + my = 1$. So its slope is $\left(-\frac{l}{m}\right)$. The slopes of the first and second lines are m_1 and m_2 respectively. The triangle formed by these three lines will be a right angled if any two of the lines be perpendicular to each other.

The first and second lines will be perpendicular if

$$m_1 m_2 = -1 \text{ i.e., } \frac{a}{b} = -1 \text{ or, } a + b = 0 \tag{4.43}$$

The first and third will be perpendicular if

$$m_1 \times \left(-\frac{l}{m}\right) = -1 \text{ i.e., } lm_1 - m = 0 \tag{4.44}$$

and similarly second and third will be perpendicular if

$$lm_2 - m = 0 \tag{4.45}$$

Combining the relations (4.43), (4.44) and (4.45) we get

$$(a+b)(lm_1 - m)(lm_2 - m) = 0$$
or $\quad (a+b)[l^2 m_1 m_2 - lm(m_1 + m_2) + m^2] = 0$
or $\quad (a+b)\left[l^2 \frac{a}{b} - lm\left(-\frac{2h}{b}\right) + m^2\right] = 0$
or $\quad (a+b)(al^2 + 2hlm + bm^2) = 0.$ $\square$

Example 4.6.11 Show that the area of the triangle formed by the straight lines $ax^2 + 2hxy + by^2 = 0$ and $lx + my + n = 0$ is $\dfrac{\sqrt{h^2 - ab}}{am^2 - 2hlm + bl^2}$.

Solution: Let the given pair represents the two lines

$$y - m_1 x = 0 \tag{4.46}$$
$$y - m_2 x = 0 \tag{4.47}$$

and the third line is

$$lx + my + n = 0 \tag{4.48}$$

Pair of Straight Lines

Then we have
$$m_1 + m_2 = -\frac{2h}{b} \quad \text{and} \quad m_1 m_2 = \frac{a}{b} \tag{4.49}$$

The coordinates of the vertices of the triangle formed by these lines (4.46), (4.47) and (4.48) are obtained by solving then by taking two at a time and these are obtained thus as

$$O(0,0),\ A\left(-\frac{n}{l+mm_1}, -\frac{m_1 n}{l+mm_1}\right),\ B\left(-\frac{n}{l+mm_2}, -\frac{m_2 n}{l+mm_2}\right).$$

Hence

$$\triangle OAB = \frac{1}{2}\left|\left\{-\frac{n}{l+mm_1}\left(-\frac{m_2 n}{l+mm_2}\right) - \frac{n}{l+mm_2}\left(+\frac{m_1 n}{l+mm_1}\right)\right\}\right|$$

$$= \frac{1}{2}\left|\frac{n^2(m_2 - m_1)}{(l+mm_1)(l+mm_2)}\right| = \frac{1}{2}\left|\frac{n^2\sqrt{(m_1+m_2)^2 - 4m_1 m_2}}{l^2 + (m_1+m_2)lm + m^2 m_1 m_2}\right|$$

$$= \frac{1}{2}\left|\frac{n^2\sqrt{\frac{4h^2}{b^2} - 4\cdot\frac{a}{b}}}{l^2 + lm\left(-\frac{2h}{b}\right) + m^2\frac{a}{b}}\right| = \frac{1}{2}\left|\frac{\sqrt{h^2 - ab}}{am^2 - 2hlm + bl^2}\right|. \qquad \square$$

Example 4.6.12 *If $ax^2 + 2hxy + by^2 + 2gx + 2fy + c = 0$ represents a pair of lines, the area of the triangle formed by their bisectors and the axis of x is $\dfrac{\sqrt{(a-b)^2 + 4h^2}}{2h} \cdot \dfrac{ca - g^2}{ab - h^2}$.*

Solution: Let (α, β) be the point of intersection of the given pair of lines represented by the equation
$$ax^2 + 2hxy + by^2 + 2gx + 2fy + c = 0 \tag{4.50}$$

Then the point (α, β) is obtained by solving any two of the equations
$$\left.\begin{array}{r} a\alpha + h\beta + g = 0 \\ h\alpha + b\beta + f = 0 \\ \text{and} \quad g\alpha + f\beta + c = 0. \end{array}\right\}$$

From first two we get $\beta = \dfrac{gh - af}{ab - h^2}$ and from the first and the last we get $\beta = \dfrac{ac - g^2}{gh - af}$. So we get

$$\beta^2 = \frac{gh - af}{ab - h^2} \times \frac{ac - g^2}{gh - af} = \frac{ac - g^2}{ab - h^2} \tag{4.51}$$

Now the equation of the bisectors of equation (4.50) is

$$\frac{(x-\alpha)^2 - (y-\beta)^2}{a-b} = \frac{(x-\alpha)(y-\beta)}{h} \qquad (4.52)$$

Let the equation (4.52) meets the x-axis at the two points $A(x_1, 0)$ and $B(x_2, 0)$. Then $AB = |x_2 - x_1|$.

Putting $y = 0$ in (4.52), we get

$$\frac{(x-\alpha)^2 - (-\beta)^2}{a-b} = \frac{(x-\alpha)(-\beta)}{h}$$

or, $h(x-\alpha)^2 - h\beta^2 + (x-\alpha)(a-b)\beta = 0$

or, $hx^2 - (2h\alpha - a\beta + b\beta)x + (\alpha^2 - \beta^2)h - (a-b)\alpha\beta = 0.$

The equation has two roots, say x_1, x_2. Then

$$x_1 + x_2 = \frac{2h\alpha - a\beta + b\beta}{h} \quad \text{and} \quad x_1 x_2 = \frac{(\alpha^2 - \beta^2)h - (a-b)\alpha\beta}{h}.$$

So, $AB = |x_2 - x_1| = \sqrt{(x_1+x_2)^2 - 4x_1x_2}$

$$= \left[\left(\frac{2h\alpha - a\beta + b\beta}{h}\right)^2 - \frac{4(\alpha^2 - \beta^2)h - 4(a-b)\alpha\beta}{h}\right]^{\frac{1}{2}}$$

$$= \frac{1}{h}[4h^2\alpha^2 + a^2\beta^2 + b^2\beta^2 - 4ah\alpha\beta + 4bh\alpha\beta - 2ab\beta^2 - 4\alpha^2 h^2$$

$$+ 4\beta^2 h^2 + 4a\alpha\beta h - 4b\alpha\beta h]^{\frac{1}{2}}$$

$$= \frac{1}{h}[a^2\beta^2 + b^2\beta^2 - 2ab\beta^2 + 4\beta^2 h^2]^{\frac{1}{2}}$$

$$= \frac{\beta}{h}\sqrt{a^2 + b^2 - 2ab + 4h^2} = \frac{\beta}{h}\sqrt{(a-b)^2 + 4h^2}.$$

Hence the required area of the triangle is

$$\frac{1}{2}AB.\beta = \frac{1}{2}|x_1 - x_2|.\beta = \frac{1}{2}\cdot\frac{\beta}{h}\sqrt{(a-b)^2 + 4h^2} \times \beta$$

$$= \frac{\beta^2}{2h}\sqrt{(a-b)^2 + 4h^2} = \frac{\sqrt{(a-b)^2 + 4h^2}}{2h}\cdot\frac{ca - g^2}{ab - h^2}. \qquad \square$$

Example 4.6.13 *If the pair of straight lines $x^2 - 2pxy - y^2 = 0$ and $x^2 - 2qxy - y^2 = 0$ be such that each pair bisects the angles between the other pair, then prove that $pq + 1 = 0$.*

Pair of Straight Lines

Solution: The equation of the bisectors of the angles between the pair of straight lines given by $x^2 - 2pxy - y^2 = 0$ is given by

$$\frac{x^2 - y^2}{1 - (-1)} = \frac{xy}{-p} \quad \text{or,} \quad px^2 + 2xy - py^2 = 0 \tag{4.53}$$

By the given condition (4.53) is same as

$$x^2 - 2qxy - y^2 = 0 \tag{4.54}$$

Comparing coefficients we get $\dfrac{p}{1} = \dfrac{2}{-2q} = \dfrac{-p}{-1}$ which gives $pq + 1 = 0$. □

Example 4.6.14 *Show that the straight lines*

$$(A^2 - 3B^2)x^2 + 8ABxy + (B^2 - 3A^2)y^2 = 0$$

form with the straight line $Ax + By + C = 0$, an equilateral triangle of area

$$\frac{C^2}{\sqrt{3}(A^2 + B^2)}.$$

Solution: The given pair of lines is

$$(A^2 - 3B^2)x^2 + 8ABxy + (B^2 - 3A^2)y^2 = 0$$

or, $\{(A + \sqrt{3}B)x + (B - \sqrt{3}A)y\}\{(A - \sqrt{3}B)x + (B + \sqrt{3}A)y\} = 0.$

So the equation of the three lines forming the triangle are given by

$$(A + \sqrt{3}B)x + (B - \sqrt{3}A)y = 0 \tag{4.55}$$
$$(A - \sqrt{3}B)x + (B + \sqrt{3}A)y = 0 \tag{4.56}$$
$$\text{and} \quad Ax + By + c = 0 \tag{4.57}$$

The coordinates of the vertex of the triangle obtained by (4.55) and (4.56) are $(0, 0)$.

The acute angle between (4.55) and (4.57) is given by

$$\tan\theta = \frac{B(A + \sqrt{3}B) - A(B - \sqrt{3}A)}{A(A + \sqrt{3}B) + B(B - \sqrt{3}A)} = \frac{\sqrt{3}(A^2 + B^2)}{A^2 + B^2} = \sqrt{3} = \tan\frac{\pi}{3}$$

$$\therefore \ \theta = \frac{\pi}{3}.$$

Similarly the acute angle made by (4.56) and (4.57) is given by

$$\tan\phi = \frac{A(B + \sqrt{3}A) - B(A - \sqrt{3}B)}{A(A - \sqrt{3}B) + B(B + \sqrt{3}A)} = \frac{\sqrt{3}(A^2 + B^2)}{A^2 + B^2} = \sqrt{3} = \tan\frac{\pi}{3}$$

$$\therefore \ \phi = \frac{\pi}{3}.$$

So the third angle is given by $\pi - \dfrac{\pi}{3} - \dfrac{\pi}{3} = \dfrac{\pi}{3}.$

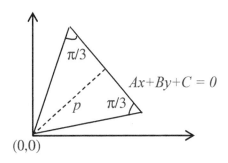

Hence the triangle is an equilateral triangle.

Let p be the length of perpendicular from the vertex $(0,0)$ of the triangle on the side $Ax+By+C = 0$. Then we get

$$p = \frac{C}{\sqrt{A^2 + B^2}}.$$

So the length of a side of the equilateral triangle $= p \csc \dfrac{\pi}{3} = \dfrac{C}{\sqrt{A^2 + B^2}} \cdot \dfrac{2}{\sqrt{3}}$.

Hence the area of the triangle

$$= \frac{1}{2} \frac{C}{\sqrt{A^2 + B^2}} \cdot \frac{2}{\sqrt{3}} p = \frac{1}{2} \frac{C}{\sqrt{A^2 + B^2}} \cdot \frac{2}{\sqrt{3}} \frac{C}{\sqrt{A^2 + B^2}} = \frac{C^2}{\sqrt{3}(A^2 + B^2)}. \qquad \square$$

Example 4.6.15 *If $ax^2 + 2hxy + by^2 + 2gx + 2fy + c = 0$ represents two straight lines equidistant from origin, then show that*

$$f^4 - g^4 = c(bf^2 - ag^2).$$

Solution: If $lx + my + n = 0$ and $l'x + m'x + n' = 0$ be the straight lines represented by the given pair then we have

$$\left.\begin{array}{ccc} ll' = a & mm' = b & nn' = c \\ lm' + l'm = 2h & ln' + l'n = 2g & mn' + m'n = 2f. \end{array}\right\}$$

Since the lines are equidistant from the origin, then

$$\frac{n}{\sqrt{l^2 + m^2}} = \frac{n'}{\sqrt{l'^2 + m'^2}} \quad \text{or,} \quad n^2(l'^2 + m'^2) = n'^2(l^2 + m^2)$$

or, $\quad n^2 l'^2 - n'^2 l^2 = m^2 n'^2 - m'^2 n^2$

or, $\quad (nl' + n'l)(nl' - n'l) = (mn' + m'n)(mn' - m'n)$

or, $\quad 2g(nl' - n'l) = 2f(mn' - m'n)$ \hfill (4.58)

Now $nl' - n'l = \sqrt{(nl' + n'l)^2 - 4nn'll'} = \sqrt{4g^2 - 4ac} = 2\sqrt{g^2 - ac}$
and $mn' - m'n = \sqrt{(mn' + m'n)^2 - 4nn'mm'} = \sqrt{4f^2 - 4bc} = 2\sqrt{f^2 - bc}$.
So we get from (4.58)

$$2g.2\sqrt{g^2 - ac} = 2f.2\sqrt{f^2 - bc}$$

or, $\quad g^2(g^2 - ac) = f^2(f^2 - bc)$

or, $\quad f^4 - g^4 = -g^2 ac + f^2 bc = c(bf^2 - ag^2).$ \hfill $\square$

Pair of Straight Lines

Example 4.6.16 *If the equation $ax^2 + 2hxy + by^2 + 2gx + 2fy + c = 0$ represents a pair of straight lines, prove that the equation of the third pair of straight lines passing through the point where these meet the axes is*

$$ax^2 - 2hxy + by^2 + 2gx + 2fy + c + \frac{4fg}{c}xy = 0.$$

Solution: The given pair of straight lines is

$$ax^2 + 2hxy + by^2 + 2gx + 2fy + c = 0 \qquad (4.59)$$

The equation of the axes is

$$xy = 0 \qquad (4.60)$$

The equation of any curve through the intersection of (4.59) and (4.60) is

$$ax^2 + 2hxy + by^2 + 2gx + 2fy + c + \lambda xy = 0 \qquad (4.61)$$

where λ is a variable parameter.

If the equation (4.61) represents a pair of straight line then its discriminant = 0, i.e.,

$$abc + 2fg\left(h + \frac{\lambda}{2}\right) - af^2 - bg^2 - c\left(h + \frac{\lambda}{2}\right)^2 = 0$$

or, $\qquad abc + 2fgh - af^2 - bg^2 - ch^2 + fg\lambda - c\lambda h - c\frac{\lambda^2}{4} = 0 \qquad (4.62)$

But since (4.59) represents a pair of straight lines so

$$abc + 2fgh - af^2 - bg^2 - ch^2 = 0$$

and hence (4.62) gives

$$fg\lambda - ch\lambda - c\frac{\lambda^2}{4} = 0 \ \ \text{or,} \ \ fg - ch - c\frac{\lambda}{4} = 0$$

or, $\qquad \frac{c\lambda}{4} = fg - ch \ \ \Rightarrow \ \ \lambda = \frac{4(fg - ch)}{c}.$

Therefore putting this value of λ in (4.61) we get the required equation of the third pair of straight lines as

$$ax^2 + 2hxy + by^2 + 2gx + 2fy + c + \frac{4(fg - ch)}{c}xy = 0$$

or, $ax^2 - 2hxy + by^2 + 2gx + 2fy + c + \dfrac{4fg}{c}xy = 0.$ □

Example 4.6.17 If the equation $ax^2 + 2hxy + by^2 + 2gx + 2fy + c = 0$ represents two straight lines, prove that the square of the distance of their point of intersection from the origin is $\dfrac{c(a+b) - f^2 - g^2}{ab - h^2}$.

Solution: Let $lx + my + n = 0$ and $l'x + m'x + n' = 0$ be the equations of the straight lines represented by the given pair then we have

$$\left.\begin{array}{ccc} ll' = a & mm' = b & nn' = c \\ lm' + l'm = 2h & ln' + l'n = 2g & mn' + m'n = 2f. \end{array}\right\}$$

Solving the equations of the lines

$$\left.\begin{array}{c} lx + my + n = 0 \\ \text{and} \quad l'x + m'y + n' = 0. \end{array}\right\}$$

By cross multiplication, the point of intersection is obtained as

$$\frac{x}{mn' - m'n} = \frac{y}{nl' - n'l} = \frac{1}{lm' - ml'}$$

or, $\dfrac{x}{\sqrt{(mn' + m'n)^2 - 4mm'nn'}} = \dfrac{y}{\sqrt{(nl' + n'l)^2 - 4nn'll'}}$

$$= \frac{1}{\sqrt{(lm' + ml')^2 - 4ll'mm'}}$$

or, $\dfrac{x}{2\sqrt{f^2 - bc}} = \dfrac{y}{2\sqrt{g^2 - ca}} = \dfrac{1}{2\sqrt{h^2 - ab}}.$

So the point of intersection is given by $\left(\sqrt{\dfrac{f^2 - bc}{h^2 - ab}}, \sqrt{\dfrac{g^2 - ac}{h^2 - ab}}\right)$.

Therefore the required square of the distance between the origin and the point of intersection =

$$\frac{f^2 - bc}{h^2 - ab} + \frac{g^2 - ac}{h^2 - ab} = \frac{f^2 + g^2 - ac - bc}{h^2 - ab} = \frac{c(a+b) - f^2 - g^2}{ab - h^2}. \quad \square$$

Example 4.6.18 The distance of the point (h, k) from each of the two given straight lines through the origin is d. Show that the equation of the pair of straight lines is $(xk - yh)^2 = d^2(x^2 + y^2)$.

Solution: Let

$$y - m_1 x = 0 \tag{4.63}$$

Pair of Straight Lines

be the equation of any one of the two given lines through the origin. Then by the condition of the problem, we have

$$d = \left|\frac{k-mh}{\sqrt{1+m^2}}\right| \quad \therefore \quad d^2 = \frac{(k-mh)^2}{1+m^2}$$

or, $(k-mh)^2 = d^2(1+m^2)$

or, $k^2 - 2mkh + m^2h^2 = d^2 + d^2m^2$

or, $(d^2 - h^2)m^2 + 2mkh + (d^2 - k^2) = 0 \quad (4.64)$

This is a quadratic equation in m. Let m_1 and m_2 be the roots of (4.64). Then

$$\left.\begin{array}{rl} m_1 + m_2 &= -\dfrac{2kh}{d^2 - h^2} \\[1em] \text{and} \quad m_1 m_2 &= \dfrac{d^2 - k^2}{d^2 - h^2} \end{array}\right\} \quad (4.65)$$

The equation of the required lines is then given by

$$y^2 - (m_1 + m_2)xy + m_1 m_2 x^2 = 0$$

or, $y^2 - \left(-\dfrac{2kh}{d^2 - h^2}\right)xy + \dfrac{d^2 - k^2}{d^2 - h^2}x^2 = 0 \quad$ [Using (4.65)]

or, $(d^2 - h^2)y^2 + 2khxy + (d^2 - k^2)x^2 = 0$

or, $d^2(x^2 + y^2) - (x^2k^2 + h^2y^2 - 2khxy) = 0$

or, $d^2(x^2 + y^2) - (xk - yh)^2 = 0$

or, $(xk - yh)^2 = d^2(x^2 + y^2)$. □

Example 4.6.19 *Prove that two of the lines represented by the equation $ax^4 + bx^3y + cx^2y^2 + dxy^3 + ay^4 = 0$ will bisect the angle between the other two, if $c + 6a = 0$ and $b + d = 0$.*

Solution: Since the bisectors are perpendicular to each other, so the given equation must represent two pairs of mutually perpendicular lines, which passes through the origin.

Let us take

$$ax^4 + bx^3y + cx^2y^2 + dxy^3 + ay^4 = (x^2 + pxy - y^2)(ax^2 + qxy - ay^2)$$

Comparing coefficients of x^3y, x^2y^2 and xy^3, we get

$$ap + q = b \quad (4.66)$$

$$-a - a + pq = c \Rightarrow 2a + c = pq \quad (4.67)$$

and $\quad -ap - q = d \Rightarrow ap + q = -d \quad (4.68)$

From (4.66) and (4.68) we get $b + d = 0$.

Now the bisectors of one pair are given by

$$\frac{x^2 - y^2}{1 - (-1)} = \frac{xy}{\frac{p}{2}} \Rightarrow x^2 - y^2 = \frac{4xy}{p} \tag{4.69}$$

which is same as the other pair $ax^2 + qxy - ay^2 = 0$ (by the given condition of the problem)

$$\Rightarrow x^2 - y^2 = \frac{-qxy}{a} \tag{4.70}$$

Comparing (4.69) and (4.70) we get $\dfrac{4}{p} = -\dfrac{q}{a} \Rightarrow pq = -4a$.

Therefore from (4.67), $-4a = 2a + c$ or, $c + 6a = 0$.

Hence $c + 6a = 0$ and $b + d = 0$ are the required conditions. $\square$

Example 4.6.20 *Prove that the straight lines joining the origin to the points of intersection of the lines $\dfrac{x}{\alpha} + \dfrac{y}{\beta} = 2$ with the curve $(x - \alpha)^2 + (y - \beta)^2 = \gamma^2$ are at right angles if $\alpha^2 + \beta^2 = \gamma^2$.*

Solution: The equation of the given line is

$$\frac{x}{\alpha} + \frac{y}{\beta} = 2 \tag{4.71}$$

Making homogeneous the equation of the curve with the help of (4.71) we get

$$x^2 + y^2 - 2(\alpha x + \beta y)\left(\frac{x}{2\alpha} + \frac{y}{2\beta}\right) + (\alpha^2 + \beta^2 - \gamma^2)\left(\frac{x}{2\alpha} + \frac{y}{2\beta}\right)^2 = 0 \tag{4.72}$$

which is the equation of the pair of straight lines joining the origin to the points of intersection of the line (4.71) and the given curve.

Now this pair of lines (4.72) will be at right angles to each other if the sum of the coefficients of x^2 and y^2 is zero.

i.e., if $\dfrac{\alpha^2 + \beta^2 - \gamma^2}{4\alpha^2} + \dfrac{\alpha^2 + \beta^2 - \gamma^2}{4\beta^2} = 0$

i.e., if $(\alpha^2 + \beta^2 - \gamma^2)\left(\dfrac{1}{\alpha^2} + \dfrac{1}{\beta^2}\right) = 0$

i.e., if $\alpha^2 + \beta^2 - \gamma^2 = 0$ i.e., if $\alpha^2 + \beta^2 = \gamma^2$.

($\because$ sum of two square terms can never be zero, so $\dfrac{1}{\alpha^2} + \dfrac{1}{\beta^2} \neq 0$.) $\square$

Example 4.6.21 *If $ax^2 + 2hxy + by^2 = 0$ be the equation of two adjacent sides of a parallelogram and $lx + my = 1$ be the equation of one of its diagonals, then show that the equation of its other diagonal is*

$$y(bl - hm) = x(am - hl).$$

Solution: Let $y = m_1 x$ and $y = m_2 x$ be the equations of the lines OA and OC represented by the pair of lines $ax^2 + 2hxy + by^2 = 0$. Then we have

$$m_1 + m_2 = -\frac{2h}{b}, \quad m_1 m_2 = \frac{a}{b} \tag{4.73}$$

If $OABC$ be the parallelogram, then, since $lx + my = 1$ does not pass

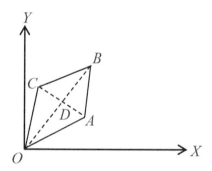

through the origin, it cannot represent the diagonal OB. So the diagonal AC is given by

$$lx + my = 1 \tag{4.74}$$

Solving $y = m_1 x$ and the line $lx + my = 1$ we get the coordinates of

$$A \equiv \left(\frac{1}{l + mm_1}, \frac{m_1}{l + mm_1} \right)$$

and similarly coordinates of

$$C \equiv \left(\frac{1}{l + mm_2}, \frac{m_2}{l + mm_2} \right).$$

The coordinates of D, the mid-point of AC is given by

$$\left(\frac{1}{2} \left(\frac{1}{l + mm_1} + \frac{1}{l + mm_1} \right), \frac{1}{2} \left(\frac{m_1}{l + mm_2} + \frac{m_2}{l + mm_2} \right) \right).$$

Now

$$\frac{1}{l+mm_1} + \frac{1}{l+mm_2} = \frac{2l+m(m_1+m_2)}{(l+mm_1)(l+mm_2)}$$

$$= \frac{2l+m\left(-\frac{2h}{b}\right)}{(l+mm_1)(l+mm_2)} = \frac{2(bl-hm)}{b(l+mm_1)(l+mm_2)}$$

and

$$\frac{m_1}{l+mm_1} + \frac{m_2}{l+mm_2} = \frac{lm_1+mm_1m_2+lm_2+mm_1m_2}{(l+mm_1)(l+mm_2)}$$

$$= \frac{l(m_1+m_2)+2mm_1m_2}{(l+mm_1)(l+mm_2)} = \frac{l\left(-\frac{2h}{b}\right)+2m\frac{a}{b}}{(l+mm_1)(l+mm_2)}$$

$$= \frac{2(am-hl)}{b(l+mm_1)(l+mm_2)}$$

So the coordinates of D is obtained as

$$\left(\frac{(bl-hm)}{b(l+mm_1)(l+mm_2)}, \frac{(am-hl)}{b(l+mm_1)(l+mm_2)}\right).$$

Now since the diagonals of a parallelogram bisect each other, so the other diagonal OB will pass through this point D and therefore its equation is given by

$$y-0 = \frac{\left(\frac{am-hl}{b(l+mm_1)(l+mm_2)}\right)}{\left(\frac{bl-hm}{b(l+mm_1)(l+mm_2)}\right)}(x-0) \text{ i.e., } y = \frac{am-hl}{bl-hm}x$$

or, $y(bl-hm) = x(am-hl)$. □

Example 4.6.22 *A triangle has the lines $ax^2 + 2hxy + by^2 = 0$ for two of its sides and the point (p,q) for its orthocentre. Show that the equation of its third side is $(a+b)(px+qy) = aq^2 - 2hpq + bp^2$.*

Solution: Clearly one vertex of the triangle is the origin O. Let the orthocentre (p,q) of the triangle be denoted by P. Then the equation of the line OP is

$$y = \frac{q}{p}x \text{ or, } qx - py = 0 \tag{4.75}$$

Let the other two vertices of the triangle be A and B. Since P is the orthocentre of the triangle, so $AB \perp OP$. So, let the equation of the third side AB be considered as

$$px + qy + k = 0 \tag{4.76}$$

Pair of Straight Lines

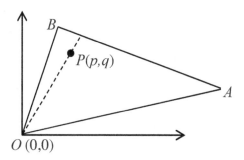

Let $y = m_1 x$ and $y = m_2 x$ be the two lines represented by the given pair $ax^2 + 2hxy + by^2 = 0$. Then we get

$$m_1 + m_2 = -\frac{2h}{b}, \quad m_1 m_2 = \frac{a}{b} \qquad (4.77)$$

The equation of any line through A, different from OA and AB can be written as

$$px + qy + k + \lambda(y - m_1 x) = 0$$
or, $\quad (p - \lambda m_1)x + (q + \lambda)y + k = 0.$

If it passes through $P(p, q)$, then

$$(p - \lambda m_1)p + (q + \lambda)q + k = 0$$
or, $\quad p^2 + q^2 - \lambda(m_1 p - q) + k = 0$
$$\Rightarrow k = \lambda(m_1 p - q) - (p^2 + q^2) \qquad (4.78)$$

The line $AP \perp OB$

$$\therefore \quad m_2 \times \frac{\lambda m_1 - p}{q + \lambda} = -1 \text{ or, } \lambda m_1 m_2 - p m_2 = -q - \lambda$$

or, $\lambda(1 + m_1 m_2) = m_2 p - q$ or, $\lambda \left(1 + \frac{a}{b}\right) = m_2 p - q$ [using (4.77)]

or, $\lambda = \dfrac{b(m_2 p - q)}{a + b}.$

Substituting this value of λ in (4.78) we get

$$\begin{aligned}
k &= \frac{b(m_2 p - q)(m_1 p - q)}{a + b} - (p^2 + q^2) \\
&= \frac{b[m_1 m_2 p^2 - (m_1 + m_2)pq + q^2]}{a + b} - (p^2 + q^2) \\
&= \frac{b[\frac{a}{b}p^2 + \frac{2h}{b}pq + q^2]}{a + b} - (p^2 + q^2)
\end{aligned}$$

$$= \frac{ap^2 + 2hpq + bq^2 - ap^2 - bq^2 - aq^2 - bp^2}{a+b}$$

$$= -\frac{aq^2 - 2hpq + bp^2}{a+b}.$$

Hence the equation (4.76) can be written as

$$px + qy - \frac{aq^2 - 2hpq + bp^2}{a+b} = 0$$

or, $(a+b)(px+qy) = aq^2 - 2hpq + bp^2$. □

Example 4.6.23 *The gradient of one of the straight lines of $ax^2 + 2hxy + by^2 = 0$ is twice that of the other. Show that $8h^2 = 9ab$.*

Solution: Let $y = m_1 x$ and $y = m_2 x$ be the two lines represented by the given pair $ax^2 + 2hxy + by^2 = 0$. Then we have

$$m_1 + m_2 = -\frac{2h}{b}, \quad m_1 m_2 = \frac{a}{b} \tag{4.79}$$

It is given that $m_2 = 2m_1$

$$\therefore \; m_1 + 2m_1 = -\frac{2h}{b} \Rightarrow m_1 = -\frac{2h}{3b} \text{ and } m_2 = -\frac{4h}{3b}$$

$$\therefore \; m_1 m_2 = \frac{a}{b} \text{ gives } \left(-\frac{2h}{3b}\right) \times \left(-\frac{4h}{3b}\right) = \frac{a}{b}$$

or, $8h^2 = 9ab$. □

Example 4.6.24 *Show that the equation of the line joining the feet of the perpendiculars from the point $(d,0)$ on the lines $ax^2 + 2hxy + by^2 = 0$ is $(a-b)x + 2hy + bd = 0$.*

Solution: Let $y = m_1 x$ and $y = m_2 x$ be the two lines represented by the given pair $ax^2 + 2hxy + by^2 = 0$. Then we have

$$m_1 + m_2 = -\frac{2h}{b}, \quad m_1 m_2 = \frac{a}{b} \tag{4.80}$$

The equation of the line through $(d,0)$ perpendicular to $y - m_1 x = 0$ is

$$m_1(y-0) + (x-d) = 0 \Rightarrow m_1 y = -(x-d) \tag{4.81}$$

Similarly the equation of the line through $(d,0)$ perpendicular to $y - m_2 x = 0$ is

$$m_2(y-0) + (x-d) = 0 \Rightarrow m_2 y = -(x-d) \tag{4.82}$$

Pair of Straight Lines

Solving (4.81) and $y = m_1 x$ we get the x-coordinates of their intersection as

$$m_1 . m_1 x = -x + d \quad \Rightarrow \quad x = \frac{d}{1 + m_1^2}.$$

Similarly solving (4.82) and $y = m_2 x$ we get the x-coordinates of their intersection as

$$m_2 . m_2 x = -x + d \quad \Rightarrow \quad x = \frac{d}{1 + m_2^2}.$$

Therefore, the point of intersection of $y = m_1 x$ and (4.81) is given by

$$\left(\frac{d}{1+m_1^2}, \frac{m_1 d}{1+m_1^2} \right)$$

and that of $y = m_2 x$ and (4.82) is

$$\left(\frac{d}{1+m_2^2}, \frac{m_2 d}{1+m_2^2} \right).$$

Let $k = \dfrac{d}{1+m_1^2}$ and $k' = \dfrac{d}{1+m_2^2}$ then the equation of the line through $(k, m_1 k)$ and $(k', m_1 k')$ is given by

$$\frac{y - m_1 k}{m_1 k - m_2 k'} = \frac{x - k}{k - k'}$$

or, $y(k - k') - m_1 k(k - k') = x(m_1 k - m_2 k') - k(m_1 k - m_2 k')$

or, $x(m_1 k - m_2 k') - y(k - k') + kk'(m_2 - m_1) = 0$

or, $x \left[m_1 \dfrac{d}{1+m_1^2} - m_2 \dfrac{d}{1+m_2^2} \right] - y \left(\dfrac{d}{1+m_1^2} - \dfrac{d}{1+m_2^2} \right)$

$$+ \frac{d^2}{(1+m_1^2)(1+m_2^2)} (m_2 - m_1) = 0$$

[Multiplying throughout by $\dfrac{(1+m_1^2)(1+m_2^2)}{d}$]

or, $x \left[(m_1 - m_2) + m_1 m_2 (m_2 - m_1) \right] - y[m_2^2 - m_1^2] + d(m_2 - m_1) = 0$

or, $x[-1 + m_1 m_2] - y(m_1 + m_2) + d = 0 \quad [\because m_2 - m_1 \neq 0]$

or, $x \left(-1 + \dfrac{a}{b} \right) - y \left(-\dfrac{2h}{b} \right) + d = 0$

or, $(a - b)x + 2hy + bd = 0$. □

Example 4.6.25 *Prove that the pair of straight lines joining the origin to the other two points of intersection of the curves $a_1x^2 + 2h_1xy + b_1y^2 + 2g_1x = 0$ and $a_2x^2 + 2h_2xy + b_2y^2 + 2g_2x = 0$ will be at right angles if $g_2(a_1 + b_1) = g_1(a_2 + b_2)$.*

Solution: The equations of the given curves are

$$a_1x^2 + 2h_1xy + b_1y^2 + 2g_1x = 0 \tag{4.83}$$
$$\text{and} \quad a_2x^2 + 2h_2xy + b_2y^2 + 2g_2x = 0 \tag{4.84}$$

Both the curves pass through the origin.

Now (4.83) × g_2 − (4.84) × g_1 gives

$$(a_1g_2 - a_2g_1)x^2 + 2(h_1g_2 - h_2g_1)xy + (b_1g_2 - b_2g_1)y^2 = 0.$$

This is a homogeneous equation of degree 2 in x, y and this equation is satisfied by the coordinates of the points which satisfy both (4.83) and (4.84). Therefore this equation represents a pair of straight lines joining origin to the other two points of intersection of the curves (4.83) and (4.84).

Now, these pair of straight lines will be at right angles

if coefficients of x^2 + coefficients of $y^2 = 0$

i.e., if $(a_1g_2 - a_2g_1) + (b_1g_2 - b_2g_1) = 0$

i.e., if $g_2(a_1 + b_1) = g_1(a_2 + b_2)$. □

4.7 Exercises

Section A: Objective Type Questions

1. For what value of k, the equation $x^2+kxy-2y^2+3y-1=0$ represents a pair of straight lines?

2. Show that the equation $x^2 + xy - 2y^2 + 2x + 4y = 0$ represents two straight lines.

3. Test whether the equation $4x^2 - 4xy + y^2 - 12x + 6y + 8 = 0$ represents a pair of straight lines.

4. Find the angle between the pair of lines $\sqrt{3}x^2 - 4xy + \sqrt{3}y^2 = 0$.

5. If the equation $6x^2 + kxy - 3y^2 + 4x + 5y - 2 = 0$ represents a pair of intersecting straight lines, find the value of k.

6. Show that the equation $6x^2 - 5xy - 6y^2 + 14x + 5y + 4 = 0$ represents a pair of straight lines which are perpendicular to each other.

7. Show that the lines passing through origin and perpendicular to the lines given by $5x^2 - 7x - 3y^2 = 0$ will be $3x^2 - 7x - 5y^2 = 0$.

8. Show that the condition for one of he straight lines given by $ax^2 + 2hxy + by^2 = 0$ may coincide with one of the straight lines given by $a'x^2 + 2hxy + b'y^2 = 0$ is $4(ah' - a'h)(hb' - h'b) = (ba' - b'a)^2$.

9. Find the area of the triangle formed by the lines $x^2 - y^2 = 0$ and $x + 2y = 1$.

10. Find the equation of the pair of straight lines joining the origin to the points of intersection of the line $x + 2y = 5$ and the parabola $y^2 = 8x$.

11. Show that the equation $4x^2 + 9y^2 + 12xy - 16x - 24y + 16 = 0$ represents a pair of coincident straight lines.

[Hints: Show $ab - h^2 = 0$, $gh - af = 0$, $g^2 - ac = 0$.]

12. Show that the straight lines given by $x^2 - 2xy - y^2 = 0$ bisect the angles between the lines $x^2 + 2xy - y^2 = 0$.

Section B: Broad Answer Type Questions

1. Deduce the condition that the general equation of second degree may represent a pair of straight lines.

2. Find an expression to give the angle between the pair of straight lines given by $ax^2 + 2hxy + by^2 + 2gx + 2fy + c = 0$.

3. Find the value of k, so that the equation $x^2 + y^2 + 2x + k = 0$ may represent a pair of straight lines.

4. Find a necessary condition that the general equation of second degree $ax^2 + 2hxy + by^2 + 2gx + 2fy + c = 0$ may represent a pair of straight lines.

5. For what value of λ, does the equation $xy + 5x + \lambda y + 15 = 0$ represents a pair of straight lines?

6. For what value of λ, does the equation $x^2 + \lambda xy - 2y^2 + 3y - 1 = 0$ represents a pair of straight lines?

7. If the equation $6x^2 + kxy - 3y^2 + 4x + 5y - 2 = 0$ represents a pair of straight lines, find the value of k.

8. Find the separate equations of the following pair of straight lines:

  (i) $x^2 - 5xy + 4y^2 + x + 2y - 2 = 0$
  (ii) $x^2 - y^2 + x - 3y - 2 = 0$
  (iii) $14x^2 + 11xy - 15y^2 - 22x + 29y - 12 = 0$.

9. Deduce the condition that the straight lines given by $ax^2 + 2hxy + by^2 + 2gx + 2fy + c = 0$ are at right angles.

10. Show that the equation $x^2 + 6xy + 9y^2 - 5x - 15y + 6 = 0$ represents a pair of parallel straight lines. Find also the distance between them.

11. Prove that the equation $8x^2 + 10xy + 3y^2 + 26x + 16y + 21 = 0$ represents a pair of straight lines. Find the coordinates of their point of intersection and the angle between them.

12. Find the angle between the pair of straight lines represented by the equation $3x^2 - 10xy + 3y^2 = 0$.

13. Find the coordinates of the centroid and the area of the triangle formed by the straight lines $2x^2 - 5xy + 2y^2 = 0$ and $x + y = 3$.

14. Find the equations of the bisectors of the angles between the pair of straight lines $ax^2 + 2hxy + by^2 = 0$.

15. Find the equations to the bisectors of the angles between the following pair of straight lines

  (i) $5x^2 - 6xy + y^2 = 0$
  (ii) $7x^2 + 6xy - 4y^2 = 0$
  (iii) $12x^2 + 7xy - 12y^2 - 31x + 17y + 7 = 0$.

16. Show that the coordinate axes are the bisectors of the angles between the pair of straight lines $x^2 - y^2 = 0$.

17. Show that the lines $y^2 + xy - x^2 = 0$ and $y^2 - 4xy - x^2 = 0$ bisect the angles between one another.

18. Show that the pair of straight lines $(a-b)(x^2 - y^2) + 4hxy = 0$ and that the lines $h(x^2 - y^2) = (a-b)xy$ are such that each bisects the angle between the other pair.

19. Prove that the pair of straight lines $a^2x^2 + 2h(a+b)xy + b^2y^2 = 0$ is equally inclined to the pair $ax^2 + 2hxy + by^2 = 0$.

Pair of Straight Lines

20. Show that the bisectors of the angles between the pair of lines $2x^2 - 7xy + 2y^2 = 0$ are equally inclined to the coordinate axes.

21. Prove that the bisectors of the angles between the pair of lines $(ax + by)^2 = 3(bx - ay)^2$ are receptively parallel and perpendicular to the line $ax + by + c = 0$.

22. Show that the equation of the straight lines bisecting the angles between the bisectors of the pair of lines $ax^2 + 2hxy + by^2 = 0$ is

$$(a - b)(x^2 - y^2) + 4hxy = 0.$$

23. Find the condition that two of the straight lines given by $ax^3 + bx^2y + cxy^2 + dy^3 = 0$ may be at right angles.
[Hints: Take $ax^3 + bx^2y + cxy^2 + dy^3 = (x^2 + kxy - y^2)(ax - by) = 0$. Then the two lines will be at right angles. Now compare the coefficients and get the condition.]

24. Find the condition that the two of the straight lines given by $ax^4 + bx^3y + cx^2y^2 + dxy^3 + ex^4 = 0$ will be at right angles if

$$(b + d)(ad + be) + (a - e)^2(a + c + e) = 0.$$

[Hints: Take $ax^4 + bx^3y + cx^2y^2 + dxy^3 + ex^4 = (x^2 + kxy - y^2)(ax^2 - k'xy - ey^2) = 0$. Then two of the lines will be at right angles. Now compare the coefficients and get the condition.]

25. Prove that the equation $ax^2 + 2hxy + by^2 + 2gx + 2fy + c = 0$ represents two parallel straight lines if $h^2 = ab$ and $bg^2 = af^2$. Also show that the distance between them is $2\sqrt{\dfrac{g^2 - ca}{a(a + b)}}$.

26. Find the equation of the pair of straight lines through the point $(2, -3)$ and parallel to the straight lines $15x^2 + xy - 6y^2 + x + 7y - 2 = 0$.

27. Find the equation of the pair of straight lines through the origin and perpendicular to the pair of straight lines $2x^2 + 5xy + 2y^2 + 10x + 5y = 0$.

28. Prove that the angle between the straight lines joining the origin to the points of intersection of the straight line $y = 3x + 2$ with the curve $x^2 + 2xy + 3y^2 + 4x + 8y - 11 = 0$ is $\tan^{-1} \dfrac{2\sqrt{2}}{3}$.

29. Prove that the equation $(x - p)^2 + 2h(x - p)(y - q) - (y - q)^2 = 0$ represents a pair of straight lines at right angles to each other.

30. Show that if the two straight lines represented by the equation $(\tan^2 \theta + \cos^2 \theta)x^2 - 2\tan\theta xy + \sin^2 \theta y^2 = 0$ make angles α, β with the axis of x, then $\tan \alpha - \tan \beta = 2$.

31. Prove that the pair of lines $a^2x^2 + 2h(a+b)xy + b^2y^2 = 0$ is equally inclined to the pair $ax^2 + 2hxy + by^2 = 0$.

Show also that the pair $ax^2 + 2hxy + by^2 + \lambda(x^2 + y^2) = 0$ is equally inclined to the same pair.

32. Prove that $x^3 - 3y^2x = 0$ represents three straight lines equally incline to one another.

33. Find the angle between the chords of the curve $x^2 + y^2 = ax + by$ obtained by joining the origin to the points of intersection of the straight line $\dfrac{x}{a} + \dfrac{y}{b} = 1$ and the curve.

34. Find the angle between the lines joining the origin to the points common to $x^2 + y^2 + 2gx + 2fy + c = 0$ and $x\cos\alpha + y\sin\alpha = p$.

Hence show that the lines represented in $x^2 + y^2 + 2gx + 2fy + c = 0$ will be at right angles if $2p^2 + 2p(g\cos\alpha + f\sin\alpha) + c = 0$.

35. Show that the angle between one of the straight lines given by $ax^2 + 2hxy + by^2 = 0$ and one of the straight lines given by $(a+\lambda)x^2 + 2hxy + (b+\lambda)y^2 = 0$ is equal to the angle between the other two straight lines of the system.

36. Prove that if all chords of the curve given by the equation $ax^2 + 2hxy + by^2 + 2gx + 2fy + c = 0$ subtend a right angle at the origin, the equation must represent two straight lines at right angles through the origin.

[*Hints:* Let $lx + my = 1$ be any chord. Then the equation to the lines joining the origin to the points of intersection of the curve and the chord is

$$ax^2 + 2hxy + by^2 + (2gx + 2fy)(lx + my) + c(lx + my)^2 = 0$$

These two lines are at right angles, if the sum of the coefficients of x^2 and y^2 is zero.

$$a + b + 2gl + 2fm + cl^2 + cm^2 = 0 \tag{4.85}$$

Since l and m are arbitrary quantities, so (4.85) will be satisfied if $a + b = 0$, $g = 0$, $f = 0$ and $c = 0$.

Hence the given equation reduces to $ax^2 + 2hxy - ay^2 = 0$ which represents two straight lines at right angles to one another.]

37. Prove that the equation to the straight lines through the origin each of which makes an angle α with the straight line $y = x$ is $x^2 - 2xy\sec 2\alpha + y^2 = 0$.

[*Hints:* The equation of the required pair is $\{y - x\tan(135° + \alpha)\}\{y - x\tan(135° + \alpha)\} = 0$. Now simplify and get the result.]

38. Find the values of a and f for which the equation $ax^2 - 12xy + 9y^2 + 20x - 2fy - 11 = 0$ represents two parallel straight lines.

39. Prove that the equation $y^3 - x^3 + 3xy(y - x) = 0$ represents three straight lines equally inclined to one another.

40. Show that the equation to the pair of lines through the origin perpendicular to the pair $ax^2 + 2hxy + by^2 = 0$ is $bx^2 - 2hxy + ay^2 = 0$.

41. Find the condition that one of the lines of $ax^2 + 2hxy + by^2 = 0$ may coincide with one of the lines of $a'x^2 + 2h'xy + b'y^2 = 0$.

42. If one of the lines given by the equation $ax^2 + 2hxy + by^2 = 0$ coincides with one of the lines given by $a'x^2 + 2h'xy + b'y^2 = 0$ and the other lines are perpendicular, then show that $\dfrac{ha'b'}{b'-a'} = \dfrac{h'ab}{b-a} = \dfrac{1}{2}\sqrt{-aa'bb'}$.

[*Hints:* See Worked our **Example 4.6.4**. Use (4.33) and (4.34) for $m_1 m_2$ with necessary changes in the coefficients, i.e., with the replacements $a_1 = a$, $h_1 = h$, $b_1 = b$ and $a_2 = a$, $h_2 = h$, $b_2 = b$ etc.]

43. Find the values of b, g and c for which the equation $x^2 - 6xy + by^2 + 2gx - 12y + c = 0$ represents a pair of coincident straight lines.

44. If one of the lines $ax^2 + 2hxy + by^2 = 0$ coincides with one of the lines given by $a'x^2 + 2h'xy + b'y^2 = 0$ and θ be the angle between the other two lines then prove that $\tan\theta = \dfrac{2(bh' - b'h)(a'h - ah')}{aa'(bh' - b'h) + bb'(a'h - ah')}$.

45. Find the conditions that the equation $ax^2 + 2hxy + by^2 + 2gx + 2fy + c = 0$ may represent a pair of straight lines (i) intersecting on the x-axis and (ii) intersecting on the y-axis.

46. If the equation $u \equiv ax^2 + 2hxy + by^2 + 2gx + 2fy + c = 0$ represents a pair of straight lines, then prove that the equation to the third pair of straight lines passing through the points where these straight lines meet the coordinate axes is $(ax^2 - 2hxy + by^2 + 2gx + 2fy + c)c + 4fgxy = 0$.

Another form: $cu + 4(fg - ch)xy = 0$.

47. If $f(x,y) \equiv ax^2 + 2hxy + by^2 + 2gx + 2fy + c = 0$ represents two straight lines, show that the product of the perpendiculars from the point (α, β) on the straight lines is

$$\dfrac{f(\alpha,\beta)}{\sqrt{(a-b)^2 + 4h^2}}.$$

48. Show that the four straight lines represented by $6x^2 - 5xy - 6y^2 = 0$ and $6(x^2 - y^2) - 5xy + x + 5y - 1 = 0$ form the sides of a square. Find the equations to the diagonals of the square.

[*Hints:* The separate lines are obtained from the given pairs by factor-

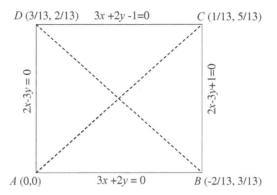

izing linearly these equations and are given by

$$AB \equiv 3x + 2y = 0 \qquad (4.86)$$
$$AD \equiv 2x - 3y = 0 \qquad (4.87)$$
$$DC \equiv 3x + 2y = 1 \qquad (4.88)$$
$$\text{and} \quad BC \equiv 2x - 3y = -1 \qquad (4.89)$$

The coordinates of the points of intersection are obtained by solving (4.86), (4.87), (4.88) and (4.89), taken twice at a time and thus we get

$$A \equiv (0,0), \ B\left(-\frac{2}{13}, \frac{3}{13}\right), \ C\left(\frac{1}{13}, \frac{5}{13}\right) \text{ and } D\left(\frac{3}{13}, \frac{2}{13}\right).$$

∴ Equation of the diagonal AC is given by

$$y - 0 = \frac{\frac{5}{13} - 0}{\frac{1}{13} - 0}(x - 0) \Rightarrow y = 5x$$

and that of BD is

$$y - \frac{3}{13} = \frac{\frac{2}{13} - \frac{3}{13}}{\frac{3}{13} + \frac{2}{13}}\left(x - \frac{3}{13}\right) \quad \text{or,} \quad y - \frac{3}{13} = -\frac{1}{5}\left(x - \frac{3}{13}\right)$$

or, $13y - 3 = -\dfrac{1}{5}(13x - 3)$

or, $65y - 15 = -13x + 3 \Rightarrow 13x + 65y = 12.$

Hence the equations of diagonals are $y = 5x$ and $13x + 65y = 12$.
Now we see that distance

$$AC = \sqrt{\left(\frac{1}{13} - 0\right)^2 + \left(\frac{5}{13} - 0\right)^2} = \frac{\sqrt{26}}{13}$$

and distance

$$BD = \sqrt{\left(\frac{3}{13}+\frac{2}{13}\right)^2 + \left(\frac{2}{13}-\frac{1}{13}\right)^2} = \frac{\sqrt{26}}{13}.$$

So, length of the diagonal AC = length of diagonal BD.
Also slope of diagonal $AC = m_1 = 5$ and slope of diagonal $BD = m_2 = -\frac{13}{65} = -\frac{1}{5}$.

Therefore product of their gradient $= m_1 m_2 = 5 \times \left(-\frac{1}{5}\right) = -1$.

So the diagonals are perpendicular to each other.

Thus we see that the diagonals AC and BD of the quadrilateral $ABCD$ formed by the lines represented by the given equations are equal and perpendicular to each other and so $ABCD$ is a square.]

49. Show that the orthocentre of the triangle formed by the straight lines $ax^2 + 2hxy + by^2 = 0$ and $lx + my = 1$ is a point (x', y') such that

$$\frac{x'}{l} = \frac{y'}{m} = \frac{a+b}{am^2 - 2hlm + bl^2}$$

and that its distance form the origin is

$$\frac{(a+b)\sqrt{l^2+m^2}}{am^2 - 2hlm + bl^2}.$$

Hence show that the locus of the orthocentre of the triangle of which two sides are given in position and whose third side goes through a fixed point (α, β) is $bx^2 - 2hxy + ay^2 = (a+b)(\alpha x + \beta y)$.

50. If each of the equation $ax^2 + 2hxy + by^2 + 2gx + 2fy + c = 0$ and $ax^2 + 2hxy + by^2 = 0$ represents a pair of straight lines, find the equations of the two diagonals of the parallelogram formed by them.

51. If each of the equations $ax^2 + 2hxy + by^2 + 2gx + 2fy + c = 0$ and $ax^2 + 2hxy + by^2 - 2gx - 2fy + c = 0$ represents a pair of intersecting straight lines, prove that the area of the parallelogram enclosed by them is

$$\frac{2c}{\sqrt{h^2 - ab}}.$$

52. Show that the four straight lines given by the equations $(y - mx)^2 = c^2(1 + m^2)$ and $(y - nx)^2 = c^2(1 + n^2)$ form a rhombus.

[*Hints:* The four lines expressed separately are

$$y - mx = \pm\sqrt{1+m^2} \qquad (4.90)$$

and $\quad y - nx = \pm\sqrt{1+n^2} \qquad (4.91)$

The two lines represented by (4.90) are parallel and the distance between them = $2c$.

The two lines represented by (4.91) are also parallel and the distance between them = $2c$.

Since the lines (4.90) are not perpendicular to the lines (4.91), so the four lines form a rhombus.]

53. The line $ax + by + c = 0$ bisects an angle between a pair of lines of which one is $lx + my + n = 0$. Show that the other line of the pair is $(lx + my + n)(a^2 + b^2) - 2(al + bm)(ax + by + c) = 0$.

54. Show that one of the bisectors of the angles between the pair of straight lines $ax^2 + 2hxy + by^2 + 2gx + 2fy + c = 0$ will pass through the point of intersection of the two straight lines $ax^2 + 2hxy + by^2 + 2gx + 2fy + c = 0$ if $h(g^2 - f^2) = fg(a - b)$.

55. The straight line joining the origin to the common points of intersection of the curve $ax^2 + 2hxy + by^2 + 2gx + 2fy + c = 0$ and the variable straight line $lx + my = 1$ are at right angles. Find the locus of the foot of the perpendicular form the origin on the line $lx + my = 1$.

[*Hints:* Similar to **Example 36** § **4.7 Section B**. Replace $'2g'$ by $'g'$, $'2f'$ by $'f'$ and $'c'$ by $' - c'$.]

ANSWERS

Section A: 1. $k = \pm 1$; **3.** Yes; **4.** $\frac{\pi}{6}$; **5.** $k = -3, -7$; **9.** $\frac{1}{3}$ square units; **10.** $8x^2 + 16xy - 5y^2 = 0$.

Section B: 1. $\triangle = 0$; **2.** $\phi = \tan^{-1}\frac{2\sqrt{h^2-ab}}{a+b}$; **3.** $k = 1$; **4.** $\triangle = 0$; **5.** $\lambda = 3$; **6.** $\lambda = \pm 1$; **7.** $k = -3, -7$; **8. (i)** $x - 4y + 2 = 0$ and $x - y - 1 = 0$; **(ii)** $x - y - 1 = 0$ and $x + y + 2 = 0$; **(iii)** $7x - 5y + 3 = 0$ and $2x + 3y - 4 = 0$; **9.** $a + b = 0$; **10.** $\frac{1}{\sqrt{10}}$ units; **11.** $(-1, -1)$, $\tan^{-1}\frac{2}{11}$; **12.** $\tan^{-1}\frac{4}{3}$; **13.** $(1, 1)$, $\frac{3}{2}$ square units; **14.** $\frac{x^2 - y^2}{a - b} = \frac{xy}{h}$; **15. (i)** $3(x^2 - y^2) + 4xy = 0$; **(ii)** $3(x^2 - y^2) - 11xy = 0$; **(iii)** $7x^2 - 48xy - 7y^2 + 34x + 62y - 48 = 0$; **23.** $a^2 + ac + bd + d^2 = 0$; **26.** $15x^2 + xy - 6y^2 - 57x - 38y = 0$; **27.** $2x^2 - 5xy + 2y^2 = 0$; **33.** $\frac{\pi}{2}$; **34.** $\tan^{-1}\left[\frac{2p\sqrt{\{f^2 + g^2 - c - (g\cos\alpha + f\sin\alpha - p)^2\}}}{2p^2 + 2p(g\cos\alpha + f\sin\alpha) + c}\right]$; **38.** $a = 4$, $f = 15$; **41.** $(ab' - a'b)^2 = 4(ha' - h'a)(bh' - b'h)$; **43.** $b = 9$, $g = 2$, $c = 4$; **45. (i)** $\frac{g}{a} = \frac{f}{h} = \frac{c}{g}$; **(ii)** $\frac{a}{h} = \frac{f}{b} = \frac{c}{f}$; **50.** $2gx + 2fy + c = 0$, $(gh - fa)x + (bg - fh)y = 0$; **55.** $a + b + gl + fm = c(l^2 + m^2)$.

Chapter 5

Tangents and Normals, Pair of Tangents, Chord of Contact & Pole and Polar

5.1 Tangent to a Curve

Definition 5.1.1 *The tangent at any point P to a given curve (vide. **Figure 5.1**) is defined as the limiting position of the secant PQ, when such a limit exists, as the point Q approaches P along the curve.*

5.1.1 To Find the Equation of the Tangent to a Conic at a Given Point on the Conic

First Method:

Let the equation of the conic be

$$f(x, y) \equiv ax^2 + 2hxy + by^2 + 2gx + 2fy + c = 0 \tag{5.1}$$

We are to find the equation of the tangent to the conic (5.1) at the given point (x_1, y_1) lies on the conic.

The equation of any line through the point (x_1, y_1) is given by

$$\frac{x - x_1}{l} = \frac{y - y_1}{m} = r \tag{5.2}$$

Any point on this line is taken as $(x_1 + lr, y_1 + mr)$, where r is the distance of any point from the point (x_1, y_1).

209

For the points of intersection of (5.1) and (5.2), we have

$$a(x_1+lr)^2 + 2h(x_1+lr)(y_1+mr) + b(y_1+mr)^2$$
$$+2g(x_1+lr) + 2f(y_1+mr) + c = 0$$

or, $(al^2 + 2hlm + bm^2)r^2 + 2\{(ax_1 + hy_1 + g)l + (hx_1 + by_1 + f)m\}r$
$$+ ax_1^2 + 2hx_1y_1 + by_1^2 + 2gx_1 + 2fy_1 + c = 0 \quad (5.3)$$

This is a quadratic equation in r giving two values of r.

Since (x_1, y_1) is a point on the conic (5.1), so we have

$$ax_1^2 + 2hx_1y_1 + by_1^2 + 2gx_1 + 2fy_1 + c = 0.$$

So one value of r is zero.

For tangency the two points must coincide. So two values of r must vanish and for this, we must have sum of the roots of (5.3) is zero, i.e.,

$$(ax_1 + hy_1 + g)l + (hx_1 + by_1 + f)m = 0 \quad (5.4)$$

Now eliminating l, m form (5.2) and (5.4) we get

$$(ax_1 + hy_1 + g)(x - x_1) + (hx_1 + by_1 + f)(y - y_1) = 0$$
or, $axx_1 + h(xy_1 + yx_1) + byy_1 + gx + fy$
$$= ax_1^2 + 2hx_1y_1 + by_1^2 + gx_1 + fy_1.$$

Adding $gx_1 + fy_1 + c$ on both sides, we get

$$axx_1 + h(xy_1 + yx_1) + byy_1 + g(x + x_1) + f(y + y_1) + c$$
$$= ax_1^2 + 2hx_1y_1 + by_1^2 + 2gx_1 + 2fy_1 + c = 0.$$

Therefore the required equation of the tangent to the given conic (5.1) at the point (x_1, y_1) on it, is given by

$$axx_1 + h(xy_1 + yx_1) + byy_1 + g(x + x_1) + f(y + y_1) + c = 0 \quad (5.5)$$

Second Method:

Let $Q(x_2, y_2)$ be a point on the conic (5.1) close to the point $P(x_1, y_1)$ on it. The equation of the line through the points $P(x_1, y_1)$ and $Q(x_2, y_2)$ is given by

$$\frac{x - x_1}{x_2 - x_1} = \frac{y - y_1}{y_2 - y_1} \quad (5.6)$$

Tangents and Normals

Since (x_1, y_1) and (x_2, y_2) both lies on the conic (5.1), we have

$$ax_1^2 + 2hx_1y_1 + by_1^2 + 2gx_1 + 2fy_1 + c$$
$$= ax_2^2 + 2hx_2y_2 + by_2^2 + 2gx_2 + 2fy_2 + c$$

or, $a(x_2^2 - x_1^2) + 2h(x_2y_2 - x_1y_1) + b(y_2^2 - y_1^2)$
$$+ 2g(x_2 - x_1) + 2f(y_2 - y_1) = 0$$

or, $a(x_2 - x_1)(x_2 + x_1) + 2h\{x_2(y_2 - y_1) + y_1(x_2 - x_1)\}$
$$+ b(y_2 - y_1)(y_2 + y_1) + 2g(x_2 - x_1) + 2f(y_2 - y_1) = 0$$

or, $(x_2 - x_1)\{a(x_1 + x_2) + 2hy_1 + 2g\} + (y_2 - y_1)$
$$\times \{2hx_2 + b(y_1 + y_2) + 2f\} = 0$$

or, $\dfrac{y_2 - y_1}{x_2 - x_1} = -\dfrac{a(x_1 + x_2) + 2hy_1 + 2g}{2hx_2 + b(y_1 + y_2) + 2f}.$

Using equation (5.6) we get

or, $\dfrac{y_2 - y_1}{x_2 - x_1} = \dfrac{y - y_1}{x - x_1} = -\dfrac{a(x_1 + x_2) + 2hy_1 + 2g}{2hx_2 + b(y_1 + y_2) + 2f}$

or, $a(x - x_1)(x_1 + x_2) + 2hy_1(x - x_1) + 2g(x - x_1)$
$$+ 2hx_2(y - y_1) + b(y - y_1)(y_1 + y_2) + 2f(y - y_1) = 0 \quad (5.7)$$

Now let the point (x_2, y_2) gradually approaching towards (x_1, y_1) and ulti-

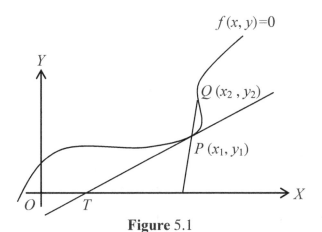

Figure 5.1

mately coincides with it. Then, in the limiting position $(x_2, y_2) \to (x_1, y_1)$, the equation (5.7) becomes the tangent at (x_1, y_1).

Hence putting $x_2 = x_1$ and $y_2 = y_1$ in the equation (5.7), we obtain

$$a(x - x_1).2x_1 + 2hy_1(x - x_1) + 2g(x - x_1) + 2hx_1(y - y_1)$$
$$+ b(y - y_1).2y_1 + 2f(y - y_1) = 0$$
$$axx_1 + h(xy_1 + yx_1) + byy_1 + g(x + x_1) + f(y + y_1) + c$$
$$= ax_1^2 + 2hx_1y_1 + by_1^2 + 2gx_1 + 2fy_1 + c = 0.$$

Therefore, the equation of the tangent at the point (x_1, y_1) to the conic (5.1) is obtained as

$$axx_1 + h(xy_1 + yx_1) + byy_1 + g(x + x_1) + f(y + y_1) + c = 0 \quad (5.8)$$

Note 5.1.1 *The formula for the equation of the tangent at the point (x_1, y_1) to a curve of the form (5.1) can be easily written by the following rule:*

The Rule: *In the equation of the conic (5.1), write*

$x.x_1$ *for* $x.x$, *i.e., for* x^2,

$y.y_1$ *for* $y.y$, *i.e., for* y^2,

$x + x_1$ *for* $x + x$, *i.e., for* $2x$,

$y + y_1$ *for* $y + y$, *i.e., for* $2y$,

$xy_1 + yx_1$ *for* $xy + yx$, *i.e., for* $2xy$,

and the constant term c remains unchanged.

Note 5.1.2 *The equation (5.8) of the tangent to the conic (5.1) may also be written in the form given below as*

$$(ax_1 + hy_1 + g)x + (hx_1 + by_1 + f)y_1 + (gx_1 + fy_1 + c) = 0.$$

This form sometimes becomes very useful in solving various problems.

5.1.2 Tangents of the Standard Equations (Conics)

The following tables, viz., **Table 5.1** and **Table 5.2** will give the equations of tangents at a glance in Cartesian and parametric form of equations of the standard conics respectively.

Table 5.1: **Tangents in Cartesian Form**

	Conics	Equations (in Cartesian Form)	Tangents at (x_1, y_1)
1.	General Circle:	$x^2 + y^2 + 2gx + 2fy + c = 0$	$xx_1 + yy_1 + g(x+x_1) + f(y+y_1) + c = 0$
2.	Circle with centre at origin:	$x^2 + y^2 = a^2$	$xx_1 + yy_1 = a^2$
3.	Parabola:	$y^2 = 4ax$	$yy_1 = 2a(x + x_1)$
4.	Ellipse:	$\frac{x^2}{a^2} + \frac{y^2}{b^2} = 1$	$\frac{xx_1}{a^2} + \frac{yy_1}{b^2} = 1$
5.	Hyperbola:	$\frac{x^2}{a^2} - \frac{y^2}{b^2} = 1$	$\frac{xx_1}{a^2} - \frac{yy_1}{b^2} = 1$

Table 5.2: **Tangents in Parametric Form**

	Conics	Parametric Co-ordinates	Tangents at 't' or 'θ'
1.	Circle: $x^2 + y^2 = a^2$	$(a\cos\theta, a\sin\theta)$	$x\cos\theta + y\sin\theta = a$
2.	Parabola: $y^2 = 4ax$	$(at^2, 2at)$	$y + tx = 2at + at^2$
3.	Ellipse: $\frac{x^2}{a^2} + \frac{y^2}{b^2} = 1$	$(a\cos\theta, b\sin\theta)$	$\frac{x}{a}\cos\theta + \frac{y}{b}\sin\theta = 1$
4.	Hyperbola: $\frac{x^2}{a^2} - \frac{y^2}{b^2} = 1$	$(a\sec\theta, b\tan\theta)$	$\frac{x}{a}\sec\theta - \frac{y}{b}\tan\theta = 1$

5.1.3 Condition for Tangency of a Straight Line to a Conic

Let us consider the conic

$$ax^2 + 2hxy + by^2 + 2gx + 2fy + c = 0 \qquad (5.9)$$

We are to find the condition that a straight line

$$lx + my + n = 0 \qquad (5.10)$$

is a tangent to the conic (5.9). The equation of the tangent to the conic (5.9) is given by

$$(ax_1 + hy_1 + g)x + (hx_1 + by_1 + f)y + (gx_1 + fy_1 + c) = 0 \qquad (5.11)$$

If the line (5.10) be a tangent to the conic (5.9), then (5.11) and (5.10) must be identical. Comparing the coefficients we get

$$\frac{ax_1 + hy_1 + g}{l} = \frac{hx_1 + by_1 + f}{m} = \frac{gx_1 + fy_1 + c}{n} = \lambda \text{ (say)}$$

from which we get

$$ax_1 + hy_1 + g - l\lambda = 0 \qquad (5.12)$$
$$hx_1 + by_1 + f - m\lambda = 0 \qquad (5.13)$$
$$\text{and } gx_1 + fy_1 + c - n\lambda = 0 \qquad (5.14)$$

Also (x_1, y_1) be a point on the line (5.10), so

$$lx_1 + my_1 + n = 0 \qquad (5.15)$$

Eliminating x_1, y_1 and λ form (5.12), (5.13), (5.14) and (5.15), we get

$$\begin{vmatrix} a & h & g & l \\ h & b & f & m \\ g & f & c & n \\ l & m & n & 0 \end{vmatrix} = 0$$

which on expansion becomes

$$Al^2 + Bm^2 + Cn^2 + 2Fmn + 2Gnl + 2Hlm = 0$$

where A, B, C, F, G, H are the respective cofactors of a, b, c, f, g, h in the determinant $\begin{vmatrix} a & h & g \\ h & b & f \\ g & f & c \end{vmatrix}$.

5.1.4 Condition of Tangency for Standard Conics of a Line

The following tables, viz., **Table 5.3** and **Table 5.4** will give the condition of tangency at a glance of the line $lx + my + n = 0$ to the standard conics.

5.1.5 Some Important Remarks

1. The equations of the tangents to the circle $x^2 + y^2 = a^2$ are $y = mx \pm a\sqrt{1+m^2}$. These lines are parallel and touch the given circle for all values of m except $m = 0$. The coordinates of the point of contact (x_1, y_1) are given by $\left(-\dfrac{a^2 m}{c}, \dfrac{a^2}{c}\right)$ where $c = a\sqrt{1+m^2}$, i.e., the point of contact is written as $\left(-\dfrac{a^2 m}{a\sqrt{1+m^2}}, \dfrac{a^2}{a\sqrt{1+m^2}}\right)$, i.e., $\left(-\dfrac{am}{\sqrt{1+m^2}}, \dfrac{a}{\sqrt{1+m^2}}\right)$.

Similarly, it can be shown that each of the lines $y - k = m(x - h) \pm a\sqrt{1+m^2}$ is always a tangent to the circle $(x-h)^2 + (y-k)^2 = a^2$, whatever

Tangents and Normals

Table 5.3: Condition of Tangency of the Straight Line $lx+my+n=0$

	Conics	Equations (in Cartesian Form)	Condition of Tangency
1.	General Circle:	$x^2+y^2+2gx+2fy+c=0$	$(lg+mf-n)^2 = (l^2+m^2)(g^2+f^2-c)$
2.	Circle with centre at origin:	$x^2+y^2=a^2$	$a^2(l^2+m^2)=n^2$ or $n=\pm a\sqrt{l^2+m^2}$
3.	Parabola:	$y^2=4ax$	$am^2=nl$
4.	Ellipse:	$\frac{x^2}{a^2}+\frac{y^2}{b^2}=1$	$a^2l^2+b^2m^2=n^2$ or $n=\pm\sqrt{a^2l^2+b^2m^2}$
5.	Hyperbola:	$\frac{x^2}{a^2}-\frac{y^2}{b^2}=1$	$a^2l^2-b^2m^2=n^2$ or $n=\pm\sqrt{a^2l^2-b^2m^2}$

Table 5.4: Condition of Tangency of the Straight Line $y=mx+c$

	Conics	Equations (in Cartesian Form)	Condition of Tangency
1.	General Circle:	$x^2+y^2+2gx+2fy+c=0$	$(mg-f-c)^2=(1+m^2)(g^2+f^2-c)$
2.	Circle with centre at origin:	$x^2+y^2=a^2$	$c=\pm a\sqrt{1+m^2}$
3.	Parabola:	$y^2=4ax$	$c=\frac{a}{m}$
4.	Ellipse:	$\frac{x^2}{a^2}+\frac{y^2}{b^2}=1$	$c=\pm\sqrt{a^2m^2+b^2}$
5.	Hyperbola:	$\frac{x^2}{a^2}-\frac{y^2}{b^2}=1$	$c=\pm\sqrt{a^2m^2-b^2}$

the value of m may be.

2. The line $y=mx+\frac{a}{m}$ is always a tangent to the parabola $y^2=4ax$ for all values of m except $m=0$ and the point of contact is given by $\left(\frac{c}{m},\frac{2a}{m}\right)$, i.e., $\left(\frac{a}{m^2},\frac{2a}{m}\right)$ since $c=\frac{a}{m}$.

3. The parallel lines $y=mx\pm\sqrt{a^2m^2+b^2}$ are always tangents to the ellipse $\frac{x^2}{a^2}+\frac{y^2}{b^2}=1$ for all values of m. The coordinates of the points of contact are $\left(-\frac{ma^2}{c},\frac{b^2}{c}\right)$, i.e., $\left(\mp\frac{a^2m}{\sqrt{a^2m^2+b^2}},\pm\frac{b^2}{\sqrt{a^2m^2+b^2}}\right)$ as $c=\pm\sqrt{a^2m^2+b^2}$.

4. The parallel lines $y = mx \pm \sqrt{a^2m^2 - b^2}$ are always tangents to the hyperbola $\dfrac{x^2}{a^2} - \dfrac{y^2}{b^2} = 1$ for all values of m. The coordinates of the points of contact are $\left(-\dfrac{ma^2}{c}, -\dfrac{b^2}{c}\right)$, i.e., $\left(\mp\dfrac{a^2m}{\sqrt{a^2m^2-b^2}}, \pm\dfrac{b^2}{\sqrt{a^2m^2-b^2}}\right)$ since $c = \pm\sqrt{a^2m^2 - b^2}$.

5.2 Normals

Definition 5.2.1 *The normal at any point of a conic is the straight line through that point drawn perpendicular to the tangent at that point.*

5.2.1 Equation of a Normal of a Conic at a Given Point

Let us consider the conic

$$ax^2 + 2hxy + by^2 + 2gx + 2fy + c = 0 \tag{5.16}$$

We have to find the equation of the normal to the conic (5.16) at the given point (x_1, y_1).

The equation of the tangent to the conic (5.16) at the point (x_1, y_1) is

$$(ax_1 + hy_1 + g)x + (hx_1 + by_1 + f)y + (gx_1 + fy_1 + c) = 0 \tag{5.17}$$

The equation of any line perpendicular to (5.17) is taken as

$$(hx_1 + by_1 + f)x - (ax_1 + hy_1 + g)y + k = 0 \tag{5.18}$$

If it passes through (x_1, y_1), then

$$(hx_1 + by_1 + f)x_1 - (ax_1 + hy_1 + g)y_1 + k = 0 \tag{5.19}$$

Subtracting (5.19) form (5.18) we get

$$(hx_1 + by_1 + f)(x - x_1) - (ax_1 + hy_1 + g)(y - y_1) = 0$$

$$\text{or,} \quad \dfrac{x - x_1}{ax_1 + hy_1 + g} = \dfrac{y - y_1}{hx_1 + by_1 + f} \tag{5.20}$$

This is the required equation of the normal.

Tangents and Normals

Table 5.5: Normals in Cartesian Form

	Conics	Equations (in Cartesian Form)	Normals at (x_1, y_1)
1.	General Circle:	$x^2+y^2+2gx+2fy+c=0$	$\frac{x-x_1}{x_1+g} = \frac{y-y_1}{y_1+f}$
2.	Circle with centre at origin:	$x^2+y^2=a^2$	$\frac{x}{x_1} = \frac{y}{y_1}$ or $xy_1 - yx_1 = 0$
3.	Parabola:	$y^2 = 4ax$	$y - y_1 = -\frac{y_1}{2a}(x-x_1)$
4.	Ellipse:	$\frac{x^2}{a^2} + \frac{y^2}{b^2} = 1$	$\frac{xx_1}{\frac{x_1^2}{a^2}} = \frac{yy_1}{\frac{y_1^2}{b^2}}$
5.	Hyperbola:	$\frac{x^2}{a^2} - \frac{y^2}{b^2} = 1$	$\frac{xx_1}{\frac{x_1^2}{a^2}} = -\frac{yy_1}{\frac{y_1^2}{b^2}}$

Table 5.6: Normals in Parametric Form

	Conics	Parametric Co-ordinates	Normals at 't' or 'θ'
1.	Circle: $x^2+y^2=a^2$	$(a\cos\theta, a\sin\theta)$	$\frac{x}{\cos\theta} = \frac{y}{\sin\theta}$ i.e., $x\sin\theta - y\cos\theta = 0$
2.	Parabola: $y^2 = 4ax$	$(at^2, 2at)$	$y + tx = 2at + at^3$
3.	Ellipse: $\frac{x^2}{a^2} + \frac{y^2}{b^2} = 1$	$(a\cos\theta, b\sin\theta)$	$\frac{ax}{\cos\theta} - \frac{by}{\sin\theta} = a^2 - b^2$
4.	Hyperbola: $\frac{x^2}{a^2} - \frac{y^2}{b^2} = 1$	$(a\sec\theta, b\tan\theta)$	$\frac{ax}{\sec\theta} + \frac{by}{\tan\theta} = a^2 + b^2$

5.2.2 Normals of the Standard Equations of Conics

The following tables, viz., **Table 5.5** and **Table 5.6** will give the equations of normals at a glance in Cartesian and parametric form of equations of the standard conics respectively.

Note 5.2.1 *Slope form of the equation of tangent to the parabola* $y^2 = 4ax$ *at* $P(at^2, 2at)$

Equation of tangent at $P(x_1, y_1)$ to the parabola $y^2 = 4ax$ is

$$yy_1 = 2a(x+x_1) \tag{5.21}$$

In (5.21) replacing y_1 by $2at$ and x_1 by at^2 we get

$$2aty = 2a(x+at^2) \Rightarrow ty = x + at^2 \tag{5.22}$$

This is the equation of the tangent at $P(t)$ or $P(at^2, 2at)$.

In (5.22) the slope of the tangent is $m = \dfrac{1}{t}$, so in (5.22) replacing t by $\dfrac{1}{m}$ we get

$$y = mx + \frac{a}{m} \qquad (5.23)$$

This is the equation of the tangent in terms of slope, i.e., the slope form of the equation of the tangent.

This is the tangent at point $\left(\dfrac{a}{m^2}, \dfrac{2a}{m}\right)$.

If the line $y = mx + c$ touches the parabola $y^2 = 4ax$, we must have $c = \dfrac{a}{m}$ (Comparing with (5.23)).

Note 5.2.2 The equation of the tangent to the parabola $(y-k)^2 = 4a(x-h)$ having slope m is $y - k = m(x - h) + \dfrac{a}{m}$.

Note 5.2.3 *Slope form of the equation of the normal to the parabola $y^2 = 4ax$ at $P(at^2, 2at)$*

Equation of the normal at $P(x_1, y_1)$ to the parabola $y^2 = 4ax$ is

$$y - y_1 = -\frac{y_1}{2a}(x - x_1) \qquad (5.24)$$

Parametric Form: Replacing x_1 by at^2 and y_1 by $2at$ in (5.24), we get

$$y - 2at = -\frac{2at}{2a}(x - at^2) \text{ or, } y - 2at = -t(x - at^2)$$

$$\text{or, } y = -tx + 2at + at^3 \qquad (5.25)$$

Slope Form: In (5.25), the slope of the normal is $m = -t$. So in (5.25), replacing t by $-m$, we get

$$y = mx - 2am - am^3 \qquad (5.26)$$

which is the equation of the normal to the parabola $y^2 = 4ax$, where m is the slope of the normal.

The coordinates of the foot of normal are $(am^2, -2am)$.

Comparing (5.26) with $y = mx + c$, we get $c = -2am - am^3$ which is the condition that $y = mx + c$ is the normal of $y^2 = 4ax$.

Note 5.2.4 *Co-normal Points:* The points on a curve at which the normals pass through a fixed point are known as **Co-normal Points**.

Tangents and Normals

5.3 Pair of Tangents

Theorem 5.3.1 *A pair of tangents can be drawn from a point not lying on the conic.*

Proof: Let (α, β) be a point not lying on the conic

$$ax^2 + 2hxy + by^2 + 2gx + 2fy + c = 0 \tag{5.27}$$

Let (x_1, y_1) be a point on the conic the tangent at which passes through (α, β). The equation of the tangent at (x_1, y_1) is

$$axx_1 + h(xy_1 + x_1y) + byy_1 + g(x + x_1) + f(y + y_1) + c = 0 \tag{5.28}$$

Since (5.28) passes through (α, β), we get

$$a\alpha x_1 + h(\alpha y_1 + x_1\beta) + b\beta y_1 + g(\alpha + x_1) + f(\beta + y_1) + c = 0 \tag{5.29}$$

and since (x_1, y_1) lies on the conic, so

$$ax_1^2 + 2hx_1y_1 + by_1^2 + 2gx_1 + 2fy_1 + c = 0 \tag{5.30}$$

Solving (5.29) and (5.30) generally two values of x_1 and y_1 are obtained. Thus there are two points of contact of tangents from (α, β) but they may not be real in all cases.

Hence, in general two tangents can be drawn from a point not lying on the conic. □

5.3.1 Equations of the chord of contact of tangents to a conic from any point outside it

Let $Q(x_2, y_2)$ and $R(x_3, y_3)$ be the points of contact of the tangents drawn from an external point $P(x_1, y_1)$ to the conic

$$ax^2 + 2hxy + by^2 + 2gx + 2fy + c = 0 \tag{5.31}$$

The equation of the tangent at (x_2, y_2) to (5.31) is

$$axx_2 + h(xy_2 + x_2y) + byy_2 + g(x + x_2) + f(y + y_2) + c = 0 \tag{5.32}$$

Since it passes through (x_1, y_1), so

$$ax_1x_2 + h(x_1y_2 + x_2y_1) + by_1y_2 + g(x_1 + x_2) + f(y_1 + y_2) + c = 0 \tag{5.33}$$

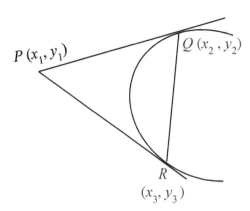

Figure 5.2

Similarly the tangent at (x_3, y_3) passes through (x_1, y_1) and we get

$$ax_1x_3 + h(x_1y_3 + x_3y_1) + by_1y_3 + g(x_1 + x_3) + f(y_1 + y_3) + c = 0 \quad (5.34)$$

These relations (5.33) and (5.34) indicates that the points $Q(x_2, y_2)$ and $R(x_3, y_3)$ lies on the straight line

$$axx_1 + h(xy_1 + x_1y) + byy_1 + g(x + x_1) + f(y + y_1) + c = 0 \quad (5.35)$$

This is, therefore, the equation of the chord of contact of tangents from the outside point $P(x_1, y_1)$.

Note 5.3.1 *The equation (5.35) is identical with the equation of the tangent at (x_1, y_1) on the conic (5.31), but in this case (x_1, y_1) does not lie on the conic.*

5.3.2 Equation of the pair of tangents from an external point to a conic

Let $P(x_1, y_1)$ be an external point to the conic

$$ax^2 + 2hxy + by^2 + 2gx + 2fy + c = 0 \quad (5.36)$$

The equation of any line through the point (x_1, y_1) is

$$\frac{x - x_1}{l} = \frac{y - y_1}{m} = r \text{ (say)} \quad (5.37)$$

where $l^2 + m^2 = 1$.

Tangents and Normals

Coordinates of any point on the line (5.37) are $(x_1 + lr, y_1 + mr)$, where r is the distance from (x_1, y_1).

If the point $(x_1 + lr, y_1 + mr)$ lies on (5.36), then

$$a(x_1 + lr)^2 + 2h(x_1 + lr)(y_1 + mr) + b(y_1 + mr)^2$$
$$+ 2g(x_1 + lr) + 2f(y_1 + mr) + c = 0$$

or, $\quad (al^2 + 2hlm + bm^2)r^2 + 2\{(ax_1 + hy_1 + g)l + (hx_1 + by_1 + f)m\}r$
$$+ ax_1^2 + 2hx_1y_1 + by_1^2 + 2gx_1 + 2fy_1 + c = 0 \qquad (5.38)$$

If the line (5.37) be a tangent to the conic (5.36), the two points of intersection will coincide and therefore both the roots of the above equation must be equal, for which the discriminant of (5.38) must vanishes giving

$$\{l(ax_1 + hy_1 + g) + m(hx_1 + by_1 + f)\}^2 = (al^2 + 2hlm + bm^2)$$
$$(ax_1^2 + 2hx_1y_1 + by_1^2 + 2gx_1 + 2fy_1 + c)$$
$$= (al^2 + 2hlm + bm^2)S_1 \qquad (5.39)$$

where $S_1 = ax_1^2 + 2hx_1y_1 + by_1^2 + 2gx_1 + 2fy_1 + c$.

Eliminating l, m, n between (5.37) and (5.39) we get

$$\{(ax_1 + hy_1 + g)(x - x_1) + (hx_1 + by_1 + f)(y - y_1)\}^2$$
$$= \{a(x - x_1)^2 + 2h(x - x_1)(y - y_1) + b(y - y_1)^2\}S_1$$

or, $\quad \{axx_1 + h(xy_1 + yx_1) + byy_1 + g(x + x_1) + f(y + y_1) + c$
$- (ax_1^2 + 2hx_1y_1 + by_1^2 + 2gx_1 + 2fy_1 + c)\}^2 = [ax^2 + 2hxy + by^2$
$+ 2gx + 2fy + c + ax_1^2 + 2hx_1y_1 + by_1^2 + 2gx_1 + 2fy_1 + c$
$- 2\{axx_1 + h(xy_1 + yx_1) + byy_1 + g(x + x_1) + f(y + y_1) + c\}]S_1$

or, $\quad (T - S_1)^2 = (S + S_1 - 2T)S_1$

where $\quad S = ax^2 + 2hxy + by^2 + 2gx + 2fy + c$
$\quad\quad\quad S_1 = ax_1^2 + 2hx_1y_1 + by_1^2 + 2gx_1 + 2fy_1 + c$
and $\quad T = axx_1 + h(xy_1 + yx_1) + byy_1 + g(x + x_1) + f(y + y_1) + c$.

So the equation of the pair of tangents is obtained as

$$T^2 - 2TS_1 + S_1^2 = SS_1 + S_1^2 - 2TS_1$$
or, $\quad SS_1 = T^2$.

5.4 Director Circle

Definition 5.4.1 *If the locus of the points of intersection of pair of perpendicular tangents to a conic is a circle, then this circle is called the **director circle** of the conic.*

5.4.1 Equation of the director circle of a conic

Let the equation of the conic be

$$S \equiv ax^2 + 2hxy + by^2 + 2gx + 2fy + c = 0 \qquad (5.40)$$

The equation of the pair of tangents from (x_1, y_1) to the conic (5.40) is

$$SS_1 = T^2$$

i.e., $(ax^2 + 2hxy + by^2 + 2gx + 2fy + c)$
$\times (ax_1^2 + 2hx_1y_1 + by_1^2 + 2gx_1 + 2fy_1 + c)$
$= \{axx_1 + h(xy_1 + yx_1) + byy_1 + g(x + x_1) + f(y + y_1) + c\}^2$
$= \{(ax_1 + hy_1 + g)x + (hx_1 + by_1 + f)y + (gx_1 + fy_1 + c)\}^2.$

If these tangents are at right angles to each other, then the sum of the coefficients of x^2 and y^2 is zero.

i.e., $(a + b)(ax_1^2 + 2hx_1y_1 + by_1^2 + 2gx_1 + 2fy_1 + c)$
$\qquad -(ax_1 + hy_1 + g)^2 - (hx_1 + by_1 + f)^2 = 0$

or, $(ab - h^2)(x_1^2 + y_1^2) + 2(bg - hf)x_1 + 2(af - gh)y_1$
$\qquad + ab + bc - g^2 - f^2 = 0$

or, $(ab - h^2)(x_1^2 + y_1^2) - 2(hf - bg)x_1 - 2(gh - af)y_1$
$\qquad + (bc - f^2) + (ac - g^2) = 0$

or, $C(x_1^2 + y_1^2) - 2Gx_1 + 2Fy_1 + A + B = 0$

where A, B, C, F, G are the respective cofactors of a, b, c, f, g in the determinant $\begin{vmatrix} a & h & g \\ h & b & f \\ g & f & c \end{vmatrix}$.

Hence the locus of (x_1, y_1) is

$$C(x^2 + y^2) - 2Gx + 2Fy + A + B = 0.$$

This is the required equation of the director circle.

Tangents and Normals

Particular Case for Circle $x^2 + y^2 = a^2$.

The equation of the pair of tangents form (x_1, y_1) to the circle is

$$(x^2 + y^2 - a^2)(x_1^2 + y_1^2 - a^2) = (xx_1 + yy_1 - a^2)^2.$$

If these lines are at right angles to each other, then

the coefficient of x^2 + the coefficient of $y^2 = 0$

or, $(x_1^2 + y_1^2 - x_1^2 - a^2) + (x_1^2 + y_1^2 - y_1^2 - a^2) = 0$

or, $x_1^2 + y_1^2 = 2a^2$.

Hence the locus of (x_1, y_1) is $x^2 + y^2 = 2a^2$ which is the required equation of the director circle.

5.4.2 Equations of Director Circles of Standard Conics

The following **Table 5.7** gives the equations of the director circles in some special cases.

Table 5.7: Director Circles for Conics

	Conics	Equations	Director Circles
1.	General Circle:	$x^2 + y^2 + 2gx + 2fy + c = 0$	$x^2 + y^2 + 2gx + 2fy - (g^2 + f^2 - 2c) = 0$
2.	Circle with centre at origin:	$x^2 + y^2 = a^2$	$x^2 + y^2 = 2a^2$
3.	Parabola:	$y^2 = 4ax$	$x + a = 0$, which is the directrix of the parabola.
4.	Ellipse:	$\frac{x^2}{a^2} + \frac{y^2}{b^2} = 1$	$x^2 + y^2 = a^2 + b^2$
5.	Hyperbola:	$\frac{x^2}{a^2} - \frac{y^2}{b^2} = 1$	$x^2 + y^2 = a^2 - b^2$
6.	General Conic:	$ax^2 + 2hxy + by^2 + 2gx + 2fy + c = 0$	$(h^2 - ab)(x^2 + y^2) + 2(hf - bg)x + 2(gh - af)y + g^2 + f^2 - ac - bc = 0$

5.5 Chords in Terms of Middle Point

Let the equation of the conic be

$$S \equiv ax^2 + 2hxy + by^2 + 2gx + 2fy + c = 0 \tag{5.41}$$

Let $P(x_1, y_1)$ be the middle point of any chord QR to the conic (5.41). The equation of any line through (x_1, y_1) is given by

$$\frac{x - x_1}{r} = \frac{y - y_1}{m} = r \text{ (say)} \tag{5.42}$$

So any point on the line may be taken as $(x_1 + lr, y_1 + mr)$. If this point lies on the conic (5.41), then

$$a(x_1 + lr)^2 + 2h(x_1 + lr)(y_1 + mr) + b(y_1 + mr)^2$$
$$+ 2g(x_1 + lr) + 2f(y_1 + mr) + c = 0$$
or, $(al^2 + 2hlm + bm^2)r^2 + 2\{(ax_1 + hy_1 + g)l + (hx_1 + by_1 + f)m\}r$
$$+ ax_1^2 + 2hx_1y_1 + by_1^2 + 2gx_1 + 2fy_1 + c = 0 \tag{5.43}$$

This is a quadratic equation in r and it has two roots, say r_1 and r_2. Since (x_1, y_1) is a middle point of the chord, so we get

$$r_1 + r_2 = 0 \Rightarrow (ax_1 + hy_1 + g)l + (hx_1 + by_1 + f)m = 0 \tag{5.44}$$

Now eliminating l, m between (5.42) and (5.44) we get

$$(ax_1 + hy_1 + g)(x - x_1) + (hx_1 + by_1 + f)(y - y_1) = 0$$
or, $axx_1 + h(xy_1 + yx_1) + byy_1 + g(x + x_1) + f(y + y_1) + c$
$$= ax_1^2 + 2hx_1y_1 + by_1^2 + 2gx_1 + 2fy_1 + c \tag{5.45}$$

which is the required equation of the chord in terms of middle point (x_1, y_1).

Note 5.5.1 *The above equation (5.45) can be expressed in the form $T = S_1$ where $T = 0$ is the equation of the tangent to the conic (5.41) at the point (x_1, y_1) and $S_1 = S(x_1, y_1)$.*

Note 5.5.2 *In case of a circle, the chord bisected at the point (x_1, y_1), i.e., the chord in terms of the middle point (x_1, y_1) is the furthest from the centre among all the chords passing through it. Also, for such chord, the length of the chord is minimum.*

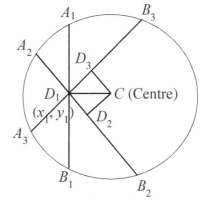

Figure 5.3

Tangents and Normals

In **Figure 5.3**, A_1B_1, A_2B_2, A_3B_3 are chords of which $D_1(x_1, y_1)$ is the middle point of only the chord A_1B_1, and then it is clear form the figure that of the diameters CD_1, CD_2 and CD_3 form the centre, CD_1 is the largest.

Also among the chords A_1B_1, A_2B_2 and A_3B_3 drawn through $D_1(x_1, y_1)$, the length of A_1B_1 is minimum.

5.6 Diameter of a Conic

Definition 5.6.1 *The locus of the middle points of a systems of parallel chords to a conic is known as a **diameter** of the conic with respect to that system of parallel chords.*

5.6.1 Equation of a Diameter

Let a system of parallel chords parallel to a straight line
$$\frac{x}{l} = \frac{y}{m} \tag{5.46}$$
be considered and let (x_1, y_1) be the middle point of one of such chords to the conic
$$ax^2 + 2hxy + by^2 + 2gx + 2fy + c = 0 \tag{5.47}$$
Then as in the previous article (§ **5.5**) we get
$$(ax_1 + hy_1 + g)l + (hx_1 + by_1 + f)m = 0$$
$$\text{or,} \quad (al + hm)x_1 + (hl + bm)y_1 + (gl + fm) = 0.$$

Therefore the locus of the middle point will be
$$(al + hm)x + (hl + bm)y + (gl + fm) = 0 \tag{5.48}$$
which is the diameter of the conic with respect to the system of chords parallel to the line $\frac{x}{l} = \frac{y}{m}$.

5.6.2 Conjugate Diameter

Definition 5.6.2 *Two diameters of a central conic are said to be **Conjugate diameters** of the conic, if each contains the middle points of the system of chords parallel to the other.*

Thus the major and minor axes of an ellipse are conjugate diameters of the ellipse, because each contains the middle points of the chord parallel to the other.

5.7 Pole and Polar

Definition 5.7.1 *The **polar** of a point with respect to a conic is the locus of the points of intersection of tangents at the extremities of the chords through that point, the point itself is called the **pole** of its polar.*

5.7.1 Equation to the polar of a point with respect to a non-singular conic

Let us consider the conic be

$$S \equiv ax^2 + 2hxy + by^2 + 2gx + 2fy + c = 0 \qquad (5.49)$$

and $P(x_1, y_1)$ be the given point.

Let a secant QR of S through $P(x_1, y_1)$ intersect the conic at $Q(x_2, y_2)$ and $R(x_3, y_3)$ (see **Figure 5.4**). The equation of the tangents at Q and R are given by

$$(ax_2 + hy_2 + g)x + (hx_2 + by_2 + f)y + (gx_2 + fy_2 + c) = 0 \qquad (5.50)$$
$$(ax_3 + hy_3 + g)x + (hx_3 + by_3 + f)y + (gx_3 + fy_3 + c) = 0 \qquad (5.51)$$

respectively. If QR does not pass through the centre of the conic (if it is a

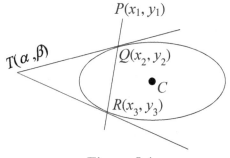

Figure 5.4

central conic), then the lines (5.50) and (5.51) will intersect. Let the point of intersection is denoted by $T(\alpha, \beta)$.

Let U be the set of all points like T corresponding to all secants through P. Now we shall show that the points of U lie on a straight line.

We observe that (5.50) and (5.51) pass through $T(\alpha, \beta)$, so

$$(ax_2 + hy_2 + g)\alpha + (hx_2 + by_2 + f)\beta + (gx_2 + fy_2 + c) = 0 \qquad (5.52)$$
$$(ax_3 + hy_3 + g)\alpha + (hx_3 + by_3 + f)\beta + (gx_3 + fy_3 + c) = 0 \qquad (5.53)$$

Tangents and Normals

From (5.52) and (5.53), it follows that the point $Q(x_2, y_2)$ and $R(x_3, y_3)$ lie on the line

$$(ax + hy + g)\alpha + (hx + by + f)\beta + (gx + fy + c) = 0 \quad (5.54)$$

Hence the equation of the line QR is given by (5.54). Since (x_1, y_1) is a point on (5.54), so

$$(ax_1 + hy_1 + g)\alpha + (hx_1 + by_1 + f)\beta + (gx_1 + fy_1 + c) = 0$$

which means that (α, β) lies on

$$(ax_1 + hy_1 + g)x + (hx_1 + by_1 + f)y + (gx_1 + fy_1 + c) = 0 \quad (5.55)$$

This represents a straight line which indicates that the points of the set U lie on the fixed straight line (5.55).

The straight line (5.55) is called the **polar** of P with respect to the conic S and the point P is called the **pole** of this line.

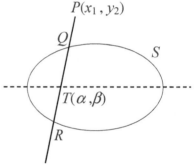

Figure 5.5

Alternative Definition

Definition 5.7.2 *Let P be a point and S be a conic. From P, a line PQR is drawn to meet the conic at Q and R. Let T be a point on QR which is the harmonic conjugate of the point P with respect to the points Q and R, i.e.,*

$$\frac{2}{PT} = \frac{1}{PQ} + \frac{1}{PR},$$

*then the locus of T is called the **polar** of the point P with respect to the conic S and the point P is called the **pole** of the polar.*

Note 5.7.1 Meaning of Harmonic Conjugate

*If P, Q, R be three points of a line g, then another point T of g different from them is said to be the **harmonic conjugate** of P with respect to Q and R, if the value of the segment PT is the harmonic mean of the values of the segments $\overline{PQ}$ and $\overline{PR}$, i.e.,*

$$\frac{1}{PQ} + \frac{1}{PR} = \frac{2}{PT}.$$

5.7.2 Deduction of the Equation of Polar from the Alternative Definition

Let the coordinates of the point T which is the harmonic conjugate of P with respect to Q and R be (α, β). Equation of any line through $P(x_1, y_1)$ is given by

$$\frac{x - x_1}{l} = \frac{y - y_1}{m} = r \text{ (say)} \tag{5.56}$$

Any point on (5.56) is taken as $(x_1 + lr, y_1 + mr)$. If this point lies on the conic

$$S \equiv ax^2 + 2hxy + by^2 + 2gx + 2fy + c = 0 \tag{5.57}$$

then we get

$$a(x_1 + lr)^2 + 2h(x_1 + lr)(y_1 + mr) + b(y_1 + mr)^2$$
$$+ 2g(x_1 + lr) + 2f(y_1 + mr) + c = 0$$

or, $(al^2 + 2hlm + bm^2)r^2 + 2\{(ax_1 + hy_1 + g)l + (hx_1 + by_1 + f)m\}r$
$$+ ax_1^2 + 2hx_1y_1 + by_1^2 + 2gx_1 + 2fy_1 + c = 0 \tag{5.58}$$

which is a quadratic equation in r. If r_1 and r_2 be its roots then we get

$$r_1 + r_2 = -\frac{2\{(ax_1 + hy_1 + g)l + (hx_1 + by_1 + f)m\}}{al^2 + 2hlm + bm^2}$$

and $\quad r_1 r_2 = \dfrac{ax_1^2 + 2hx_1y_1 + by_1^2 + 2gx_1 + 2fy_1 + c}{al^2 + 2hlm + bm^2}.$

Let us take $PQ = r_1$, $PR = r_2$ and $PT = \rho$, then

$$\frac{\alpha - x_1}{l} = \frac{\beta - y_1}{m} = \rho, \text{ i.e., } l\rho = \alpha - x_1 \text{ and } m\rho = \beta - y_1.$$

Now T is the harmonic conjugate of P with respect to P and Q

$$\therefore \quad \frac{2}{PT} = \frac{1}{PQ} + \frac{1}{PR} \text{ or, } \frac{2}{\rho} = \frac{1}{r_1} + \frac{1}{r_2} = \frac{r_1 + r_2}{r_1 r_2}$$

$$\therefore \quad \frac{2}{\rho} = -\frac{2\{(ax_1 + hy_1 + g)l + (hx_1 + by_1 + f)m\}}{ax_1^2 + 2hx_1y_1 + by_1^2 + 2gx_1 + 2fy_1 + c}$$

or, $(ax_1 + hy_1 + g)l\rho + (hx_1 + by_1 + f)m\rho$
$$= -(ax_1^2 + 2hx_1y_1 + by_1^2 + 2gx_1 + 2fy_1 + c).$$

Tangents and Normals

Putting the values of $l\rho$ and $m\rho$ we get

$$(ax_1 + hy_1 + g)(\alpha - x_1) + (hx_1 + by_1 + f)(\beta - y_1)$$
$$= -(ax_1^2 + 2hx_1y_1 + by_1^2 + 2gx_1 + 2fy_1 + c)$$
$$\text{or,} \quad a\alpha x_1 + h(\alpha y_1 + \beta x_1) + b\beta y_1 + g(x_1 + \alpha) + f(y_1 + \beta) + c = 0$$

Hence the locus of (α, β) will be

$$axx_1 + h(xy_1 + yx_1) + byy_1 + g(x_1 + x) + f(y_1 + y) + c = 0 \quad (5.59)$$

which is the required equation of the polar of the point (x_1, y_1) with respect to the given conic, which can also be written as

$$(ax_1 + hy_1 + g)x + (hx_1 + by_1 + f)y + (gx_1 + fy_1 + c) = 0$$

and thus equation is same as deduced earlier.

Remarks:

1. The *polar of a point* with respect to a conic *coincides with the chord of contact of tangents form the point* to the conic *when the point does not lie on the conic*.

2. If (x_1, y_1) is a point *on the conic*, its polar *coincides with the tangent* to the conic at (x_1, y_1).

3. If (x_1, y_1) lies *within the conic*, its polar is the line through the points of contact of the *two imaginary tangents to the conic* form (x_1, y_1).

4. If P be the mid point of the chord QR, then r_1, r_2 are equal in magnitude but opposite in sign. In such case

$$\frac{2}{\rho} = \frac{1}{r_1} + \frac{1}{r_2} = \frac{1}{r_1} - \frac{1}{r_1} = 0 \Rightarrow \rho = \infty.$$

This indicates that the harmonic conjugate point of P is for away from the conic. Therefore, there will be no real polar of a point which is mid point of all chords through it. Hence for **a central conic the centre has no real polar**.

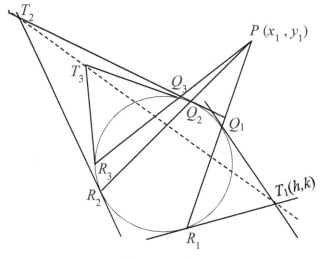

Figure 5.6

5.7.3 Particular Cases

1. Equation of the polar of a given point with respect to the circle

The equation of the given circle be

$$x^2 + y^2 + 2gx + 2fy + c = 0 \tag{5.60}$$

Let PQ_1R_1 be any chord drawn through the given exterior point $P(x_1, y_1)$ with respect to the circle (*vide.* **Figure 5.6**). If the point $P(x_1, y_1)$ lies interior of the circle then the chord will be Q_1PR_1 (*vide.* **Figure 5.7**). Let the tangents at Q_1 and R_1 meet in the point $T_1(h, k)$. Therefore, PQ_1R_1 is the chord of contact of the tangents drawn for the point $T_1(h, k)$. Hence its equation is

$$xh + yk + g(x + h) + f(y + k) + c = 0.$$

This line passes through (x_1, y_1).

$$\therefore \quad x_1 h + y_1 k + g(x_1 + h) + f(y_1 + k) + c = 0.$$

Hence the locus of (h, k) is given by

$$xx_1 + yy_1 + g(x + x_1) + f(y + y_1) + c = 0 \tag{5.61}$$

which is the required equation of the polar. The point (x_1, y_1) is the pole of the polar (5.61).

Tangents and Normals

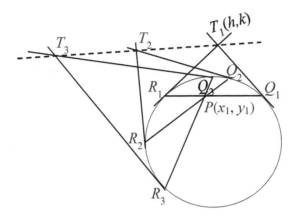

Figure 5.7

Corollary 5.7.1 *The polar of the point (x_1, y_1) with respect to the circle $x^2 + y^2 = a^2$ is $xx_1 + yy_1 = a^2$.*

Corollary 5.7.2 *The line joining the point (x_1, y_1) and the centre of the given circle (5.60) is*

$$y - y_1 = \frac{y_1 + f}{x_1 + g}(x - x_1).$$

So the gradient of this line $= \dfrac{y_1 + f}{x_1 + g}.$

And the gradient of the polar (5.61) is $-\dfrac{x_1 + g}{y_1 + f}.$

[Since (5.61) can be written as $(x_1 + g)x + (y_1 + f)y + gx_1 + fy_1 + c = 0$.]
Therefore, the product of the two gradients $= -1$.

Hence the polar of a point (x_1, y_1) with respect to a circle is perpendicular to the line joining the point to the centre of the circle.

2. Polar of a given point with respect to a parabola

The equation of the parabola be

$$y^2 = 4ax \tag{5.62}$$

Let the tangents at the extremities of a chord drawn through the point (x_1, y_1) intersect at (h, k).

The equation of the chord of contact of the tangents form (h, k) to the parabola (5.62) is $yk = 2a(x + h)$.

Thus the locus of the point (h, k) is $yy_1 = 2a(x + x_1)$, which is the required equation of the polar of the point (x_1, y_1) with respect to the parabola $y^2 = 4ax$.

3. Polar of a given point with respect to an ellipse

The equation of the ellipse be

$$\frac{x^2}{a^2} + \frac{y^2}{b^2} = 1 \tag{5.63}$$

Let the tangents at the extremities of a chord drawn through the point (x_1, y_1) intersect at (h, k).

The equation of the chord of contact of the tangents form (h, k) to the ellipse (5.63) is

$$\frac{xh}{a^2} + \frac{yk}{b^2} = 1.$$

Since it passes through (x_1, y_1)

$$\therefore \quad \frac{hx_1}{a^2} + \frac{ky_1}{b^2} = 1.$$

Hence the locus of the point (h, k) is

$$\frac{xx_1}{a^2} + \frac{yy_1}{b^2} = 1,$$

which is the required equation of the polar of the point (x_1, y_1) with respect to the ellipse (5.63).

Corollary 5.7.3 *In a similar way, the equation of the polar of a point (x_1, y_1) with respect to the hyperbola*

$$\frac{x^2}{a^2} - \frac{y^2}{b^2} = 1$$

will be obtained as

$$\frac{xx_1}{a^2} - \frac{yy_1}{b^2} = 1.$$

Tangents and Normals

5.8 Pole of a Polar with respect to a Conic

5.8.1 Pole of a Given Straight Line with respect to a Given Conic

Let the equation of the given conic be

$$ax^2 + 2hxy + by^2 + 2gx + 2fy + c = 0 \tag{5.64}$$

and the given straight line be

$$lx + my + n = 0 \tag{5.65}$$

Also let (x_1, y_1) be the required pole. Then the equation of the polar of (x_1, y_1) with respect to the conic (5.64) is

$$(ax_1 + hy_1 + g)x + (hx_1 + by_1 + f)y + gx_1 + fy_1 + c = 0 \tag{5.66}$$

If this line is identical to the line (5.65) then comparing the equations we get

$$\frac{ax_1 + hy_1 + g}{l} = \frac{hx_1 + by_1 + f}{m} = \frac{gx_1 + fy_1 + c}{n}.$$

From the equality of the first two ratios, we get

$$(am - hl)x_1 + (hm - bl)y_1 + gm - fl = 0 \tag{5.67}$$

Again from the equality of the second and the third ratios, we get

$$(hn - gm)x_1 + (bn - fm)y_1 + fn - cm = 0 \tag{5.68}$$

Solving (5.67) and (5.68) we get

$$x_1 = \frac{(hm - bl)(fn - cm) - (gm - fl)(bn - fm)}{(am - hl)(bn - fm) - (hn - gm)(hm - bl)}$$

and $$y_1 = \frac{(gm - fl)(hn - gm) - (am - hl)(fn - cm)}{(am - hl)(bn - fm) - (hn - gm)(hm - bl)}$$

or, $$x_1 = \frac{Alm + Hm^2 + Gmn}{Gln + Fm^2 + Cmn} \tag{5.69}$$

and $$y_1 = \frac{Hlm + Bm^2 + Fmn}{Gln + Fm^2 + Cmn} \tag{5.70}$$

where A, B, C, F, G, H are the respective cofactors of a, b, c, f, g, h in the determinant $\Delta = \begin{vmatrix} a & h & g \\ h & b & f \\ g & f & c \end{vmatrix}$.

The required pole is given by (5.69) and (5.70).

Particular Cases

1. Pole of a given line with respect to a given circle

Let the equation of the given circle be

$$x^2 + y^2 + 2gx + 2fy + c = 0 \tag{5.71}$$

and the given straight line be

$$lx + my + n = 0 \tag{5.72}$$

Also let (x_1, y_1) be the required pole. Then the equation of the polar of (x_1, y_1) with respect to the circle (5.71) is

$$xx_1 + yy_1 + g(x + x_1) + f(y + y_1) + c = 0$$
$$\text{or,} \quad (x_1 + g)x + (y_1 + f)y + gx_1 + fy_1 + c = 0 \tag{5.73}$$

Therefore this must be identical with the line (5.72). By comparing (5.72) and (5.73) we get

$$\frac{x_1 + g}{l} = \frac{y_1 + f}{m} = \frac{gx_1 + fy_1 + c}{n}$$

$$= \frac{g(x_1 + g) + f(y_1 + f) - (gx_1 + fy_1 + c)}{gl + fm - n}$$

$$= \frac{g^2 + f^2 - c}{gl + fm - n}$$

$$\therefore x_1 = \frac{l(g^2 + f^2 - c)}{gl + fm - n} - g = \frac{lg^2 + lf^2 - lc - lg^2 - fgm + gn}{gl + fm - n}$$

$$= \frac{l(f^2 - c) - gfm + gn}{gl + fm - n}$$

and $y_1 = \dfrac{m(g^2 + f^2 - c)}{gl + fm - n} - f = \dfrac{mg^2 + mf^2 - cm - glf - mf^2 + fn}{gl + fm - n}$

$$= \frac{m(g^2 - c) - glf + fn}{gl + fm - n}.$$

Alternatively,

Comparing the equation of the circle (5.71) with the general conic given in § **5.8.1**, we get

$$a = 1 \quad b = 1 \quad c = c$$
$$h = 0 \quad g = g \quad f = f.$$

So we get

$$\Delta = \begin{vmatrix} a & h & g \\ h & b & f \\ g & f & c \end{vmatrix} = \begin{vmatrix} 1 & 0 & g \\ 0 & 1 & f \\ g & f & c \end{vmatrix}$$

$$\therefore \quad A = c - f^2 \quad B = c - g^2 \quad C = 1$$
$$H = gf \quad G = -g \quad F = -f.$$

So if (x_1, y_1) be the pole of the line $lx + my + n = 0$ with respect to the circle $x^2 + y^2 + 2gx + 2fy + c = 0$, then by the formulae of the last article § 5.8, we get

$$x_1 = \frac{Alm + Hm^2 + Gmn}{Gln + Fm^2 + Cmn} = \frac{(c-f^2)lm + gfm^2 - gmn}{-glm - fm^2 + mn}$$
$$= \frac{(c-f^2)l + gfm - gn}{-gl - fm + n} = \frac{(f^2 - c)l - gfm + gn}{gl + fm - n}$$

and $\quad y_1 = \frac{Hlm + Bm^2 + Fmn}{Gln + Fm^2 + Cmn} = \frac{gflm + (c-g^2)m^2 - fmn}{-glm - fm^2 + mn}$
$$= \frac{fgl + (c-g^2)m - fn}{-gl - fm + n} = \frac{(g^2 - c)m - gfl + fn}{gl + fm - n}.$$

So the required pole is given by

$$\left(\frac{(f^2 - c)l - gfm + gn}{gl + fm - n}, \frac{(g^2 - c)m - gfl + fn}{gl + fm - n} \right).$$

Note 5.8.1 If $gl + fm - n = 0$, i.e., the line $lx + my + n = 0$ passes through the centre $(-g, -f)$, then the corresponding pole does not exist.

Corollary 5.8.1 Proceeding in a similar manner it can be established that the pole of the line $lx + my + n = 0$ with respect to the circle $x^2 + y^2 = a^2$ will be obtained as $\left(-\frac{la^2}{n}, -\frac{ma^2}{n} \right)$.

Note 5.8.2 If $n = 0$, i.e., if the line passes through the origin (centre of the circle, in this case), the corresponding pole does not exist. Thus in a circle the diameters have no poles.

2. Pole of a given line with respect to a given parabola

Let the equation of the given parabola be

$$y^2 = 4ax \tag{5.74}$$

and the given straight line be

$$lx + my + n = 0 \tag{5.75}$$

Also let (x_1, y_1) be the required pole. Then the equation of the polar of (x_1, y_1) with respect to the parabola (5.74) is

$$yy_1 = 2a(x + x_1) \tag{5.76}$$

Therefore (5.75) and (5.76) must be identical and so comparing we get

$$\frac{2a}{l} = \frac{-y_1}{m} = \frac{2ax_1}{n} \Rightarrow x_1 = \frac{n}{l}, \; y_1 = -\frac{2am}{l}.$$

Thus the required pole is $\left(\dfrac{n}{l}, -\dfrac{2am}{l}\right)$.

Note 5.8.3 *If $l = 0$, i.e., the line is parallel to the axis of the parabola, the corresponding pole does not exist.*

3. Pole of a given line with respect to a given ellipse

Let the equation of the given ellipse be

$$\frac{x^2}{a^2} + \frac{y^2}{b^2} = 1 \tag{5.77}$$

and the given straight line be

$$lx + my + n = 0 \tag{5.78}$$

Also let (x_1, y_1) be the required pole. Then the equation of the polar of (x_1, y_1) with respect to the ellipse (5.77) is

$$\frac{xx_1}{a^2} + \frac{yy_1}{b^2} = 1 \tag{5.79}$$

Therefore (5.78) and (5.79) must be identical and so comparing we get

$$\frac{\frac{x_1}{a^2}}{l} = \frac{\frac{y_1}{b^2}}{m} = \frac{1}{-n} \Rightarrow x_1 = -\frac{a^2 l}{n}, \; y_1 = -\frac{b^2 m}{n}.$$

Thus the required pole is $\left(-\dfrac{l}{n}a^2, -\dfrac{m}{n}b^2\right)$.

Note 5.8.4 *If $n = 0$, i.e., the line passes through the centre of the ellipse (origin, in this case), the corresponding pole does not exist.*

Tangents and Normals

Corollary 5.8.2 *In the same manner, it can be shown that the pole of the line $lx + my + n = 0$ with respect to the hyperbola*
$$\frac{x^2}{a^2} - \frac{y^2}{b^2} = 1 \text{ is } \left(-\frac{l}{n}a^2, \frac{m}{n}b^2\right).$$

Note 5.8.5 *If $n = 0$, i.e., the line passes through the centre of the hyperbola (origin, in this case also), the corresponding pole does not exist.*

Corollary 5.8.3 *The pole of the chord joining two points (x_1, y_1) and (x_2, y_2) on a conic is the point of intersection of the tangents at that point.*

5.9 Conjugate Points and Conjugate Lines

Definition 5.9.1 *Two points are said to be **Conjugate Points** with respect to a conic, if each point lies on the polar of the other.*

Definition 5.9.2 *Two lines are said to be **Conjugate Lines** to any conic, if the pole of one line lies on the other line.*

5.10 Properties of Pole and Polar

1. If the polar of a point P passes through a point Q, then the polar of Q passes through P

Let the conic be the circle
$$x^2 + y^2 + 2gx + 2fy + c = 0 \tag{5.80}$$
and let (x_1, y_1) and (x_2, y_2) be the two points P and Q respectively.
  The polar of $P(x_1, y_1)$ with respect to (5.80) is
$$xx_1 + yy_1 + g(x + x_1) + f(y + y_1) + c = 0 \tag{5.81}$$
Since the line (5.81) passes through $Q(x_2, y_2)$
$$\therefore \quad x_2 x_1 + y_2 y_1 + g(x_2 + x_1) + f(y_2 + y_1) + c = 0 \tag{5.82}$$
From the relation (5.82), it follows that $P(x_1, y_1)$ lies on the line
$$xx_2 + yy_2 + g(x + x_2) + f(y + y_2) + c = 0 \tag{5.83}$$
But this equation (5.83) is the polar of $Q(x_2, y_2)$ with respect to (5.80).
  Hence the polar of $Q(x_2, y_2)$ passes through $P(x_1, y_1)$.

Note 5.10.1 *The above two points P and Q are said to be conjugate points with respect to the circle.*

2. The polar of the point of intersection of any two lines is the line joining their poles

Let O be the point of intersection of the two lines AB and CD. Let P and Q be their respective poles.

Since the polar of P, i.e., AB passes through O, the polar of O passes through P.

Similarly the polar of O passes through Q. Hence PQ is the polar of the point O.

3. If the pole of a line $L_1 = 0$ lies on another line $L_2 = 0$ then the pole of the line $L_2 = 0$ will lie on $L_1 = 0$

Let the conic be the circle

$$x^2 + y^2 = a^2 \tag{5.84}$$

and let the poles of $L_1 = 0$ and $L_2 = 0$ with respect to (5.84) be (x_1, y_1) and (x_2, y_2). Then the equation of the line $L_1 = 0$ is

$$xx_1 + yy_1 = a^2 \tag{5.85}$$

and that of the line $L_2 = 0$ is

$$xx_2 + yy_2 = a^2 \tag{5.86}$$

Now the pole of $L_1 = 0$ lies on $L_2 = 0$. This means that the point (x_1, y_1) lies on (5.86) which implies $x_1 x_2 + y_1 y_2 = a^2$.

This shows that the point (x_2, y_2) lies on the line (5.85), i.e., the point (x_2, y_2) lies on $L_1 = 0$.

Hence the pole of $L_2 = 0$ lies on the line $L_1 = 0$.

Note 5.10.2 *The above proposition can be established by following the same procedure for any conic.*

Note 5.10.3 *The above two lines $L_1 = 0$ and $L_2 = 0$ are called conjugate lines with respect to the circle.*

5.10.1 Self-polar Triangle

Definition 5.10.1 *A triangle $\triangle ABC$ is said to be a **Self-Conjugate** or **Self-polar Triangle** if the points A, B, C are such that every two of them are conjugate points with respect to a conic.*

Tangents and Normals

Since the polar of A passes through B and C, BC is the polar of A. Similarly CA is the polar of B and AB is the polar of C. Therefore each side of the $\triangle ABC$ is the polar of the opposite vertex.

5.11 Conditions for Conjugate Lines

To Find the Condition that Two Lines will be Conjugate Lines to a Conic

Let the equation of the given conic be

$$ax^2 + 2hxy + by^2 + 2gx + 2fy + c = 0 \tag{5.87}$$

and the given straight lines

$$l_1 x + m_1 y + n_1 = 0 \tag{5.88}$$
$$\text{and} \quad l_2 x + m_2 y + n_2 = 0 \tag{5.89}$$

be conjugate lines for the conic (5.87). The polar of (x_1, y_1) with respect to the conic is given by

$$(ax_1 + hy_1 + g)x + (hx_1 + by_1 + f)y + (gx_1 + fy_1 + c) = 0 \tag{5.90}$$

Now (5.88) and (5.90) will be identical, so

$$\frac{ax_1 + hy_1 + g}{l_1} = \frac{hx_1 + by_1 + f}{m_1} = \frac{gx_1 + fy_1 + c}{n_1} = \lambda \text{ (say)}$$

So
$$ax_1 + hy_1 + g - \lambda l_1 = 0 \tag{5.91}$$
$$hx_1 + by_1 + f - \lambda m_1 = 0 \tag{5.92}$$
$$\text{and} \quad gx_1 + fy_1 + c - \lambda n_1 = 0 \tag{5.93}$$

Again, since lines (5.88) and (5.89) are conjugate lines, so the pole of the line (5.88) will lie on the line (5.89), i.e.,

$$l_2 x + m_2 y + n_2 = 0 \tag{5.94}$$

Eliminating x_1, y_1 and λ between (5.91), (5.92), (5.93) and (5.94) we get

$$\begin{vmatrix} a & h & g & l_1 \\ h & b & f & m_1 \\ g & f & c & n_1 \\ l_1 & m_1 & n_1 & 0 \end{vmatrix} = 0$$

which is the required condition.

Particular Cases:

1. To find the condition that the given straight lines be conjugate with respect to the circle $x^2 + y^2 = a^2$

Let (x_1, y_1) and (x_2, y_2) be the poles of the straight lines

$$l_1 x + m_1 y + n_1 = 0 \qquad (5.95)$$

$$\text{and} \quad l_2 x + m_2 y + n_2 = 0 \qquad (5.96)$$

respectively. The equation of the polar of the point (x_1, y_1) with respect to the given circle

$$x^2 + y^2 = a^2 \qquad (5.97)$$

is

$$x x_1 + y y_1 = a^2 \qquad (5.98)$$

This is identical to the equation (5.95). Comparing (5.95) and (5.98) we get

$$\frac{x_1}{l_1} = \frac{y_1}{m_1} = -\frac{a^2}{n_1} \Rightarrow x_1 = -\frac{a^2 l_1}{n_1}, \ y_1 = -\frac{a^2 m_1}{n_1}.$$

Therefore, pole of the line (5.95) is given by $\left(-\dfrac{a^2 l_1}{n_1}, -\dfrac{a^2 m_1}{n_1}\right)$.

Since the two lines (5.95) and (5.96) are conjugate, the pole of (5.95) lie on (5.96)

i.e., $\quad l_2\left(-\dfrac{a^2 l_1}{n_1}\right) + m_2\left(-\dfrac{a^2 m_1}{n_1}\right) + n_2 = 0$

or, $\quad (l_1 l_2 + m_1 m_2) a^2 - n_1 n_2 = 0$

which is the required condition that the lines (5.95) and (5.96) are conjugate lines.

2. Proceeding in a similar manner it can be shown that the straight lines $l_1 x + m_1 y + n_1 = 0$ **and** $l_1 x + m_1 y + n_1 = 0$ **will be conjugate lines with respect to the**

(i) parabola $y^2 = 4ax$ if $l_2 n_1 + l_1 n_2 = 2a m_1 m_2$,

(ii) ellipse $\dfrac{x^2}{a^2} + \dfrac{y^2}{b^2} = 1$ if $a^2 l_1 l_2 + b^2 m_1 m_2 = n_1 n_2$,

(ii) hyperbola $\dfrac{x^2}{a^2} - \dfrac{y^2}{b^2} = 1$ if $a^2 l_1 l_2 - b^2 m_1 m_2 = n_1 n_2$.

Tangents and Normals

5.12 Worked Out Examples

Example 5.12.1 *Find the equation of the tangent and normal at the point $(1,1)$ to the hyperbola $x^2 - 2y^2 + x + y = 1$.*

Solution: Clearly the point $(1,1)$ lies on the given hyperbola. Therefore the equation of the tangent at $(1,1)$ to it is

$$x(1) - 2y(1) + \frac{1}{2}(x+1) + \frac{1}{2}(y+1) = 1 \;\;\Rightarrow\;\; x - y = 0.$$

Since the normal is perpendicular to the tangent at $(1,1)$, so the required equation of the normal is

$$\frac{x-1}{1+\frac{1}{2}} + \frac{y-1}{1+\frac{1}{2}} = 0 \;\;\Rightarrow\;\; x + y = 2. \qquad \square$$

Example 5.12.2 *Find the equations of the tangents to the circle $x^2 + y^2 + 8x + 10y - 4 = 0$ which are parallel to the line $x + 2y + 3 = 0$.*

Solution: The equation of any line parallel to $x + 2y + 3 = 0$ is given by

$$x + 2y + k = 0 \;\;\Rightarrow\;\; x = -(2y + k).$$

Let this line be a tangent to the given circle. Putting the value of $x = -(2y + k)$ in the equation of the circle, we get

$$(2y+k)^2 + y^2 - 8(2y+k) + 10y - 4 = 0$$
$$\text{or,} \;\; 5y^2 + 2(2k-3)y + (k^2 - 8k - 4) = 0.$$

Since the given line is a tangent, so the above equation in y must have equal roots

$$\therefore \;\; 4(2k-3)^2 - 4.5(k^2 - 8k - 4) = 0$$
$$\text{or,} \;\; 4(4k^2 - 12k + 9) - 20(k^2 - 8k - 4) = 0$$
$$\text{or,} \;\; -4k^2 + 112k + 116 = 0$$
$$\text{or,} \;\; k^2 - 28k - 29 = 0$$
$$\text{or,} \;\; (k+1)(k-29) = 0 \;\;\Rightarrow\;\; k = -1,\; 29.$$

Thus the equation of the tangents to the given circle parallel to the line $x + 2y + 3 = 0$ are given by

$$x + 2y - 1 = 0 \text{ and } x + 2y + 29 = 0.$$

Alternative Method:

The equation of any line parallel to the line $x + 2y + 3 = 0$ is

$$x + 2y + k = 0 \tag{5.99}$$

The equation of the given circle is

$$(x + 4)^2 + (y + 5)^2 = (\sqrt{45})^2 \tag{5.100}$$

If the line (5.99) be tangent to the circle (5.100) then its distance for the centre $(-4, -5)$ = radius of the circle

i.e., $\dfrac{-4 - 10 + k}{\pm\sqrt{1 + 4}} = \sqrt{45}$

or, $-14 + k = \pm 15 \Rightarrow k = -1, 29.$

Hence the tangents are $x + 2y - 1 = 0$ and $x + 2y + 29 = 0$. □

Note 5.12.1 *The alternative method can be applied only to the case of a circle. In case of a conic, only first method is to be followed.*

Example 5.12.3 *Prove that the locus of a point from which two perpendicular tangents can be drawn to the circle $x^2 + y^2 = a^2$ is the circle $x^2 + y^2 = 2a^2$.*

Solution: Let (x_1, y_1) be any position of the point. The equation to the pair of tangents drawn from (x_1, y_1) to the given circle $S \equiv x^2 + y^2 - a^2 = 0$ is given by $SS_1 = T^2$

i.e., $(x^2 + y^2 - a^2)(x_1^2 + y_1^2 - a^2) = (xx_1 + yy_1 - a^2)^2.$

The tangents represented by the above equation will be perpendicular if

coefficient of x^2 + coefficient of $y^2 = 0$

i.e., $(x_1^2 + y_1^2 - a^2 - x_1^2) + (x_1^2 + y_1^2 - a^2 - y_1^2) = 0$

or, $x_1^2 + y_1^2 - 2a^2 = 0.$

Hence the equation of the locus of (x_1, y_1) is a circle $x^2 + y^2 = a^2$.

Alternative Method:

We know that the straight line $y = mx + a\sqrt{1 + m^2}$ always touches the circle $x^2 + y^2 = a^2$ for all values of m. Let this tangent passes through the point $P(x_1, y_1)$, then

$$y_1 = mx_1 + a\sqrt{1 + m^2}$$

or, $(x_1^2 - a^2)m^2 - 2mx_1y_1 + y_1^2 - a^2 = 0$

Tangents and Normals

which is a quadratic equation in m giving two values, say m_1 and m_2 denoting the gradients of the tangents. If the tangents be mutually perpendicular to each other, then $m_1 m_2 = -1$ giving

$$\frac{y_1^2 - a^2}{x_1^2 - a^2} = -1 \text{ or, } x_1^2 + y_1^2 = 2a^2.$$

Hence the locus of the point (x_1, y_1) is given by $x^2 + y^2 = 2a^2$ which represents a circle. □

Example 5.12.4 *If $y = x \sin\alpha + a \sec\alpha$ be a tangent to the circle $x^2 + y^2 = a^2$, then show that $\cos^2\alpha = 1$.*

Solution: The condition that the line $y = mx + c$ will be a tangent to the circle $x^2 + y^2 = a^2$ is $c^2 = a^2(1 + m^2)$.

Thus if $y = x \sin\alpha + a \sec\alpha$ be a tangent to the circle $x^2 + y^2 = a^2$, then

$$a^2 \sec^2\alpha = a^2(1 + \sin^2\alpha) \text{ or, } \sec^2\alpha = 1 + \sin^2\alpha$$

or, $\quad \dfrac{1}{\cos^2\alpha} - 1 - \sin^2\alpha = 0 \;$ or, $\; \dfrac{1 - \cos^2\alpha}{\cos^2\alpha} - (1 - \cos^2\alpha) = 0$

or, $\quad (1 - \cos^2\alpha)\left(\dfrac{1}{\cos^2\alpha} - 1\right) = 0 \;$ or, $\; (1 - \cos^2\alpha)^2 = 0$

or, $\quad 1 - \cos^2\alpha = 0 \;\; \therefore \; \cos^2\alpha = 1.$ □

Example 5.12.5 *Prove that the straight line $lx + my + n = 0$ touches the parabola $y^2 - 4px + 4pq = 0$ if $l^2 q + ln - pm^2 = 0$.*

Solution: Let the given line

$$lx + my + n = 0 \tag{5.101}$$

touches the parabola

$$y^2 - 4px + 4pq = 0 \tag{5.102}$$

at the point (x_1, y_1).

The equation of the tangent to the parabola (5.102) at (x_1, y_1) is

$$yy_1 - 2p(x + x_1) + 4pq = 0 \text{ or, } 2px - yy_1 + 2p(x_1 - 2q) = 0 \tag{5.103}$$

So the lines (5.101) and (5.103) must be identical. Comparing the coefficients we get

$$\frac{l}{2p} = \frac{m}{-y_1} = \frac{n}{2p(x_1 - 2q)} \Rightarrow x_1 = 2q + \frac{n}{l} \text{ and } y_1 = -\frac{2pm}{l}.$$

Since the point (x_1, y_1) lies on the parabola (5.102), so the point (x_1, y_1) will satisfy (5.102) and we get

$$\left(-\frac{2pm}{l}\right)^2 - 4p\left(2q + \frac{n}{l}\right) + 4pq = 0$$

or, $\quad \dfrac{4p^2 m^2}{l^2} - 8pq - \dfrac{4pn}{l} + 4pq = 0$

or, $\quad -\dfrac{4p}{l^2}(-pm^2 + ql^2 + ln) = 0$ i.e., $l^2 q + ln - pm^2 = 0$ ($\because p \neq 0$)

which is the required condition. $\square$

Example 5.12.6 *Show that the locus of the intersection of a pair of perpendicular tangents to the parabola $x^2 = 4by$ is the directrix.*

Solution: Let (x_1, y_1) be the point of intersection of the pair of tangents. The equation of the pair of tangents are thus given by

$$(x^2 - 4by)(x_1^2 - 4by_1) = \{xx_1 - 2b(y + y_1)\}^2 \quad [SS_1 = T^2].$$

If this pair is perpendicular to each other, so

coefficient of x^2 + coefficient of $y^2 = 0$

i.e., $\quad x_1^2 - 4by_1 - x_1^2 - 4b^2 = 0$ or, $4b(y_1 + b) = 0$ or, $y_1 + b = 0$.

Hence the locus of (x_1, y_1) is the straight line $y + b = 0$ which is the directrix of the given parabola. $\square$

Example 5.12.7 *Find the equation to the common tangent to the circle $x^2 + y^2 = 4ax$ and the parabola $y^2 = 4ax$.*

Solution: Let $y = mx + c$ be the common tangent. Since it touches the parabola $y^2 = 4ax$

$$\therefore \quad c = \frac{a}{m} \tag{5.104}$$

Again, since it touches the given circle, so its distance from the centre $(2a, 0)$ is equal to the radius of it, i.e.,

$$\frac{2am + c}{\pm\sqrt{1 + m^2}} = 2a \quad \text{or,} \quad \left(2am + \frac{a}{m}\right)^2 = 4a^2(1 + m^2) \quad [\text{using (5.104)}]$$

or, $\quad 4a^2 m^2 + 4a^2 + \dfrac{a^2}{m^2} = 4a^2 + 4a^2 m^2$

or, $\quad \dfrac{a^2}{m^2} = 0 \quad \therefore \quad \dfrac{1}{m} = 0. \quad [\because a \neq 0]$

Hence the common tangent is $x = \dfrac{y}{m} - \dfrac{c}{m} = y.0 - c.0 = 0$. $\square$

Tangents and Normals

Example 5.12.8 *Fing the angle between the two tangents from an external point (x_1, y_1) to the circle $x^2 + y^2 = a^2$.*

Solution: The equation of the pair of tangents drawn form (x_1, y_1) to the given circle $x^2 + y^2 = a^2$ is

$$(xx_1 + yy_1 - a^2)^2 = (x^2 + y^2 - a^2)(x_1^2 + y_1^2 - a^2) \quad [\because SS_1 = T^2]$$

or, $(x_1^2 - x_1^2 - y_1^2 + a^2)x^2 + (y_1^2 - x_1^2 - y_1^2 + a^2)y^2 + 2x_1 y_1 xy$
$$-2a^2 x_1 x - 2a^2 y_1 y + a^2(x_1^2 + y_1^2) = 0$$

or, $(a^2 - y_1^2)x^2 + (a^2 - x_1^2)y^2 + 2x_1 y_1 xy - 2a^2 x_1 x$
$$-2a^2 y_1 y + a^2(x_1^2 + y_1^2) = 0$$

or, $Ax^2 + By^2 + 2Hxy + 2Gx + 2Fy + C = 0$

where $A = a^2 - y_1^2, B = a^2 - x_1^2, H = x_1 y_1, G = -a^2 x_1, F = -a^2 y_1$ and $C = a^2(x_1^2 + y_1^2)$.

Now, if α be the angle between the tangents, then

$$\tan \alpha = \frac{2\sqrt{H^2 - AB}}{A + B} = \frac{2\sqrt{x_1^2 y_1^2 - (a^2 - y_1^2)(a^2 - x_1^2)}}{2a^2 - (x_1^2 + y_1^2)}. \qquad \square$$

Example 5.12.9 *Find the equation of the locus of the points of intersection of the tangents to a parabola (i) which are at right angles and (ii) which include an angle β.*

Solution: The straight line $y = mx + \dfrac{a}{m}$ is always a tangent to the parabola $y^2 = 4ax$, $m \neq 0$.

Let this tangent passes through an external point (h, k).

$$\therefore \quad k = mh + \frac{a}{m}$$

or, $hm^2 - km + a = 0 \qquad (5.105)$

Let m_1 and m_2 be the roots of (5.105), then m_1, m_2 denote the gradients of the tangents to the parabola drawn from the external point (h, k). Then we have

$$m_1 + m_2 = \frac{k}{h} \qquad (5.106)$$

and $m_1 m_2 = \dfrac{a}{h} \qquad (5.107)$

Now (i) When the tangents intersect at right angles, $m_1 m_2 = -1$

$$\therefore \quad \frac{a}{h} = -1 \text{ i.e., } h = -a.$$

So the equation of the locus of (h, k) is $x + a = 0$, which represents the directrix of the parabola.

(ii) When β is the angle between the tangents, then

$$\tan \beta = \frac{m_2 - m_1}{1 + m_1 m_2} = \frac{\sqrt{(m_2 + m_1)^2 - 4m_1 m_2}}{1 + m_1 m_2}$$

$$= \frac{\sqrt{\left(\frac{k}{h}\right)^2 - 4\frac{a}{h}}}{1 + \frac{a}{h}} = \frac{\sqrt{k^2 - 4ah}}{a + h}$$

or, $(a + h)^2 \tan^2 \beta = k^2 - 4ah$.

Hence the required locus is $y^2 - 4ax = (a + x)^2 \tan^2 \beta$. □

Example 5.12.10 *A tangent to the parabola $y^2 + 4bx = 0$ meets the parabola $y^2 = 4ax$ at P and Q. Prove that the locus of the mid-point of PQ is $y^2(2a + b) = 4a^2 x$.*

Solution: Let (h, k) be the mid-point of PQ. Any line through the mid-point (h, k) is

$$y - k = m(x - h) \tag{5.108}$$

where m is the gradient of the line.

Let it meets the parabola $y^2 = 4ax$ at the points whose ordinates are y_1 and y_2.

Eliminating x between the equation of the parabola $y^2 = 4ax$ and the equation of the line (5.108), we get

$$y - k = m\left(\frac{y^2}{4a} - h\right) \text{ or, } my^2 - 4ay + 4ak - 4amh = 0.$$

If the roots of the equation be y_1 and y_2 then

$$y_1 + y_2 = \frac{4a}{m}, \text{ but } \frac{y_1 + y_2}{2} = k \text{ i.e., } y_1 + y_2 = 2k.$$

So we get

$$\frac{4a}{m} = 2k \text{ i.e., } m = \frac{4a}{2k}.$$

Putting this value of m in (5.108), we get

$$y - k = \frac{4a}{2k}(x - h) \text{ i.e., } y = \frac{2a}{k}x + \frac{k^2 - 2ah}{k}.$$

Tangents and Normals

Now if it be tangent to the parabola $y^2 = -4bx$, then from the condition of tangency, we have

$$c = -\frac{b}{m} \text{ i.e., } \frac{k^2 - 2ah}{k} = -\frac{b}{\frac{2a}{k}}$$

or, $2a(k^2 - 2ah) + bk^2 = 0$ or, $k^2(2a + b) = 4a^2 h$.

Hence the locus of (h, k) is $(2a + b)y^2 = 4a^2 x$. □

Example 5.12.11 *If a circle be so drawn that it always touches a given line and a given circle, prove that the locus of its centre is a parabola.*

Solution: Without loss of generality, let us consider the given line as the x-axis and let the given circle be

$$x^2 + y^2 = a^2 \tag{5.109}$$

Let the equation of the circle which touches the x-axis be taken as

$$(x - h)^2 + (y - k)^2 = k^2 \tag{5.110}$$

The circle (5.110) touches the circle (5.109), if the sum of the radii of the two circles be equal to the distance between their centers,

i.e., $a + k = \sqrt{(h-0)^2 + (k-0)^2}$

or, $(a+k)^2 = h^2 + k^2$ or, $a^2 + 2ak + k^2 = h^2 + k^2$

or, $h^2 = a(a + 2k) = 2a\left(k + \frac{a}{2}\right)$.

Hence the locus of (h, k) is given by $x^2 = 2a\left(y + \frac{a}{2}\right)$ which is clearly a parabola. □

Example 5.12.12 *Find the point of intersection of the tangents at the points $(at_1^2, 2at_1)$ and $(at_2^2, 2at_2)$ to the parabola $y^2 = 4ax$.*

Solution: Let (x_1, y_1) be the point of intersection of two tangents. The equation of the chord of contact of the tangents from (x_1, y_1) is

$$yy_1 = 2a(x + x_1) \quad [\text{by } T = 0].$$

Since it passes through the points $(at_1^2, 2at_1)$ and $(at_2^2, 2at_2)$ we get

$$\left. \begin{array}{l} 2at_1 y_1 = 2a(at_1^2 + x_1) \Rightarrow t_1 y_1 = x_1 + at_1^2 \\ \text{and } 2at_2 y_1 = 2a(at_2^2 + x_1) \Rightarrow t_2 y_1 = x_1 + at_2^2. \end{array} \right\}$$

Solving these we get $x_1 = at_1 t_2$ and $y = a(t + t_1)$.

Hence the point of intersection is $(at_1 t_2, a(t_1 + t_2))$. □

Example 5.12.13 *If the normal at the point $(at_1^2, 2at_1)$ on the parabola $y^2 = 4ax$ meets it again at the point $(at^2, 2at)$, then prove that $t = -t_1 - \dfrac{2}{t_1}$.*

Solution: The equation of the normal to the parabola $y^2 = 4ax$ at the point $(at_1^2, 2at_1)$ is $y = -t_1 x + 2at_1 + at_1^3$ [See **Table 5.6**].

Since it passes through $(at^2, 2at)$,

$$\therefore \quad 2at = -at^2 t_1 + 2at_1 + at_1^3 \text{ or, } 2a(t - t_1) = -at_1(t^2 - t_1^2)$$

or, $2 = -t_1(t + t_1)$ $[\because t_1 \ne t_2]$

or, $t + t_1 = -\dfrac{2}{t_1}$ $\therefore t = -t_1 - \dfrac{2}{t_1}$. □

Example 5.12.14 *Show that the condition that the normals at (x_1, y_1), (x_2, y_2) and (x_3, y_3) on the ellipse $\dfrac{x^2}{a^2} + \dfrac{y^2}{b^2} = 1$, may be concurrent if*

$$\begin{vmatrix} x_1 & y_1 & x_1 y_1 \\ x_2 & y_2 & x_2 y_2 \\ x_3 & y_3 & x_3 y_3 \end{vmatrix} = 0.$$

Solution: The equation of the normal at (x_1, y_1) to the ellipse $\dfrac{x^2}{a^2} + \dfrac{y^2}{b^2} = 1$ is

$$\dfrac{x - x_1}{\frac{x_1}{a^2}} = \dfrac{y - y_1}{\frac{y_1}{b^2}} \quad [vide \textbf{ Table 5.5}]$$

$$\Rightarrow b^2 x_1 y - a^2 x y_1 + (a^2 - b^2) x_1 y_1 = 0 \quad (5.111)$$

Similarly the normals at (x_1, y_1) and (x_2, y_2) are obtained as

$$b^2 x_2 y - a^2 x y_2 + (a^2 - b^2) x_2 y_2 = 0 \quad (5.112)$$

and $\quad b^2 x_3 y - a^2 x y_3 + (a^2 - b^2) x_3 y_3 = 0 \quad (5.113)$

Eliminating $b^2 y, -a^2 x$ and $(a^2 - b^2)$ from (5.111), (5.112) and (5.113) we get

$$\begin{vmatrix} x_1 & y_1 & x_1 y_1 \\ x_2 & y_2 & x_2 y_2 \\ x_3 & y_3 & x_3 y_3 \end{vmatrix} = 0$$

which is the required condition. □

Example 5.12.15 *Show that the straight line $lx + my = n$ is a normal to the ellipse $\dfrac{x^2}{a^2} + \dfrac{y^2}{b^2} = 1$ if $\dfrac{a^2}{l^2} + \dfrac{b^2}{m^2} = \dfrac{(a^2 - b^2)^2}{n^2}.$*

Tangents and Normals

Solution: The equation of the normal to the ellipse $\dfrac{x^2}{a^2} + \dfrac{y^2}{b^2} = 1$ at any point (x_1, y_1) is

$$\frac{a^2}{x_1}x - \frac{b^2}{y_1}y = a^2 - b^2 \tag{5.114}$$

If the line

$$lx + my = n \tag{5.115}$$

be a normal to the ellipse at the point (x_1, y_1), then (5.115) must be identical to (5.114).

$$\therefore \quad \frac{l}{\frac{a^2}{x_1}} = \frac{m}{-\frac{b^2}{y_1}} = \frac{n}{a^2 - b^2}$$

giving $\quad x_1 = \dfrac{a^2 n}{l(a^2 - b^2)}, \quad y_1 = -\dfrac{b^2 n}{m(a^2 - b^2)}$

and since (x_1, y_1) lies on the ellipse, so

$$\frac{x_1^2}{a^2} + \frac{y_1^2}{b^2} = 1 \text{ i.e., } \frac{a^2 n^2}{l^2(a^2 - b^2)^2} + \frac{b^2 n^2}{m^2(a^2 - b^2)^2} = 1$$

or, $\quad \dfrac{a^2}{l^2} + \dfrac{b^2}{m^2} = \dfrac{(a^2 - b^2)^2}{n^2}.$ $\quad\square$

Example 5.12.16 *The normals to the ellipse $\dfrac{x^2}{a^2} + \dfrac{y^2}{b^2} = 1$ at the ends of the chords $lx + my = 1$ and $l'x + m'y = 1$ will be concurrent if*

$$a^2 ll' = b^2 mm' = -1.$$

Solution: Two points of the four concurrent normals lie on the straight line

$$lx + my = 1 \tag{5.116}$$

Let the other two points lie on the line

$$l'x + m'y = 1 \tag{5.117}$$

Hence the four feet lie on $(lx + my - 1)(l'x + m'y - 1) = 0$

or, $ll'x^2 + mm'y^2 + (lm' + l'm)xy - (l + l')x - (m + m')y + 1 = 0$ (5.118)

Now the four feet of normals lie on the ellipse

$$\frac{x^2}{a^2} + \frac{y^2}{b^2} = 1 \tag{5.119}$$

So (5.118) and (5.119) must be identical. Comparing we get

$$\frac{ll'}{\frac{1}{a^2}} = \frac{mm'}{\frac{1}{b^2}} = \frac{1}{-1} \Rightarrow a^2 ll' = b^2 mm' = -1. \qquad \square$$

Example 5.12.17 *Tangents are drawn from any point on the ellipse $\frac{x^2}{a^2} + \frac{y^2}{b^2} = 1$ to the circle $x^2 + y^2 = r^2$. Prove that the chords of contact are tangents to the ellipse $a^2 x^2 + b^2 y^2 = r^4$.*

Solution: Let $P(a\cos\theta, b\sin\theta)$ be any point on the given ellipse

$$\frac{x^2}{a^2} + \frac{y^2}{b^2} = 1.$$

The equation of the chord of contact of the tangents drawn from the point $P(a\cos\theta, b\sin\theta)$ to the given circle $x^2 + y^2 = r^2$ is

$$ax\cos\theta + by\sin\theta = r^2$$

or, $\quad ax \dfrac{1 - \tan^2 \frac{\theta}{2}}{1 + \tan^2 \frac{\theta}{2}} + by \dfrac{2\tan \frac{\theta}{2}}{1 + \tan^2 \frac{\theta}{2}} = r^2$

or, $\quad ax \dfrac{1 - t^2}{1 + t^2} + by \dfrac{2t}{1 + t^2} = r^2, \quad$ where $t = \tan \dfrac{\theta}{2}$

or, $\quad (r^2 + ax)t^2 - 2byt + (r^2 - ax) = 0 \tag{5.120}$

We know that an envelope is a curve which touches each member of the family of curves and each point of the curve is touched by some member of the family. In this case, the envelope of the family of chords of contact (5.120) will be the required curve.

Hence the required envelope of the family of straight line (5.120) is given by

$$4b^2 y^2 = 4(r^2 + ax)(r^2 - ax) \text{ or, } a^2 x^2 + b^2 y^2 = r^4. \qquad \square$$

Note 5.12.2 *The envelope of the family of curves*

$$A(x,y)\lambda^2 + B(x,y)\lambda + C(x,y) = 0$$

where λ is the parameter, is given by $B^2 = 4AC$.

Tangents and Normals

Example 5.12.18 *Show that the locus of the point of intersection of tangents at two points on the ellipse $\dfrac{x^2}{a^2} + \dfrac{y^2}{b^2} = 1$, where the difference of their eccentric angels is 2α is $\dfrac{x^2}{a^2} + \dfrac{y^2}{b^2} = \sec^2 \alpha$.*

Solution: Let (h, k) be the point of intersection of the two tangents at the two points on the given ellipse $\dfrac{x^2}{a^2} + \dfrac{y^2}{b^2} = 1$ whose eccentric angles are ϕ_1 and ϕ_2 and $\phi_1 - \phi_2 = 2\alpha$.

The equation of the chord of contact of tangents drawn form the point (h, k) to the ellipse $\dfrac{x^2}{a^2} + \dfrac{y^2}{b^2} = 1$ is

$$\frac{hx}{a^2} + \frac{ky}{b^2} = 1 \tag{5.121}$$

The equation of the chord of the ellipse, joining the two points ϕ_1 and ϕ_2 is

$$\frac{x}{a} \cos \frac{1}{2}(\phi_1 + \phi_2) + \frac{y}{b} \sin \frac{1}{2}(\phi_1 + \phi_2) = \cos \frac{1}{2}(\phi_1 - \phi_2) = \cos \alpha \tag{5.122}$$

By the condition of the problem, equations (5.121) and (5.122) represent the same straight line. Comparing the coefficients of (5.121) and (5.122), we get

$$\frac{\frac{h}{a^2}}{\frac{1}{a} \cos \frac{1}{2}(\phi_1 + \phi_2)} = \frac{\frac{k}{b^2}}{\frac{1}{b} \sin \frac{1}{2}(\phi_1 + \phi_2)} = \frac{1}{\cos \alpha}$$

$$\therefore \; h = \frac{a}{\cos \alpha} \cos \frac{1}{2}(\phi_1 + \phi_2), \; k = \frac{b}{\cos \alpha} \sin \frac{1}{2}(\phi_1 + \phi_2)$$

$$\therefore \; \frac{h^2}{a^2} + \frac{k^2}{b^2} = \frac{1}{\cos^2 \alpha} = \sec^2 \alpha.$$

Hence the locus of (h, k) is $\dfrac{x^2}{a^2} + \dfrac{y^2}{b^2} = \sec^2 \alpha$. □

Example 5.12.19 *Find the equation of the common tangent to the parabolas $y^2 = 32x$ and $x^2 = 108y$.*

Solution: The straight line

$$y = mx + \frac{8}{m} \tag{5.123}$$

is always a tangent to the parabola $y^2 = 32x$ for all non-zero values of m.

The abscissa of the points of intersection of the line (5.123) with the parabola $x^2 = 108y$ are given by

$$x^2 = 108\left(mx + \frac{8}{m}\right)$$

or, $\quad mx^2 - 108m^2 x + 8 \times 108 = 0 \qquad (5.124)$

The straight line (5.123) will be tangent to the second parabola $x^2 = 108y$ if the roots of (5.124) are equal, which requires

$$(108m^2)^2 - 4.m.8.108 = 0$$

or, $\quad m\left(m^3 + \dfrac{8}{27}\right) = 0 \quad \therefore \ m = -\dfrac{2}{3} \ (\because m \neq 0).$

Hence the equation of the common tangent is

$$y = -\frac{2}{3}x - \frac{8 \times 3}{2} \quad \text{or,} \ 2x + 3y + 36 = 0. \qquad \square$$

Example 5.12.20 *The polar of the point P with respect to the circle $x^2 + y^2 = a^2$ touches in circle $4x^2 + 4y^2 = a^2$. Show that the locus of P is the circle $x^2 + y^2 = 4a^2$.*

Solution: Let (x_1, y_1) be the coordinates of the point P. Thus the equation of the polar of the point $P(x_1, y_1)$ with respect to the circle $x^2 + y^2 = a^2$ is

$$xx_1 + yy_1 = a^2 \quad \text{or,} \ y = -\frac{x_1}{y_1}x + \frac{a^2}{y_1} \qquad (5.125)$$

If this line (5.125) be tangent to the circle $x^2 + y^2 = \dfrac{a^2}{4}$, then we get

$$\frac{a^2}{y_1} = \pm\frac{a}{2}\sqrt{1 + \left(-\frac{x_1}{y_1}\right)^2} \quad \text{or,} \ \frac{4a^2}{y_1^2} = \frac{y_1^2 + x_1^2}{y_1^2}$$

i.e., $\quad x_1^2 + y_1^2 = 4a^2.$

Hence the locus of (x_1, y_1) is $x^2 + y^2 = 4a^2$. $\qquad \square$

Example 5.12.21 *Show that the polar of the point $(-1, 5)$ with respect to the parabola $y^2 = 4x$ passes through the focus.*

Solution: The equation of the polar of the point $(-1, 5)$ with respect to the parabola $y^2 = 4x$ is $5y = 2(x - 1)$, i.e., $2x - 5y - 2 = 0$ which evidently passes through $(1, 0)$, the focus of the parabola. $\qquad \square$

Tangents and Normals

Example 5.12.22 *If the pole of the straight line with respect to the circle $x^2 + y^2 = a^2$ lies on $x^2 + y^2 = k^2 a^2$, then prove that the straight line will touch the circle $x^2 + y^2 = \dfrac{a^2}{k^2}$.*

Solution: Let (x_1, y_1) be a point on the circle

$$x^2 + y^2 = k^2 a^2 \tag{5.126}$$

The polar of (x_1, y_1) with respect to the circle $x^2 + y^2 = a^2$ is $xx_1 + yy_1 = a^2$. If it touches the circle $x^2 + y^2 = \dfrac{a^2}{k^2}$, then

its perpendicular distance form $(0,0)$ = the radius $\dfrac{a}{k}$

i.e., if $\dfrac{a^2}{\sqrt{x_1^2 + y_1^2}} = \pm \dfrac{a}{k}$

i.e., if $x_1^2 + y_1^2 = a^2 k^2$ which is true by (5.126). □

Example 5.12.23 *Find the locus of the poles with respect to the circle $x^2 + y^2 = a^2$ of the tangents to the circle $x^2 + y^2 = 2ax$.*

Solution: Equation of the circle can be written as $(x-a)^2 + y^2 = a^2$.
Its tangent is given by

$$y = m(x-a) + a\sqrt{1+m^2}$$
$$\text{or,} \quad mx - y + a(\sqrt{1+m^2} - m) = 0 \tag{5.127}$$

Let (α, β) be the pole of this tangent line with respect to the circle $x^2 + y^2 = a^2$.

The equation of the polar of (α, β) with respect to the circle $x^2 + y^2 = a^2$ is

$$\alpha x + \beta y = a^2 \tag{5.128}$$

Since (5.127) and (5.128) represent the same straight line, so comparing we get

$$\dfrac{\alpha}{m} = \dfrac{\beta}{-1} = \dfrac{-a^2}{a(\sqrt{1+m^2}-m)} = \dfrac{-a}{\sqrt{1+m^2}-m}$$

$$\therefore \quad m = -\dfrac{\alpha}{\beta} \quad \text{and} \quad \sqrt{1+m^2} - m = \dfrac{a}{\beta}.$$

Eliminating m we get

$$\sqrt{1+\left(-\frac{\alpha}{\beta}\right)^2}+\frac{\alpha}{\beta}=\frac{a}{\beta}$$

or, $\sqrt{\alpha^2+\beta^2}+\alpha=a$ i.e., $(a-\alpha)^2=\alpha^2+\beta^2$
or, $a^2-2a\alpha+\alpha^2=\alpha^2+\beta^2$ or, $\beta^2=a^2-2a\alpha$.

Hence the required locus is $y^2+2ax=a^2$ which represents a parabola. □

Example 5.12.24 *Show that the polar of any point on the circle $x^2+y^2-2ax-3a^2=0$ with respect to the circle $x^2+y^2+2ax-3a^2=0$ will be tangent to the parabola $y^2=-4ax$.*

Solution: Let (α,β) be coordinates of any point P on the circle

$$x^2+y^2-2ax-3a^2=0 \tag{5.129}$$

Then $\alpha^2+\beta^2-2a\alpha-3a^2=0 \tag{5.130}$

The equation of the polar of $P(\alpha,\beta)$ with respect to the circle

$$x^2+y^2+2ax-3a^2=0 \tag{5.131}$$

is $\quad \alpha x+\beta y+a(x+\alpha)-3a^2=0$

or $\quad (\alpha+a)x+\beta y+a(\alpha-3a)=0$

i.e., $y=-\dfrac{a+\alpha}{\beta}x-\dfrac{a(\alpha-3a)}{\beta}=-\dfrac{a+\alpha}{\beta}x-\dfrac{a(\alpha-3a)\beta}{\beta^2}$

$\qquad =-\dfrac{a+\alpha}{\beta}x+\dfrac{a(\alpha-3a)\beta}{3a^2+2a\alpha-\alpha^2}$ [from (5.130)]

or $\quad y=-\dfrac{a+\alpha}{\beta}x+\dfrac{a\beta}{a+\alpha}=-mx+\dfrac{a}{m}$ where $m=\dfrac{a+\alpha}{\beta}$.

Evidently, this straight line, i.e., the polar of $P(\alpha,\beta)$ is a tangent to the parabola $y^2=-4ax$. □

Example 5.12.25 *If the polar of a point with respect to the parabola $y^2=4ax$ touches the parabola $x^2=4by$, then show that the locus of the point is the hyperbola $xy+2ab=0$.*

Solution: Equation of the polar of $P(\alpha,\beta)$ with respect to the parabola $y^2=4ax$ is

$$\beta y=2a(x+\alpha) \text{ or } 2ax=\beta y-2a\alpha \tag{5.132}$$

Now the straight line
$$x = my + \frac{b}{m} \qquad (5.133)$$
is a tangent to the parabola $x^2 = 4by$.

Evidently (5.132) and (5.133) are identical, so comparing we get
$$\frac{1}{2a} = \frac{m}{\beta} = \frac{\frac{b}{m}}{-2a\alpha} \quad \therefore \ m = \frac{\beta}{2a} = \frac{-2ab}{2a\alpha}$$
i.e., $\alpha\beta = -2ab$.

Hence the equation of the locus of $P(\alpha, \beta)$ is $xy + 2ab = 0$ which represents a hyperbola. □

A. Special Problems on Tangents and Normals; Chord of Contacts; Pair of Tangents

Example 5.12.26 TP and TQ are tangents to the parabola $y^2 = 4ax$ drawn form a variable point T. If TP and TQ are always perpendicular to each other, show that the locus of the point of intersection of the normals to the parabola at P and Q is the parabola $y^2 = a(x - 3a)$.

Solution: Let $(at_1^2, 2at_1)$ and $(at_2^2, 2at_2)$ be the coordinates of the points P and Q respectively. Then normals at P and Q are respectively given by
$$y - 2at_1 = -t_1(x - at_1^2) \qquad (5.134)$$
$$\text{and} \quad y - 2at_2 = -t_2(x - at_2^2) \qquad (5.135)$$
Now since the tangents at P and Q are at right angles to each other, so also the normals thereat
$$\therefore \ t_1 t_2 = -1 \ \text{ i.e., } \ t_2 = -\frac{1}{t_1}.$$
Let they intersect at $L(\alpha, \beta)$
$$\therefore \quad \beta - 2at_1 = -t_1(\alpha - at_1^2) \qquad (5.136)$$
$$\text{and} \quad \beta - 2at_2 = -t_2(\alpha - at_2^2) \qquad (5.137)$$

From (5.136), $\quad at_1^3 + (2a - \alpha)t_1 - \beta = 0 \ \Rightarrow \ t_1^3 = \dfrac{\beta - (2a - \alpha)t_1}{a}.$

From (5.137), $\quad \beta + \dfrac{2a}{t_1} = \dfrac{1}{t_1}\left(\alpha - \dfrac{a}{t_1^2}\right) \ $ or $\ \beta t_1^3 + (2a - \alpha)t_1^2 + a = 0$

$$\therefore \quad \beta \cdot \frac{\beta - (2a - \alpha)t_1}{a} + (2a - \alpha)t_1^2 + a = 0$$
$$\text{or} \quad a(2a - \alpha)t_1^2 - (2a - \alpha)\beta t_1 + (\beta^2 + a^2) = 0.$$

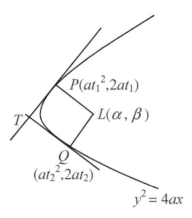

Similarly, form (5.137) we get

$$a(2a - \alpha)t_2^2 - (2a - \alpha)\beta t_2 + (\beta^2 + a^2) = 0.$$

Therefore we see that t_1 and t_2 are the roots of the equation

$$a(2a - \alpha)t^2 - (2a - \alpha)\beta t + (\beta^2 + a^2) = 0 \text{ giving } t_1 t_2 = \frac{\beta^2 + a^2}{a(2a - \alpha)}.$$

Again since $t_1 t_2 = -1$, so we get

$$\frac{\beta^2 + \alpha^2}{a(2a - \alpha)} = -1 \text{ or, } \beta^2 + a^2 = -2a^2 + a\alpha \text{ or, } \beta^2 = a(\alpha - 3a).$$

Hence the locus of $L(\alpha, \beta)$ is $y^2 = a(x - 3a)$. □

Example 5.12.27 *If three normals from a point to the parabola $y^2 = 4ax$ cut the axis in points whose distance from the vertex are in arithmetic progression (A.P.), show that the point lies on the curve $27ay^2 = 2(x - 2a)^3$.*

Solution: The equation of the normal to the parabola $y^2 = 4ax$ at any point $(at^2, 2at)$ on it for all values of t is given by

$$y - 2at = -t(x - at^2) \text{ i.e., } y + tx = 2at + at^3 \tag{5.138}$$

Let this normal passes through the point (x_1, y_1), then we get

$$y_1 + tx_1 = 2at + at^3 \text{ or, } at^3 + (2a - x_1)t - y_1 = 0 \tag{5.139}$$

If t_1, t_2 and t_3 be the roots of (5.139), then
$$t_1 + t_2 + t_3 = 0 \tag{5.140}$$
$$t_1 t_2 + t_2 t_3 + t_3 t_1 = \frac{2a - x_1}{a} \tag{5.141}$$
and $\quad t_1 t_2 t_3 = \dfrac{y_1}{a} \tag{5.142}$

The normals cut the x-axis at the points $(at_1^2 + 2a, 0)$, $(at_2^2 + 2a, 0)$ and $(at_3^2 + 2a, 0)$ [obtained by putting $y = 0$ in (5.138)] and so their distances from the vertex are $at_1^2 + 2a$, $at_2^2 + 2a$ and $at_3^2 + 2a$.

Now by condition of the given problem, since the distances are in A.P., we have
$$at_1^2 + 2a + at_3^2 + 2a = 2(at_2^2 + 2a) \text{ or, } t_1^2 + t_3^2 = 2t_2^2$$
or, $(t_1 + t_3)^2 - 2t_1 t_3 = 2t_2^2$ or, $t_2^2 - 2t_1 t_3 = 2t_2^2$ [by (5.140)]

or, $\quad t_1 t_3 = -\dfrac{1}{2} t_2^2 \tag{5.143}$

or, $\quad t_2^3 = -\dfrac{2y_1}{a}$ [by (5.142)] $\tag{5.144}$

Again form (5.141) we get
$$t_2(t_1 + t_3) + t_1 t_3 = \frac{2a - x_1}{a}$$
or, $\quad t_2(-t_2) - \dfrac{1}{2} t_2^2 = \dfrac{2a - x_1}{a}$ [using (5.140) and (5.143)]

or, $\quad -\dfrac{3}{2} t_2^2 = \dfrac{2a - x_1}{a}$ i.e., $t_2^2 = \dfrac{2(x_1 - 2a)}{3 a} \tag{5.145}$

Eliminating t_2 form (5.144) and (5.145) we get
$$\left(-\frac{2y_1}{a}\right)^2 = \left(\frac{2}{3} \times \frac{x_1 - 2a}{a}\right)^3 \text{ or, } \frac{4y_1^2}{a^2} = \frac{8}{27}\left(\frac{x_1 - 2a}{a}\right)^3$$
or, $\quad 27 a y_1^2 = 2(x_1 - 2a)^3$.

Hence the locus of (x_1, y_1) is $27 a y^2 = 2(x - 2a)^3$. $\quad\square$

Example 5.12.28 *Any ordinate NP of the ellipse $\dfrac{x^2}{a^2} + \dfrac{y^2}{b^2} = 1$ meets the auxiliary circle at Q. Prove that the locus of the point of intersection of the normals at P and Q is the circle $x^2 + y^2 = (a + b)^2$.*

Solution: The equation of the auxiliary circle of the ellipse $\dfrac{x^2}{a^2} + \dfrac{y^2}{b^2} = 1$ is given by $x^2 + y^2 = a^2$ ($\because$ centre of the auxiliary circle is the same as that of the ellipse and its radius is equal to the semi-major axis).

Let ϕ be the eccentric angle of the point P on the ellipse, i.e., the point P is given by $(a\cos\phi, b\sin\phi)$. Then the corresponding point Q on the auxiliary circle is given by $(a\cos\phi, a\sin\phi)$.

The normal at P to the ellipse is

$$y - b\sin\phi = \frac{a\sin\phi}{b\cos\phi}(x - a\cos\phi)$$

or, $ax\sin\phi - by\cos\phi =$
$(a^2 - b^2)\sin\phi\cos\phi$ \hfill (5.146)

The normal at $Q(a\cos\phi, a\sin\phi)$ to the auxiliary circle is

$$y - a\sin\phi = \frac{\sin\phi}{\cos\phi}(x - a\cos\phi)$$

or, $x\sin\phi - y\cos\phi = 0$ \hfill (5.147)

At $\phi = \frac{\pi}{2}$, the normals at P and Q are same, so $\cos\phi \neq 0$.
So, we get from (5.146),

$$\tan\phi = \frac{y}{x} \qquad (5.148)$$

and from (5.146), $ax\tan\phi - by = (a^2 - b^2)\sin\phi$

or, $ay - by = (a^2 - b^2)\sin\phi$ [using (5.148)]

or, $\sin\phi = \dfrac{y}{a+b}$ and hence $\cos\phi = \dfrac{x}{y}\sin\phi = \dfrac{x}{a+b}$.

Eliminating ϕ we get the point of intersection of the two normals as

$$\sin^2\phi + \cos^2\phi = \left(\frac{y}{a+b}\right)^2 + \left(\frac{x}{a+b}\right)^2 = 1$$

or, $x^2 + y^2 = (a+b)^2$. □

Example 5.12.29 *If the normals at any end of a latus rectum of an ellipse passes through one end of the minor axis, then prove that $e^4 + e^2 = 1$, e being the eccentricity.*

Solution: Let a, b be the lengths of the semi-major and semi-minor axes of the ellipse $\dfrac{x^2}{a^2} + \dfrac{y^2}{b^2} = 1$.

Tangents and Normals

The equation of the normal at end of the latus-rectum $\left(ae, \dfrac{b^2}{a}\right)$ is given by

$$y - \dfrac{b^2}{a} = \dfrac{1}{e}(x - ae) \quad \text{or,} \quad \dfrac{x}{e} - a = y - \dfrac{b^2}{a}.$$

This passes through one end of the minor axis $(0, -b)$.

$$\therefore \quad -a = -b - \dfrac{b^2}{a} = -b - \dfrac{a^2(1-e^2)}{a} \quad [\because b^2 = a^2(1-e^2)]$$

or, $a^2 = ab + a^2 - a^2 e^2$ or, $a^2 e^2 = ab$ i.e., $ae^2 = b$

$\Rightarrow a^2 e^4 = b^2 = a^2(1 - e^2) \Rightarrow e^4 = 1 - e^2 \Rightarrow e^4 + e^2 = 1.$ □

Example 5.12.30 Show that $6x^2 + 4xy + y^2 + 2y = 0$ is the equation of the conic with centre at the point $(1, -3)$ which passes through the point $(0, -2)$ and touches the x-axis at the origin.

Solution: Here $a = 6$, $b = 1$, $c = 0$, $h = 2$, $g = 0$, $f = 1$.

$$\therefore \quad ab - h^2 = 6 \cdot 1 - 2^2 = 2 \neq 0.$$

So the conic is a central conic.

Its centre is given by

$$\left(\dfrac{hf - bg}{ab - h^2}, \dfrac{gh - af}{ab - h^2}\right) \equiv \left(\dfrac{2 \times 1 - 1 \times 0}{2}, \dfrac{0 \times 2 - 6 \times 1}{2}\right) \equiv (1, -3).$$

Since the point $(0, -2)$ satisfies the equation of the conic, so the conic passes through this point.

The tangent at the origin is obtained by equating the lowest degree term to zero and is given by $2y = 0$ i.e., $y = 0$, i.e., the x-axis. □

Example 5.12.31 If the normal to the rectangular hyperbola $xy = c^2$ at $\left(ct_1, \dfrac{c}{t_1}\right)$ meets the curve at $\left(ct_2, \dfrac{c}{t_2}\right)$, then prove that $t_1^3 t_2 = -1$.

Solution: The equation of the tangent to the rectangular hyperbola $xy = c^2$ at the point $\left(ct_1, \dfrac{c}{t_1}\right)$ is given by

$$\dfrac{1}{2}\left(\dfrac{xc}{t_1} + yct_1\right) = c^2 \quad \text{or,} \quad y = -\dfrac{x}{t_1^2} + \dfrac{2c}{t_1}.$$

So the equation of the normal at this point is given by

$$y - \frac{c}{t_1} = t_1^2(x - ct_1).$$

If it passes through $\left(ct_2, \dfrac{c}{t_2}\right)$, then

$$\frac{c}{t_2} - \frac{c}{t_1} = t_1^2(ct_2 - ct_1) \text{ or, } \frac{t_1 - t_2}{t_1 t_2} = -t_1^2(t_1 - t_2)$$

or, $(t_1 - t_2)\left(\dfrac{1}{t_1 t_2} + t_1^2\right) = 0 \Rightarrow \dfrac{1}{t_1 t_2} + t_1^2 = 0 \; [\because t_1 \neq t_2]$

which gives $t_1^3 t_2 = -1$. □

Example 5.12.32 *If α be the angle between a pair of tangents to the parabola $y^2 = 4ax$, show that the locus of their point of intersection is*

$$(x + a)^2 \tan^2 \alpha = y^2 - 4ax.$$

Solution: Let (h, k) be the point of intersection. Equation of the pair of tangents form (h, k) to the parabola $y^2 = 4ax$ is

$$(y^2 - 4ax)(k^2 - 4ah) = \{yk - 2a(x + h)\}^2 \; [\text{by rule } SS_1 = T^2]$$

or, $y^2 k^2 - 4ahy^2 - 4axk^2 + 16a^2 hx = y^2 k^2 - 4a(x + h)yk$
$$+ 4a^2(x^2 + 2xh + h^2)$$

or, $-4ahy^2 - 4axk^2 + 8a^2 hx = -4axyk - 4ahyk + 4a^2 x^2 + 4a^2 h^2$

or, $a^2 x^2 - akxy + ahy^2 + ak^2 x - 2a^2 hx - ahky + a^2 h^2 = 0$

or, $ax^2 - kxy + hy^2 + (k^2 - 2ah)x - hky + ah^2 = 0.$

It is given that α is the angle between these pair of lines,

$$\therefore \quad \tan \alpha = \frac{2\sqrt{\left(-\frac{k}{2}\right)^2 - ah}}{a + h} \text{ or, } (h + a)^2 \tan^2 \alpha = k^2 - 4ah$$

Therefore, the locus of (h, k) is $(x + a)^2 \tan^2 \alpha = y^2 - 4ax.$ □

B. Special Problems on Poles and Polars

Example 5.12.33 *Show that the locus of the pole with respect to the ellipse $\dfrac{x^2}{a^2} + \dfrac{y^2}{b^2} = 1$ of any tangent to the director circle of the ellipse $\dfrac{x^2}{c^2} + \dfrac{y^2}{d^2} = 1$ is $\dfrac{x^2}{a^4} + \dfrac{y^2}{b^4} = \dfrac{1}{c^2 + d^2}.$*

Solution: The equation of the director circle of the ellipse $\dfrac{x^2}{c^2} + \dfrac{y^2}{d^2} = 1$ is

$$x^2 + y^2 = c^2 + d^2 \tag{5.149}$$

Let (α, β) be the pole of a tangent to (5.149) with respect to the ellipse $\dfrac{x^2}{a^2} + \dfrac{y^2}{b^2} = 1$. Then the polar is

$$\dfrac{x\alpha}{a^2} + \dfrac{y\beta}{b^2} = 1 \text{ or, } b^2\alpha x + a^2\beta y - a^2 b^2 = 1 \tag{5.150}$$

Now (5.150) is a tangent to (5.149),

$\therefore$ its distance form the centre $(0,0)$ of (5.150) = radius of (5.149)

i.e., $\left| \dfrac{a^2 b^2}{\sqrt{b^4 \alpha^2 + a^4 \beta^2}} \right| = \sqrt{c^2 + d^2}$ or, $\dfrac{b^4 \alpha^2 + a^4 \beta^2}{a^4 b^4} = \dfrac{1}{c^2 + d^2}$

or, $\dfrac{\alpha^2}{a^4} + \dfrac{\beta^2}{b^4} = \dfrac{1}{c^2 + d^2}.$

Hence the locus of (α, β) is $\dfrac{x^2}{a^4} + \dfrac{y^2}{b^4} = \dfrac{1}{c^2 + d^2}.$ □

Example 5.12.34 *Show that the locus of the poles of chords of the ellipse* $\dfrac{x^2}{a^2} + \dfrac{y^2}{b^2} = 1$, *which subtend a right angle at the centre is*

$$\dfrac{x^2}{a^4} + \dfrac{y^2}{b^4} = \dfrac{1}{a^2} + \dfrac{1}{b^2}.$$

Solution: Let (α, β) be the pole of a chord of the given ellipse. The polar of (α, β) with respect to the given ellipse will be

$$\dfrac{\alpha x}{a^2} + \dfrac{\beta y}{b^2} = 1 \tag{5.151}$$

This will be the equation of the chord. It is given that it subtends a right angle at the origin, i.e., at the centre of the ellipse. So making the equation of the given ellipse homogeneous with the help of (5.151), we get

$$\dfrac{x^2}{a^2} + \dfrac{y^2}{b^2} = \left(\dfrac{\alpha x}{a^2} + \dfrac{\beta y}{b^2} \right)^2.$$

Since these lines are perpendicular to each other, so

$$\text{coefficient of } x^2 + \text{coefficient of } y^2 = 0$$

or, $\left(\dfrac{\alpha^2}{a^4} - \dfrac{1}{a^2}\right) + \left(\dfrac{\beta^2}{b^4} - \dfrac{1}{b^2}\right) = 0$ or, $\dfrac{\alpha^2}{a^4} + \dfrac{\beta^2}{b^2} = \dfrac{1}{a^2} + \dfrac{1}{b^2}.$

Hence the locus of (α, β) is given by $\dfrac{x^2}{a^4} + \dfrac{y^2}{b^4} = \dfrac{1}{a^2} + \dfrac{1}{b^2}.$ □

Example 5.12.35 *Prove that the locus of the poles of normal chords of the hyperbola $\dfrac{x^2}{a^2} - \dfrac{y^2}{b^2} = 1$ with respect to the hyperbola is the curve*

$$a^6 y^2 - b^6 x^2 = (a^2 + b^2)^2 x^2 y^2.$$

Solution: Equation of the normal at any point $(a\sec\theta, b\tan\theta)$ on the hyperbola $\dfrac{x^2}{a^2} - \dfrac{y^2}{b^2} = 1$ is

$$\dfrac{ax}{\sec\theta} + \dfrac{by}{\tan\theta} = a^2 + b^2 \quad (\textit{vide } \textbf{Table 5.6}) \tag{5.152}$$

i.e., the normal chord at $(a\sec\theta, b\tan\theta)$ lies along this line (5.152).

If (h, k) be the pole of (5.152) with respect to the hyperbola $\dfrac{x^2}{a^2} - \dfrac{y^2}{b^2} = 1$ then (5.152) will be identical to

$$\dfrac{xh}{a^2} - \dfrac{yk}{b^2} = 1 \tag{5.153}$$

So comparing the coefficients of (5.152) and (5.153) we get

$$\dfrac{\frac{a}{\sec\theta}}{\frac{h}{a^2}} = \dfrac{\frac{b}{\tan\theta}}{-\frac{k}{b^2}} = \dfrac{a^2 + b^2}{1}$$

or, $\dfrac{a^3}{h}\cos\theta = -\dfrac{b^3}{k}\cot\theta = a^2 + b^2$

$\therefore \quad \cos\theta = \dfrac{(a^2 + b^2)h}{a^3}$

and $\cot\theta = -\dfrac{a^2 + b^2}{b^3}k \quad \therefore \quad \dfrac{\cos\theta}{\sin\theta} = -\dfrac{a^2 + b^2}{b^3}k$

$\therefore \quad \sin\theta = -\dfrac{b^3}{(a^2+b^2)k}\cos\theta = -\dfrac{b^3}{(a^2+b^2)k} \times \dfrac{(a^2+b^2)h}{a^3} = -\dfrac{b^3 h}{a^3 k}.$

Tangents and Normals

Eliminating θ we get

$$1 = \cos^2\theta + \sin^2\theta = \frac{(a^2+b^2)^2 h^2}{a^6} + \frac{b^6 h^2}{a^6 k^2}$$

or, $a^6 k^2 = (a^2+b^2)^2 h^2 k^2 + b^6 h^2$.

Hence the locus of (h,k) is $a^6 y^2 - b^6 x^2 = (a^2+b^2)^2 x^2 y^2$. □

Example 5.12.36 *Prove that the locus of the poles of tangents to the parabola $y^2 = 4ax$ with respect to the parabola $y^2 = 4bx$ is the parabola $y^2 = 4\dfrac{b^2}{a}x$.*

Solution: Let (h,k) be the pole. Then the equation of the polar with respect to $y^2 = 4bx$ is

$$yk = 2b(x+h) \quad \text{or,} \quad y = \frac{2b}{k}x + \frac{2bh}{k}.$$

If it is a tangent to the parabola $y^2 = 4ax$, then we get

$$\frac{2bh}{k} = \frac{a}{\frac{2b}{k}} \quad [\text{ by the condition of tangency } c = \frac{a}{m}]$$

$$\Rightarrow k^2 = \frac{4b^2 h}{a}.$$

Hence the locus of (h,k) is $y^2 = 4\dfrac{b^2}{a}x$. □

Example 5.12.37 *If the sum of the abscissas of two points on the ellipse $\dfrac{x^2}{a^2} + \dfrac{y^2}{b^2} = 1$ be 'a', show that the locus of the pole of the chord which joins them is $b^2 x^2 + a^2 y^2 = 2ab^2 x$.*

Solution: Let $P(a\cos\theta, b\sin\theta)$ and $Q(a\cos\phi, b\sin\phi)$ be the two points on the given ellipse. Then by the condition of the problem

$$a\cos\theta + a\cos\phi = a \quad \Rightarrow \quad \cos\theta + \cos\phi = 1$$

or, $\quad 2\cos\dfrac{\theta+\phi}{2}\cdot\cos\dfrac{\theta-\phi}{2} = 1 \qquad (5.154)$

The equation of the chord PQ is

$$y - b\sin\theta = \frac{b\sin\phi - b\sin\theta}{a\cos\phi - a\cos\theta}(x - a\cos\theta)$$

or, $b(\sin\theta - \sin\phi)x + a(\cos\phi - \cos\theta)y$
$= ab(-\sin\phi\cos\theta + \sin\theta\cos\theta + \sin\theta\cos\phi - \sin\theta\cos\theta)$
$= ab(\sin\theta\cos\phi - \cos\theta\sin\phi) = ab\sin(\theta - \phi) \qquad (5.155)$

Let its pole be (x_1, y_1). Equation of polar of (x_1, y_1) with respect to the ellipse $\dfrac{x^2}{a^2} + \dfrac{y^2}{b^2} = 1$ is

$$\frac{xx_1}{a^2} + \frac{yy_1}{b^2} = 1 \tag{5.156}$$

Comparing (5.155) and (5.156), we get

$$\frac{a^2 b(\sin\theta - \sin\phi)}{x_1} = \frac{ab^2(\cos\phi - \cos\theta)}{y_1} = ab\sin(\theta - \phi)$$

$$\therefore \; x_1 = \frac{a(\sin\theta - \sin\phi)}{\sin(\theta - \phi)}, \quad y_1 = \frac{a(\cos\phi - \cos\theta)}{\sin(\theta - \phi)}$$

$$\therefore \; b^2 x_1^2 + a^2 y_1^2 = a^2 b^2 \frac{(\sin\theta - \sin\phi)^2}{\sin^2(\theta - \phi)} + b^2 a^2 \frac{(\cos\phi - \cos\theta)^2}{\sin^2(\theta - \phi)}$$

$$= \frac{a^2 b^2}{\sin^2(\theta - \phi)}[2 - 2\cos(\theta - \phi)]$$

$$= \frac{a^2 b^2}{\sin^2(\theta - \phi)} \cdot 2 \cdot 2\sin^2\frac{\theta - \phi}{2}$$

$$= \frac{a^2 b^2}{\cos^2 \frac{\theta - \phi}{2}} = \frac{a^2 b^2 \cdot 2\cos\frac{\theta+\phi}{2}}{\cos\frac{\theta-\phi}{2}} \quad [\text{using (5.154)}]$$

$$\therefore \; b^2 x_1^2 + a^2 y_1^2 = 2a^2 b^2 \cdot \frac{2\cos\frac{\theta+\phi}{2} \cdot \sin\frac{\theta-\phi}{2}}{2\sin\frac{\theta-\phi}{2} \cdot \cos\frac{\theta-\phi}{2}} = 2a^2 b^2 \cdot \frac{\sin\theta - \sin\phi}{\sin(\theta - \phi)} = 2ab^2 x_1.$$

Hence the locus of (x_1, y_1) is $b^2 x^2 + a^2 y^2 = 2ab^2 x$. □

Example 5.12.38 *Find the condition that the lines $lx + my = 1$ and $px + qy = 1$ will be a conjugate with respect to the conic $ax^2 + 2hxy + by^2 = 1$.*

Solution: Let (x_1, y_1) be the pole of $lx + my = 1$ with respect to the given conic, then $lx + my = 1$ and $axx_1 + h(x_1 y + xy_1) + byy_1 = 1$ will be identical. Comparing we get,

$$\frac{ax_1 + hy_1}{l} = \frac{hx_1 + by_1}{m} = 1 \quad \therefore \; x_1 = \frac{lb - hm}{ab - h^2}, \quad y_1 = \frac{ma - lh}{ab - h^2}.$$

As the given lines are conjugate, so

$$px_1 + qy_1 = 1 \text{ or, } p\left(\frac{lb - hm}{ab - h^2}\right) + q\left(\frac{ma - lh}{ab - h^2}\right) = 1$$

$$\text{or, } p(lb - hm) + q(ma - lh) = ab - h^2$$

which is the required condition. □

5.13 Exercises

Section A: Objective Type Questions

1. If $y = x \sin \alpha + a \sec \alpha$ be a tangent to a circle $x^2 + y^2 = a^2$, show that $\cos^2 \alpha = 1$.

2. Find the equation of the tangent to the ellipse $\dfrac{x^2}{16} + \dfrac{y^2}{12} = 1$ at the point $(1, 2)$.

3. If the line $lx + my = 1$ touches the circle $x^2 + y^2 = a^2$, then (l, m) lies on a certain circle. Find its equation.

4. Find the equation of the tangents to the circle $x^2 + y^2 + 8x + 10y - 4 = 0$ which are parallel to the straight line $x + 2y + 3 = 0$.

5. Find the equation of the common tangent to the parabolas $y^2 = 32x$ and $x^2 = 108y$.

6. Find the equation of the normal at the point $(1, 4)$ to the ellipse $3x^2 + 7y^2 = 115$.

7. Prove that the straight line $4ax + 3by = 12c$ will be a normal to the ellipse $\dfrac{x^2}{a^2} + \dfrac{y^2}{b^2} = 1$, if $5c = a^2 e^2$, e being the eccentricity.

8. Prove that the straight line $2x + 4y = 9$ is a normal to the parabola $y^2 = 8x$.

9. Find the coordinates of the pole of the straight line $3x + 4y + 1 = 0$ with respect to the circle $x^2 + y^2 + 6x + 4y - 3 = 0$.

10. Find the polar of the point $(-1, 5)$ with respect to the parabola $y^2 = 4x$.

11. Find the locus of the point P, if its polar with respect to the hyperbola $\dfrac{x^2}{a^2} - \dfrac{y^2}{b^2} = 1$ be equally inclined to the coordinate axes.

12. Show that the locus of the poles of tangents to the parabola $y^2 = 4ax$ with respect to the parabola $y^2 = 4bx$ is $y^2 = \dfrac{4b^2}{a} x$.

13. Find the length of the tangent from $(-2, 2)$ to the conic $4x^2 + y^2 = 4$.

14. Find the length of the tangent from $(-1, 1)$ to the circle $x^2 + y^2 - 4x - 4y + 4 = 0$.

15. Find the length of the tangent from $(0, 0)$ to the circle $x^2 + y^2 - 4x - 4y + 4 = 0$.

16. Find the equation of the chord of contact to the point $(6, 5)$ with respect to the ellipse $4x^2 + 9y^2 = 36$.

17. Find the equation of the normal at $(1, -1)$ to the conic $y^2 - xy - 2x^2 - 5y + x - 6 = 0$.

18. Show that the locus of the poles of tangents to the parabola $y^2 = 4ax$ with respect to the parabola $y^2 = 4bx$ is the parabola $y^2 = \dfrac{4b^2}{a}x$.

19. Find the equation of the polar of the point $(1, 2)$ with respect to the conic $x^2 + y^2 - 4x - 6y - 12 = 0$.

20. Find the equation of the polar of the point $(2, 3)$ with respect to the conic $y^2 - 4x = 0$.

21. Chords of the ellipse $\dfrac{x^2}{a^2} + \dfrac{y^2}{b^2} = 1$ touch the circle $x^2 + y^2 = e^2$. Find the locus of their poles.

22. If the normal at any end of a latus rectum of an ellipse passes through one end of the minor axis, then prove that $e^4 + e^2 = 1$ where e is the eccentricity of the ellipse.

Section B: Broad Answer Type Questions

1. (i) Find the equation of the tangent to the parabola $y^2 = 16x$ at the point $(1, 4)$.

(ii) Find the equation of the tangent to the circle $x^2 + y^2 = 3$, which makes an angle of $60°$ with the x-axis.

(iii) Find the equation of the tangents to the parabola $y^2 = 4x + 5$ which are parallel to the straight line $y = 2x + 1$.

2. Find the equation of the tangents to the conic $x^2 + 4xy + 3y^2 - 5x - 6y + 3 = 0$ which are parallel to the straight line $x + 4y = 0$.

3. Two circles both touch the axis of y and intersect at the points $(1, 0)$ and $(2, -1)$. Show that they touch the line $y + 2 = 0$.

4. Prove that the straight lines $4x - 2y + 3 = 0$ touches the parabola $y^2 = 12x$ and find the coordinates of the point of contact.

5. If tangents be drawn from any point on the line $x + 4a = 0$ to the parabola $y^2 = 4ax$, then show that the chord of contact will subtends a right angle at the vertex.

6. Find the common tangents to the circles $x^2 - 22xy + y^2 + 4y + 100 = 0$ and $x^2 - 22xy + y^2 - 4y - 100 = 0$.

7. Find the coordinates of the point of intersection of the tangents at the points $(at_1^2, 2at_1)$ and $(at_2^2, 2at_2)$ to the parabola $y^2 = 4ax$.

8. If the angle between the two tangents drawn form the point (h, k) to the ellipse $\dfrac{x^2}{a^2} + \dfrac{y^2}{b^2} = 1$ be θ, then prove that

$$(h^2 + k^2 - a^2 - b^2)\tan\theta = 2ab\sqrt{\dfrac{h^2}{a^2} + \dfrac{k^2}{b^2} - 1}.$$

9. For the parabola $y^2 = 8x$ form the equations of two tangents which pass through the point $\left(-2, \dfrac{16}{3}\right)$. Also find the angle between them.

10. If 2α be the angle between the two tangents drawn form an external point to the circle $x^2 + y^2 = a^2$, then show that $2\alpha = 2\sin^{-1}\dfrac{a}{\sqrt{h^2+y^2}}$.

11. A tangent to a parabola $y^2 + 4bx = 0$ meets the parabola $y^2 = 4ax$ at P and Q. Prove that the locus of the mid-point of PQ is $y^2(2a+b) = 4a^2x$.

12. Show that the locus of the foot of perpendicular from the centre on any tangent to the ellipse $\dfrac{x^2}{a^2} + \dfrac{y^2}{b^2} = 1$ is the curve $a^2x^2 + b^2y^2 = (x^2+y^2)^2$.

13. The product of the tangents drawn form a point P to the parabola $y^2 = 4ax$ is equal to the product of the focal distance of P and the latus rectum. Show that the locus of P is the parabola $y^2 = 4a(x+a)$.

14. Two straight lines are at right angles to one another, one of them touches the parabola $y^2 = 4a(x+a)$ and other touches the parabola $y^2 = 4b(x+b)$. Show that the point of intersection of the straight lines will lie on the straight line $x + a + b = 0$.

15. Prove that the straight line $lx + my + n = 0$ touches

   (i) the circle $x^2 + y^2 = a^2$ if $n^2 = a^2(l^2 + m^2)$

   (ii) the parabola $y^2 = 4ax$ if $ln = am^2$

   (iii) the ellipse $\dfrac{x^2}{a^2} + \dfrac{y^2}{b^2} = 1$ if $n^2 = a^2l^2 + b^2m^2$

   (iv) the hyperbola $\dfrac{x^2}{a^2} - \dfrac{y^2}{b^2} = 1$ if $n^2 = a^2l^2 - b^2m^2$.

16. If the tangent $y = mx + \sqrt{a^2m^2 - b^2}$ touches the hyperbola $\dfrac{x^2}{a^2} - \dfrac{y^2}{b^2} = 1$ at the point $(a\sec\phi, b\tan\phi)$, prove that $\phi = \sin^{-1}\left(\dfrac{b}{am}\right)$.

17. Show that the straight line $x\cos\alpha + y\sin\alpha = p$ touches the hyperbola $\dfrac{x^2}{a^2} - \dfrac{y^2}{b^2} = 1$ if $a^2\cos^2\alpha - b^2\sin^2\alpha = p^2$.

18. Tangents are drawn from the point (h, k) to the circle $x^2 + y^2 = a^2$. Prove that the area of the triangle formed by them and the straight line joining their point of contact is $\dfrac{a(h^2 + k^2 - a^2)^{\frac{1}{2}}}{h^2 + k^2}$.

19. Prove that the length of the tangent drawn from any point on the circle $x^2 + y^2 + 2gx + 2fy + c = 0$ to the circle $x^2 + y^2 + 2gx + 2fy + c' = 0$ is $(c' - c)^{\frac{1}{2}}$.

20. Show that the locus of the point of intersection of mutually perpendicular tangents of a parabola is the directrix of the parabola.

21. Find the locus of the point of intersection of tangents drawn t an ellipse which include an angle θ.

22. Find the equation of the normal at the point $(1, 3)$ to the parabola $y^2 = 9x$.

23. Prove that the normals at the ends of a focal chord of a parabola $y^2 = 4ax$ intersect on the parabola $y^2 = a(x - 3a)$.

24. Prove that the straight line $lx + my + n = 0$ is a normal to

(i) the ellipse $\dfrac{x^2}{a^2} + \dfrac{y^2}{b^2} = 1$ if $\dfrac{a^2}{l^2} + \dfrac{b^2}{m^2} = \dfrac{(a^2 - b^2)^2}{n^2}$

(ii) the hyperbola $\dfrac{x^2}{a^2} - \dfrac{y^2}{b^2} = 1$ if $\dfrac{a^2}{l^2} - \dfrac{b^2}{m^2} = \dfrac{(a^2 + b^2)^2}{n^2}$.

25. The normal to the parabola $y^2 = 4ax$ at the point $(at^2, 2at)$ meets the parabola again at the point $(at_1^2, 2at_1)$. Prove that $t^2 + tt_1 + 2 = 0$.

26. Find the locus of the point of intersection of the two normals to a parabola which are at right angles.

27. Prove that the normal to the circle $x^2 + y^2 - 5x + 2y = 48$ at the point $(5, 6)$ is a tangent to the parabola $5y^2 + 448x = 0$.

28. Show that the locus of poles of tangents to the parabola $ay^2 + 2b^2 x = 0$ with respect to the ellipse $\dfrac{x^2}{a^2} + \dfrac{y^2}{b^2} = 1$ is the parabola $ay^2 - 2b^2 x = 0$.

29. The polar of a point P with respect to the circle $x^2 + y^2 = a^2$ touches the circle $(x - \alpha)^2 + (y - \beta)^2 = b^2$. Show that the locus of P is the curve given by the equation $(\alpha x + \beta y - a^2)^2 = b^2(x^2 + y^2)$.

30. Chords of the ellipse $\dfrac{x^2}{a^2} + \dfrac{y^2}{b^2} = 1$ touch the circle $x^2 + y^2 = c^2$. Find the locus of their poles.

31. Find the locus of P if its polar with respect to the hyperbola $\dfrac{x^2}{a^2} - \dfrac{y^2}{b^2} = 1$ be equally inclined to the coordinate axes.

Tangents and Normals

32. Prove that

   (i) the locus of the poles of the normal chords of the parabola $y^2 = 4ax$ is the curve $y^2(x+2a) + 4a^3 = 0$.

   (ii) the locus of the poles of the chords of the parabola $y^2 = 4ax$ which subtend a right angle at the vertex is $x + 4a = 0$.

33. Find the locus of the poles of the normal chords of the ellipse $\dfrac{x^2}{a^2} + \dfrac{y^2}{b^2} = 1$.

34. The polar of the point P with respect to the circle $x^2 + y^2 = a^2$ touches the circle $4(x^2 + y^2) = a^2$. Show that locus of P is the circle $x^2 + y^2 = 4a^2$.

35. Show that the normal $y = mx - 2am - am^3$ of the parabola $y^2 = 4ax$ intersects the parabola again at an angle $\tan^{-1}\left(\dfrac{1}{2}m\right)$.

36. A circle is described on a focal chord of $y^2 = 4ax$ as diameter. If m be the tangent of inclination of the chord to the axis, prove that the equation of the circle is $x^2 + y^2 - 2ax\left(1 + \dfrac{2}{m^2}\right) - \dfrac{4ay}{m} - 3a^2 = 0$.

37. If the normals at two points A and B of a parabola intersect on the curve, then show that the straight line AB passes through a fixed point.

38. Show that the locus of the middle points of normal chords of the rectangular hyperbola $x^2 - y^2 = a^2$ is $(y^2 - x^2)^3 = 4a^2x^2y^2$.

39. The normals at the ends of the latus rectum of the parabola $y^2 = 4ax$ meet the curve again in Q and Q'. Prove that $QQ' = 12a$.

40. The tangents at two points P and Q of a parabola whose focus is S, meet at T, prove that

   (i) TP and TQ subtend equal angles at the focus,

   (ii) $SP.SQ = ST^2$.

and (iii) the triangle $\triangle SPT$ and $\triangle STQ$ are similar.

41. If the tangent and normal to an ellipse meet the major axis at the points T and G respectively, then prove that $CG.CT = CS^2$, where C is the centre and S is the nearer focus to T.

42. If the normals at the points (x_1, y_1), (x_2, y_2), (x_3, y_3) and (x_4, y_4) on the rectangular hyperbola $xy = c^2$ meet at (α, β) then show that $\alpha = x_1 + x_2 + x_3 + x_4$, $\beta = y_1 + y_2 + y_3 + y_4$ and $x_1 x_2 x_3 x_4 = y_1 y_2 y_3 y_4 = -c^4$.

43. Show that the points $(1, 2)$ and $(8, -6)$ are conjugate with respect to the conic $x^2 + xy + y^2 = 1$.

44. Find the condition that the two straight lines $l_1 x + m_1 y = 1$ and

$l_2x + m_2y = 1$ will be conjugate with respect to the conic $\dfrac{x^2}{a^2} + \dfrac{y^2}{b^2} = 1$.

45. If a point lies on the ellipse $\dfrac{x^2}{a'^2} + \dfrac{y^2}{b'^2} = 1$, then prove that its polar with respect to the ellipse $\dfrac{x^2}{a^2} + \dfrac{y^2}{b^2} = 1$ touches the ellipse $\dfrac{a'^2 x^2}{a^4} + \dfrac{b'^2 y^2}{b^4} = 1$.

46. Show that the locus of the poles of the chords of the parabola $y^2 = 4ax$ which is bisected by a fixed straight line $lx + my + n = 0$ is $l(y^2 - 2ax) + 2a(my + n) = 0$.

47. Show that the locus of the poles, with respect to the parabola $y^2 = 4ax$, of tangents to the hyperbola $x^2 - y^2 = a^2$ is the ellipse $4x^2 + y^2 = 4a^2$.

48. If the pole of the normal at P on an ellipse lies on the normal at Q, then show that the pole of the normal at Q lies on the normal at P.

ANSWERS

Section A: 2. $x + 2y = 8$; **3.** $x^2 + y^2 = 1/a^2$; **4.** $x + 2y - 1 = 0$ and $x + 2y + 29 = 0$; **5.** $2x + 3y + 36 = 0$; **6.** $28x - 3y - 16 = 0$; **9.** $(0, 2)$; **10.** $2x - 5y - 2 = 0$; **11.** $b^2 x = a^2 y$; **12.** 4 units; **13.** $\sqrt{6}$ units; **14.** 2 units; **15.** $8x + 15y = 12$; **16.** $4x - y - 5 = 0$; **18.** $x + 2y + 20 = 0$; **19.** $2x - 3y + 4 = 0$; **20.** $\frac{x^2}{a^4} + \frac{y^2}{b^4} = \frac{1}{c^2}$.

Section B: 1. (i) $2x - y - 2 = 0$; (ii) $y = \sqrt{3}x \pm 2\sqrt{3}$; (iii) $y = 2x + 3$; **2.** $x + 4y - 5 = 0$ and $x + 4y - 8 = 0$; **4.** $\left(\frac{3}{4}, 3\right)$; **6.** $7x - 24y - 250 = 0$, $3x + 4y - 50 = 0$; **7.** $(at_1 t_2, a(t_1 + t_2))$; **9.** $x - 3y + 18 = 0$, $9x + 3y + 2 = 0$, $90°$; **21.** $(x^2 + y^2 - a^2 - b^2)^2 \tan^2 \theta = 4a^2 b^2 \left(\frac{x^2}{a^2} + \frac{y^2}{b^2} - 1\right)$; **22.** $2x + 3y = 11$; **26.** $y^2 = a(x - 3a)$; **31.** $b^2 x = a^2 y$; **33.** $\frac{a^4}{x^2} + \frac{b^4}{y^2} = (a^2 - b^2)^2$; **44.** $a^2 l_1 l_2 + b^2 m_1 m_2 - n_1 n_2 = 0$.

Chapter 6

Diameters and Conjugate Diameters

6.1 Diameters

Definition 6.1.1 *The locus of the middle points of a system of parallel chords to a conic is known as a diameter of the conic.*

Note 6.1.1 *For the equation of the chord of a conic in terms of middle point see § **5.5** from **Chapter 5**.*

Note 6.1.2 *Different system of parallel chords have different diameters.*

6.2 Important Theorems

Theorem 6.2.1 *The locus of the middle points of a system of parallel chords of a parabola is a straight line parallel to the axis of the parabola $y^2 = 4ax$.*

Proof: Let $V(x_1, y_1)$ be a middle point of one of the the system of parallel chords of the parabola $y^2 = 4ax$ parallel to the straight line

$$y = mx + c \qquad (6.1)$$

The equation of the chord of the parabola $y^2 = 4ax$ in terms of the middle point (x_1, y_1) is $yy_1 - 2a(x + x_1) = y_1^2 - 4ax_1$.

Since it is parallel to (6.1), we get $\dfrac{2a}{y_1} = m$, i.e., $y_1 = \dfrac{2a}{m}$.

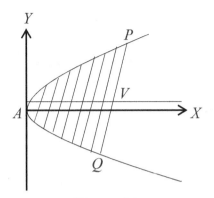

Figure 6.1

Hence the locus of $V(x_1, y_1)$ is $y = \dfrac{2a}{m}$.

This is a straight line parallel to $y = 0$, the axis of the parabola. □

Now, any line drawn parallel to the axis of the parabola is called a *diameter*.

Thus the diameter bisects each of the system of parallel chords which is called the *ordinate* of that diameter.

Note 6.2.1 *The diameter meets the curve at a point, the tangent at which is parallel to the given system.*

Theorem 6.2.2 *The locus of the middle points of a system of parallel chords of an ellipse is a straight line passing through the centre of the ellipse.*

Proof: Let the equation of the ellipse be

$$\frac{x^2}{a^2} + \frac{y^2}{b^2} = 1 \tag{6.2}$$

Let the system of chords be parallel to the straight line

$$y = mx + c \tag{6.3}$$

Let PQ be any one of the chords of the system and let V, the middle point of PQ be the point with coordinates (x_1, y_1).

Therefore the equation of PQ is

$$\frac{xx_1}{a^2} + \frac{yy_1}{b^2} = \frac{x_1^2}{a^2} + \frac{y_1^2}{b^2} \tag{6.4}$$

Diameters & Conjugate Diameters

The gradient of (6.4) is same as that of (6.3), therefore comparing the coefficients we get

$$-\frac{b^2 x_1}{a^2 y_1} = m \text{ i.e., } y_1 = -\frac{b^2}{a^2 m} x_1.$$

Hence the locus of (x_1, y_1) is $y = -\dfrac{b^2}{a^2 m} x.$

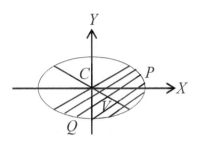

Figure 6.2

This is a straight line passing through the origin, i.e., the centre of the ellipse.

Hence, the locus of the middle points of a system of parallel chords of an ellipse is a straight line passing through its centre. □

Note 6.2.2 *This straight line is called a diameter with respect to the system of parallel chords.*

Theorem 6.2.3 *The locus of the middle points of a system of parallel chords of a hyperbola is a straight line passing through the centre of the hyperbola.*

Proof: Let the equation of the hyperbola be

$$\frac{x^2}{a^2} - \frac{y^2}{b^2} = 1 \tag{6.5}$$

Let the system of chords be parallel to the straight line

$$y = mx + c \tag{6.6}$$

The equation of the chord of the hyperbola (6.5) in terms of its middle point (x_1, y_1) is

$$\frac{xx_1}{a^2} - \frac{yy_1}{b^2} = \frac{x_1^2}{a^2} - \frac{y_1^2}{b^2} \tag{6.7}$$

Since (6.7) and (6.6) are same, we get $\dfrac{b^2 x_1}{a^2 y_1} = m$ i.e., $y_1 = \dfrac{b^2}{a^2 m} x_1.$

Therefore the locus of (x_1, y_1) is $y = \dfrac{b^2}{a^2 m} x$ which evidently passes through the origin, i.e., the centre of the hyperbola. Hence, the proposition. □

Note 6.2.3 *This straight line is called a diameter of the hyperbola with respect to the system of parallel chords.*

Remark: Any straight line passing through the centre of a central conic is a *diameter* of the conic. The length of segment of this line intercepted by the central conic is called the *length of the diameter*.

6.3 Conjugate Diameter

Definition 6.3.1 *Two diameters of a central conic are said to be conjugate diameter of the conic if each contains the middle points of the system of chords parallel to the other.*

Thus the major and minor axes of an ellipse are conjugate diameters of the ellipse for each contains the middle points of the chords parallel to the other as is evident from the adjoining **Figure 6.3**.

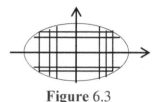

Figure 6.3

6.4 Condition that two given Straight Lines may be Conjugate Diameters of an Ellipse

Let $y = mx$ and $y = m'x$ be two conjugate diameters of the ellipse

$$\frac{x^2}{a^2} + \frac{y^2}{b^2} = 1.$$

The equation of the diameter bisecting all chords parallel to the line $y = mx$ is

$$y = -\frac{b^2}{a^2 m} x.$$

Since this line is same as

$$y = m'x \quad \therefore \quad m' = -\frac{b^2}{a^2 m} \quad \text{or,} \quad mm' = -\frac{b^2}{a^2}.$$

The symmetry of the result shows that if $y = m'x$ bisects all chords parallel to $y = mx$, then the line $y = mx$ will also bisect all chords parallel to $y = m'x$.

Therefore the lines $y = mx$ and $y = m'x$ will be conjugate diameters of the ellipse $\frac{x^2}{a^2} + \frac{y^2}{b^2} = 1$ if $mm' = -\frac{b^2}{a^2}$.

Corollary 6.4.1 *In a similar way, we can prove that the condition that the two straight lines $y = mx$ and $y = m'x$ will be a pair of conjugate diameters of the hyperbola $\frac{x^2}{a^2} - \frac{y^2}{b^2} = 1$ if $mm' = \frac{b^2}{a^2}$.*

6.5 Properties of Conjugate Diameters of an Ellipse

Property 1: *The eccentric angles of the extremities of a pair of conjugate diameters of an ellipse differ by an odd multiple of $\frac{\pi}{2}$.*

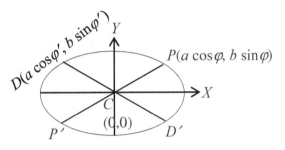

Figure 6.4

Let PCP' and DCD' be a pair of conjugate diameters of an ellipse. Let ϕ and ϕ' be the eccentric angles of P and D respectively and let the coordinates of P and D are respectively given by $(a\cos\phi, b\sin\phi)$ and $(a\cos\phi', b\sin\phi')$.

The slope of CP is $\dfrac{b\sin\phi}{a\cos\phi}$ and that of CD is $\dfrac{b\sin\phi'}{a\cos\phi'}$.

Since CP and CD are conjugate diameters

$$\therefore \text{ slope of } CP \times \text{ slope of } CD = -\frac{b^2}{a^2}$$

or, $\dfrac{b\sin\phi}{a\cos\phi} \times \dfrac{b\sin\phi'}{a\cos\phi'} = -\dfrac{b^2}{a^2} \Rightarrow \dfrac{\sin\phi\sin\phi'}{\cos\phi\cos\phi'} = -1$

$\Rightarrow \cos\phi\cos\phi' + \sin\phi\sin\phi' = 0$

$\Rightarrow \cos(\phi \sim \phi') = 0 = \cos(2n+1)\dfrac{\pi}{2}$ for $n \in \mathbb{N}$

$\therefore \phi \sim \phi' =$ an odd multiple of $\dfrac{\pi}{2}$.

Corollary 6.5.1 *If the eccentric angle of P be ϕ, then the eccentric angles of the points D, P' and D' are respectively given by $\dfrac{\pi}{2} + \phi$, $\dfrac{2\pi}{2} + \phi$, i.e., $\pi + \phi$ and $\dfrac{3\pi}{2} + \phi$. That is if the coordinates of the point P be $(a\cos\phi, b\sin\phi)$, then the coordinates of D, P' and D' are respectively $(-a\sin\phi, b\cos\phi)$, $(-a\cos\phi, -b\sin\phi)$ and $(a\sin\phi, -b\cos\phi)$.*

Property 2: *The sum of the squares of any two semi-conjugate diameters of an ellipse is constant.*

Let PCP' and DCD' be a pair of conjugate diameters of an ellipse $\dfrac{x^2}{a^2} + \dfrac{y^2}{b^2} = 1$. (*ref.* **Figure 6.4**)

Then, if ϕ be the eccentric angles of the extremity P, the coordinates of P, D, P', D' are respectively given by $(a\cos\phi, b\sin\phi)$, $(-a\sin\phi, b\cos\phi)$, $(-a\cos\phi, -b\sin\phi)$ and $(a\sin\phi, -b\cos\phi)$.

$$\therefore\ CP^2 = a^2\cos^2\phi + b^2\sin^2\phi,\ \ CD^2 = a^2\sin^2\phi + b^2\cos^2\phi$$
$$\therefore\ CP^2 + CD^2 = a^2(\cos^2\phi + \sin^2\phi) + b^2(\sin^2\phi + \cos^2\phi) = a^2 + b^2$$

which is a constant.

Property 3: *The tangents at the ends of a pair of conjugate diameters of an ellipse form a parallelogram of constant area.*

Let PCP' and DCD' be a pair of conjugate diameters of an ellipse $\dfrac{x^2}{a^2} + \dfrac{y^2}{b^2} = 1$.

If $(a\cos\phi, b\sin\phi)$ be the coordinates of the point P, then the coordinates of D, P' and D' are respectively given by $(-a\sin\phi, b\cos\phi)$, $(-a\cos\phi, -b\sin\phi)$ and $(a\sin\phi, -b\cos\phi)$.

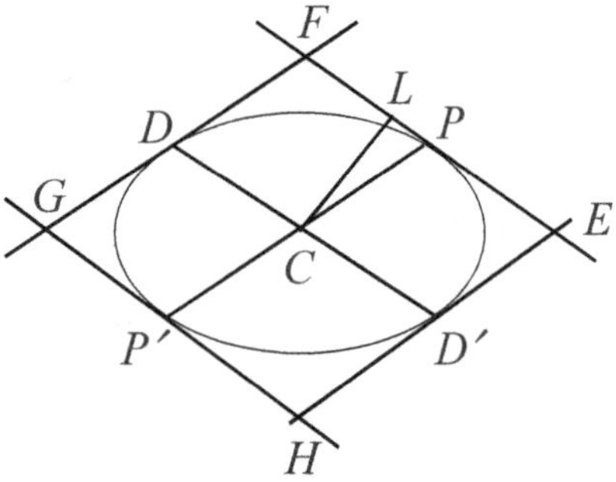

Then equation of the tangents at these points P, D, P', D' are respectively given by

Figure 6.5

$$\frac{x\cos\phi}{a} + \frac{y\sin\phi}{b} = 1,\ \ \frac{x\sin\phi}{a} - \frac{y\cos\phi}{b} = -1,$$
$$\frac{x\cos\phi}{a} + \frac{y\sin\phi}{b} = -1\ \text{ and }\ \frac{x\sin\phi}{a} - \frac{y\cos\phi}{b} = 1$$

so that the tangents at P, P' and D, D' are respectively parallel to each other. Therefore, the tangents at these points P, D, P' and D', i.e., at the end points of a pair of diameters of an ellipse form a parallelogram.

Now the gradient of the tangent at P is $-\dfrac{b\cos\phi}{a\sin\phi}$ and the gradient of CD is also $-\dfrac{b\cos\phi}{a\sin\phi}$.

Hence the tangent is parallel to DCD'. Similarly, the tangent at D is parallel to PCP'.

Therefore, the area of the parallelogram $EFGH$
$$= 4 \times \text{area of the parallelogram } CPFD. \text{ (See \textbf{Figure 6.5})}$$

Let CL be the perpendicular from the centre C upon the tangent at P. Then

$$CL = \frac{1}{\sqrt{\frac{\cos^2 \phi}{a^2} + \frac{\sin^2 \phi}{b^2}}} = \frac{ab}{\sqrt{b^2 \cos^2 \phi + a^2 \sin^2 \phi}} = \frac{ab}{CD}$$

so, area of the parallelogram $CPFD = CD \times CL = CD \times \dfrac{ab}{CD} = ab$.

∴ area of the parallelogram $EFGH$
$$= 4 \times \text{area of the parallelogram } CPFD = 4ab = \text{a constant}.$$

Note 6.5.1 *The area of the parallelogram $CPFD$ can be obtained alternatively as follows:*

$$\text{Area } CPFD = \frac{1}{2} \times \triangle CPD = \frac{1}{2} \begin{vmatrix} 0 & 0 & 1 \\ a \cos \phi & b \sin \phi & 1 \\ -a \sin \phi & b \cos \phi & 1 \end{vmatrix}$$

$$= \frac{1}{2} ab(\cos^2 \phi + \sin^2 \phi) = \frac{1}{2} ab.$$

Corollary 6.5.2 *Since $CL.CD = ab = \text{constant}$, so the **product of the length of perpendicular from the centre of the ellipse on the tangent at one end of a diameter and the semi-conjugate diameter is constant** [i.e., $p.CD = ab = \text{constant}$ where p is the said length of perpendicular.]*

Property 4: *The product of the focal distances of a point on an ellipse is equal to the square on the semi-diameter parallel to the tangent at P.*

Let $P(a \cos \theta, b \sin \theta)$ be a point on the ellipse $\dfrac{x^2}{a^2} + \dfrac{y^2}{b^2} = 1$. Then the coordinates of D are $(-a \sin \theta, b \cos \theta)$.

The coordinates of the foci S and S' are respectively $(ae, 0)$ and $(-ae, 0)$. We have

$$\begin{aligned} SP &= \sqrt{(ae - a \cos \theta)^2 + b^2 \sin^2 \theta} \\ &= \sqrt{a^2 e^2 - 2a^2 e \cos \theta + a^2 \cos^2 \theta + b^2 \sin^2 \theta} \\ &= \sqrt{a^2 e^2 - 2a^2 e \cos \theta + a^2 \cos^2 \theta + a^2(1 - e^2) \sin^2 \theta} \\ &= \sqrt{a^2 e^2 - 2a^2 e \cos \theta + a^2 \cos^2 \theta + a^2 \sin^2 \theta - a^2 e^2 \sin^2 \theta} \end{aligned}$$

$$= \sqrt{a^2e^2 - 2a^2 e \cos\theta + a^2 + a^2 e^2 \sin^2\theta}$$
$$= \sqrt{a^2 e^2 \cos^2\theta - 2a^2 e \cos\theta + a^2} = \sqrt{(a - ae\cos\theta)^2} = a - ae\cos\theta$$

and similarly, $S'P = \sqrt{(ae + a\cos\theta)^2 + b^2 \sin^2\theta} = a + ae\cos\theta.$

$$\therefore \quad SP.S'P = a^2 - a^2 e^2 \cos^2\theta = a^2 + (b^2 - a^2)\cos^2\theta$$
$$= a^2 \sin^2\theta + b^2 \cos^2\theta = CD^2.$$

6.6 Properties of Conjugate Diameters of a Hyperbola

Property 1: *If a pair of diameters be conjugate with respect to a hyperbola, one of them meets the hyperbola in real points and the other meets the conjugate hyperbola in real points.*

Let $y = m_1 x$ and $y = m_2 x$ be a pair of conjugate diameters with respect to the hyperbola $\dfrac{x^2}{a^2} - \dfrac{y^2}{b^2} = 1$. Then $m_1 m_2 = \dfrac{b^2}{a^2}$.

Let $|m_1| < \dfrac{b}{a}$, hence $|m_2| > \dfrac{b}{a}$.

If the straight line $y = m_1 x$ meets the hyperbola $\dfrac{x^2}{a^2} - \dfrac{y^2}{b^2} = 1$ then for the abscissas of the points of intersection we get

$$\frac{x^2}{a^2} - \frac{m_1^2 x^2}{b^2} = 1 \quad \Rightarrow \quad x^2 = \frac{a^2 b^2}{b^2 - a^2 m_2^2} \quad \Rightarrow \quad x = \pm \frac{ab}{\sqrt{b^2 - a^2 m_1^2}}$$

which is real if $b^2 - a^2 m_1^2 > 0 \Rightarrow |m_1| < \dfrac{b}{a}$. It is true according to our assumption. Therefore the straight line $y = m_1 x$ meets the hyperbola in real points.

The abscissa of points of intersection of the line $y = m_2 x$ and the conjugate hyperbola $\dfrac{y^2}{b^2} - \dfrac{x^2}{a^2} = 1$ are given by $x^2 \left(\dfrac{m_2^2}{b^2} - \dfrac{1}{a^2} \right) = 1$, i.e., $x^2 = \dfrac{a^2 b^2}{a^2 m_2^2 - b^2}.$

Since $m_2^2 > \dfrac{b^2}{a^2}$, these abscissas are real. Therefore the line $y = m_2 x$ meets the conjugate hyperbola in real points.

Diameters & Conjugate Diameters

Property 2: *If a pair of diameters be conjugate with respect to a hyperbola, then they are also conjugate with respect to its conjugate hyperbola.*

If two straight line $y = m_1 x$ and $y = m_2 x$ be a pair of conjugate diameters with respect to the hyperbola $\dfrac{x^2}{a^2} - \dfrac{y^2}{b^2} = 1$. Then

$$m_1 m_2 = \dfrac{b^2}{a^2} \qquad (6.8)$$

The equation of the conjugate hyperbola is $\dfrac{y^2}{b^2} - \dfrac{x^2}{a^2} = 1$, i.e., $\dfrac{x^2}{-a^2} - \dfrac{y^2}{-b^2} = 1$.

The lines $y = m_1 x$ and $y = m_2 x$ be a pair of conjugate diameters of the second hyperbola if $m_1 m_2 = \dfrac{-b^2}{-a^2} = \dfrac{b^2}{a^2}$. This is true from the above condition (6.8).

Hence the two straight lines are a pair of conjugate diameters of the conjugate hyperbola also.

Property 3: *If a pair of conjugate diameters meet the hyperbola and its conjugate in P and D, then*

(a) $CP^2 - CD^2 = a^2 - b^2$ and

(b) *the tangents at P, D and the other ends of the diameters through them from a parallelogram whose area is constant.*

(a) Let $P(a \sec\theta, b \tan\theta)$ be any point on the hyperbola $\dfrac{x^2}{a^2} - \dfrac{y^2}{b^2} = 1$ whose centre C is at $(0, 0)$.

The equation of the diameter CP is

$$y = \dfrac{b \tan\theta}{a \sec\theta} x \text{ or, } y = \dfrac{b}{a} \sin\theta . x \qquad (6.9)$$

Therefore the equation of CD is

$$y = \dfrac{b}{a \sin\theta} x \quad \left(\because m_1 m_2 = \dfrac{b^2}{a^2} \right) \qquad (6.10)$$

The abscissas of the points of intersection of the line (6.10) and the conjugate hyperbola $\dfrac{y^2}{b^2} - \dfrac{x^2}{a^2} = 1$ are given by the equation

$$x^2 \left(\dfrac{1}{a^2 \sin^2\theta} - \dfrac{1}{a^2} \right) = 1 \text{ or, } \dfrac{x^2}{a^2} \left(\dfrac{1 - \sin^2\theta}{\sin^2\theta} \right) = 1$$

or, $x^2 = a^2 \tan^2\theta$ i.e., $x = \pm a \tan\theta$, so, $y = \pm b \sec\theta$.

Hence the coordinates of D are $(a\tan\theta, b\sec\theta)$. So, we have
$$CP^2 = a^2\sec^2\theta + b^2\tan^2\theta \quad \text{and} \quad CD^2 = a^2\tan^2\theta + b^2\sec^2\theta$$
$$\therefore \quad CP^2 - CD^2 = (a^2\sec^2\theta + b^2\tan^2\theta) - (a^2\tan^2\theta + b^2\sec^2\theta)$$
$$= a^2(\sec^2\theta - \tan^2\theta) - b^2(\sec^2\theta - \tan^2\theta) = a^2 - b^2.$$

(b) The tangents at P and D to the hyperbola and its conjugate hyperbola are respectively

$$\frac{x\sec\theta}{a} - \frac{y\tan\theta}{b} = 1 \qquad (6.11)$$

and $\quad \dfrac{x\tan\theta}{a} - \dfrac{y\sec\theta}{b} = -1 \qquad (6.12)$

From **Figure 6.6**, it is obvious that the tangents at P and D are parallel to CD and CP respectively. Similarly the tangents at P' and D' are parallel to CD and CP respectively. Hence the tangents at P, P', D, D' from a parallelogram.

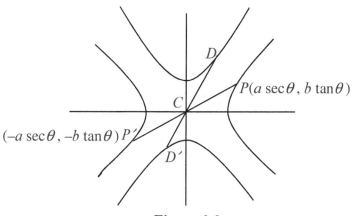

Figure 6.6

Now the area of this parallelogram $= 4 \times CD.p$,

where p be the length of perpendicular form C to the tangent at P

$$= \frac{1}{\sqrt{\frac{1}{a^2}\sec^2\theta + \frac{1}{b^2}\tan^2\theta}} = \frac{1}{\sqrt{a^2\tan^2\theta + b^2\sec^2\theta}} = \frac{ab}{CD}$$

[since D is the point $(a\tan\theta, b\sec\theta)$ (from above)].

So the area of this parallelogram $= 4 \times CD \times p = 4 \times CD \times \dfrac{ab}{CD}$

$$= 4ab = \text{a constant}.$$

Note 6.6.1 *If p be the length of perpendicular form the centre of the hyperbola to the tangent at any point P, then $p = \dfrac{ab}{CD}$.*

6.7 Equi-conjugate Diameters of an Ellipse

Definition 6.7.1 *When two conjugate diameters of an ellipse are equal, they are said to be equi-conjugate diameters of the ellipse.*

Let P and D be the extremities of the two equi-conjugate diameters of the ellipse $\dfrac{x^2}{a^2} + \dfrac{y^2}{b^2} = 1$ whose centre C is at $(0,0)$, i.e., $CP^2 = CD^2$.

Now if $(a\cos\theta, b\sin\theta)$ be the coordinates of P, then we have

$$a^2\cos^2\theta + b^2\sin^2\theta = a^2\sin^2\theta + b^2\cos^2\theta$$
or, $\quad (a^2 - b^2)\cos^2\theta = (a^2 - b^2)\sin^2\theta$
or, $\quad \tan^2\theta = 1$ or, $\tan\theta = \pm 1 \Rightarrow \theta = 45°$ or $135°$.

Hence the equation of CP is $y = \dfrac{b}{a}\tan\theta.x$ or $y = \pm\dfrac{b}{a}x \quad\quad (6.13)$

and the equation of CD is $y = \dfrac{b}{a}\cot\theta.x = \mp\dfrac{b}{a}x \quad\quad (6.14)$

From (6.13) and (6.14) it is clear that the two equi-conjugate diameters CP and CD are equally inclined to the axis of x in opposite senses.

Note 6.7.1 *If a rectangle be formed by the tangents at A, A', B, B' the two diameters (6.13) and (6.14) are clearly its diagonals.*

Also the length of each equi-conjugate semi-diameters

$$= CP = \sqrt{a^2\cos^2\theta + b^2\sin^2\theta} = \sqrt{\dfrac{a^2 + b^2}{2}} \quad \left[\because \sin^2\theta = \cos^2\theta = \dfrac{1}{2}\right].$$

Property: *The acute angle between a pair of conjugate diameters of an ellipse is least when they are equi-conjugate.*

Proof: Let PCP' and DCD' be a pair of conjugate diameters of the ellipse $\dfrac{x^2}{a^2} + \dfrac{y^2}{b^2} = 1$ and let θ be the angle between the pair of semi conjugate diameters CP and CD (see **Figure 6.5**).

We have $CP.CD.\sin\theta = ab$.

Since the perpendicular distance form the centre C to the tangent at P is equal to the perpendicular distance form P to the chord DCD' which is again equal to $CP.\sin\theta$.

Clearly, $\sin\theta$ is least when the product $(CP.CD)$ is maximum. Again, by **Property 2** of § 6.5 we get $CP^2 + CD^2 = a^2 + b^2$. Hence $(CP.CD)$ is maximum when $CP = CD$, i.e., when the conjugate diameters are equi-conjugate then the acute angle between them is least. □

6.8 Worked Out Examples

Example 6.8.1 *Find the diameter of the ellipse $3x^2 + 4y^2 = 5$ conjugate to the line $y + 3x = 0$.*

Solution: The given equation of the ellipse is written as $\dfrac{x^2}{\frac{5}{3}} + \dfrac{y^2}{\frac{5}{4}} = 1$. Let $y = mx$ be the diameter which is conjugate to $y = -3x$.

$$\therefore\ m(-3) = \dfrac{-\frac{5}{4}}{\frac{5}{3}} \text{ or, } m = \dfrac{1}{4}.$$

Hence the equation to the diameter of the given ellipse conjugate to the line $y + 3x = 0$ is $y = \dfrac{1}{4}x$ or $4y - x = 0$. □

Example 6.8.2 *CP and CQ are two conjugate semi-diameters of the ellipse $2x^2 + 3y^2 = 14$. If the coordinates of P be $(1,2)$, find the length of CQ.*

Solution: Here, the equation of the ellipse is

$$\dfrac{x^2}{7} + \dfrac{y^2}{\frac{14}{3}} = 1, \quad \therefore\ a^2 = 7,\ b^2 = \dfrac{14}{3}.$$

$$\therefore\ CP^2 + CQ^2 = a^2 + b^2 = 7 + \dfrac{14}{3} = \dfrac{35}{3} \qquad (6.15)$$

Again, $CP^2 = 1 + 4 = 5$ [$\because$ C is the origin.] So, form (6.15), we get

$$CQ^2 = \dfrac{35}{3} - 5 = \dfrac{20}{3} \quad \therefore\ CQ = \dfrac{2\sqrt{5}}{\sqrt{3}} = \dfrac{2\sqrt{15}}{3}.$$ □

Example 6.8.3 *For the ellipse $2x^2 + 3y^2 = 24$, find the pairs of semi-conjugate diameters inclined at an angle $\tan^{-1} 7$.*

Diameters & Conjugate Diameters   285

Solution: The equation of the ellipse is $\dfrac{x^2}{12} + \dfrac{y^2}{8} = 1$.

If the required equation of the conjugate diameters be $y = m_1 x$ and $y = m_2 x$, then $m_1 m_2 = -\dfrac{b^2}{a^2} = -\dfrac{8}{12} = -\dfrac{2}{3}$.

If θ be the angle between these conjugate diameters, then

$$\tan\theta = 7 = \frac{m_1 \sim m_2}{1 + m_1 m_2} = \frac{m_1 - m_2}{1 - \frac{2}{3}} \quad \therefore \quad m_1 - m_2 = \frac{7}{3}.$$

Now $(m_1 + m_2)^2 = (m_1 - m_2)^2 + 4 m_1 m_2 = \dfrac{49}{9} - \dfrac{8}{3} = \dfrac{25}{9}$

$$\therefore \quad m_1 + m_2 = \pm \frac{5}{3}$$

So we get, $m_1 + m_2 = \pm \dfrac{5}{3}$ and $m_1 - m_2 = \pm \dfrac{7}{3}$

whence we get, $m_1 = 2, \dfrac{1}{3}$ and $m_2 = -\dfrac{1}{3}, -2$.

Hence the required conjugate diameters are
$$\left.\begin{array}{l} y = 2, \quad y = -\dfrac{1}{3}x \\ y = \dfrac{1}{3}x, \quad y = -2x. \end{array}\right\} \quad \square$$

Example 6.8.4 *Prove that the locus of the middle points of the chords of the ellipse $\dfrac{x^2}{a^2} + \dfrac{y^2}{b^2} = 1$ which pass through a fixed point (α, β) is given by $b^2 x(x - \alpha) + a^2 y(y - \beta) = 0$.*

Solution: Let (h, k) be the middle point of a chord of the ellipse $\dfrac{x^2}{a^2} + \dfrac{y^2}{b^2} = 1$.
The equation of the chord whose middle point is (h, k) is

$$\frac{xh}{a^2} + \frac{yk}{b^2} = \frac{h^2}{a^2} + \frac{k^2}{b^2}.$$

Since this chord passes through (α, β), so

$$\frac{\alpha h}{a^2} + \frac{\beta k}{b^2} = \frac{h^2}{a^2} + \frac{k^2}{b^2}.$$

Hence the locus of the middle point (h, k) is

$$\frac{\alpha x}{a^2} + \frac{\beta y}{b^2} = \frac{x^2}{a^2} + \frac{y^2}{b^2}$$

or, $b^2 x^2 + a^2 y^2 = a^2 \beta y + b^2 \alpha x$

or, $b^2 x(x - \alpha) + a^2 y(y - \beta) = 0$. $\square$

Example 6.8.5 *Find the locus of the middle points of the chords of the ellipse* $\dfrac{x^2}{a^2} + \dfrac{y^2}{b^2} = 1$, *which subtend a right angle at the centre.*

Solution: Let the middle point of a chord of the ellipse $\dfrac{x^2}{a^2} + \dfrac{y^2}{b^2} = 1$ be (h, k).

Therefore the equation of the chord of the ellipse in terms of middle point (h, k) is

$$\frac{xh}{a^2} + \frac{yk}{b^2} = \frac{h^2}{a^2} + \frac{k^2}{b^2} \tag{6.16}$$

The equation of the pair of lines joining the centre (i.e., the origin) to the points of intersection of the ellipse and the chord given in (6.16) is obtained by making the equation of the ellipse homogeneous with the help of (6.16) and is given by

$$\frac{x^2}{a^2} + \frac{y^2}{b^2} = \left(\frac{\frac{xh}{a^2} + \frac{yk}{b^2}}{\frac{h^2}{a^2} + \frac{k^2}{b^2}} \right)^2$$

or, $\left(\dfrac{x^2}{a^2} + \dfrac{y^2}{b^2}\right)\left(\dfrac{h^2}{a^2} + \dfrac{k^2}{b^2}\right)^2 = \left(\dfrac{xh}{a^2} + \dfrac{yk}{b^2}\right)^2$ (6.17)

Since the chord (6.16) subtends a right angle at the centre, the lines represented by (6.17) are perpendicular to each other if the sum of the coefficients of x^2 and y^2 is zero,

i.e., $\left(\dfrac{h^2}{a^2} + \dfrac{k^2}{b^2}\right)^2 \left(\dfrac{1}{a^2} + \dfrac{1}{b^2}\right) - \left(\dfrac{h^2}{a^4} + \dfrac{k^2}{b^4}\right) = 0$

or, $(b^2 h^2 + a^2 k^2)^2 (a^2 + b^2) = a^2 b^2 (b^4 h^2 + a^4 k^2).$

Hence the locus of (h, k) is $(b^2 x^2 + a^2 y^2)^2 (a^2 + b^2) = a^2 b^2 (b^4 x^2 + a^4 y^2).$ □

Example 6.8.6 *Prove that the locus of the middle points of normal chords of the parabola* $y^2 = 4ax$ *is* $\dfrac{y^2}{2a} + \dfrac{4a^3}{y^2} = x - 2a.$

Solution: Equation of any normal at the point $'t'$ to the given parabola is

$$y + tx = 2at + at^3 \tag{6.18}$$

Let (α, β) be the middle point of the chord. Then the equation of the chord in terms of the middle point (α, β) is

$$\beta y - 2a(x + \alpha) = \beta^2 - 4a\alpha \;\Rightarrow\; \beta y - 2ax = \beta^2 - 2a\alpha \tag{6.19}$$

Diameters & Conjugate Diameters 287

Equations (6.18) and (6.19) are identical

$$\therefore \quad \frac{t}{-2a} = \frac{1}{\beta} = \frac{2at + at^3}{\beta^2 - 2a\alpha}$$

$$\therefore \quad t = -\frac{2a}{\beta} \text{ and } 2at + at^3 = \frac{\beta^2 - 2a\alpha}{\beta}.$$

Eliminating 't', we get

$$2a\left(-\frac{2a}{\beta}\right) + a\left(-\frac{2a}{\beta}\right)^3 = \frac{\beta^2 - 2a\alpha}{\beta}$$

or, $-4a^2\beta^2 - 8a^4 = \beta^4 - 2a\alpha\beta^2$

or, $\beta^4 + 8a^4 = 2a\alpha\beta^2 - 4a^2\beta^2 = 2a\beta^2(\alpha - 2a)$

or, $\dfrac{\beta^2}{2a} + \dfrac{4a^3}{\beta^2} = \alpha - 2a.$

Hence the required locus of (α, β) is $\dfrac{y^2}{2a} + \dfrac{4a^3}{y^2} = x - 2a.$ □

Example 6.8.7 PCP' and QCQ' are two conjugate diameters of the hyperbola $3x^2 - 2y^2 = 1$. If the coordinates of P be $(1,1)$ and θ be the angle between the diameters, then show that $\sin\theta = \dfrac{1}{\sqrt{26}}$.

Solution: The equation of the given hyperbola be $\dfrac{x^2}{\frac{1}{3}} - \dfrac{y^2}{\frac{1}{2}} = 1$, i.e., we get $a^2 = \dfrac{1}{3}, b^2 = \dfrac{1}{2}.$

$$\therefore \quad CP^2 - CQ^2 = a^2 - b^2 = \frac{1}{3} - \frac{1}{2} = -\frac{1}{6} \tag{6.20}$$

Since the coordinates of P are $(1,1)$, $CP^2 = 1 + 1 = 2$ (6.21)

Hence form (6.20) we get $CQ^2 = 2 + \dfrac{1}{6} = \dfrac{13}{6}$ (6.22)

Now $CP.CQ\sin\theta = ab$ or, $\sqrt{2}.\dfrac{\sqrt{13}}{\sqrt{6}}\sin\theta = \dfrac{1}{\sqrt{3}}\dfrac{1}{\sqrt{2}}$ [by (6.21) and (6.22)]

Hence $\sin\theta = \dfrac{1}{\sqrt{3}}.\dfrac{1}{\sqrt{2}}.\dfrac{1}{\sqrt{2}}.\dfrac{\sqrt{6}}{\sqrt{13}} = \dfrac{1}{\sqrt{26}}.$ □

Example 6.8.8 *Find the condition that the pair of straight lines $Ax^2 + 2Hxy + By^2 = 0$ may be conjugate diameters of the ellipse $\dfrac{x^2}{a^2} + \dfrac{y^2}{b^2} = 1$.*

Solution: Let $y = m_1 x$ and $y = m_2 x$ be the pair of lines represented by the given equation $Ax^2 + 2Hxy + By^2 = 0$. Then $m_1 m_2 = \dfrac{A}{B}$. Again if these lines became conjugate diameters of the given ellipse $\dfrac{x^2}{a^2} + \dfrac{y^2}{b^2} = 1$, then $m_1 m_2 = -\dfrac{b^2}{a^2}$.

Therefore the required condition is $\dfrac{A}{B} = -\dfrac{b^2}{a^2}$ or, $a^2 A + b^2 B = 0$. □

Example 6.8.9 *Find the eccentric angles corresponding to two equal conjugate diameters of the ellipse $\dfrac{x^2}{a^2} + \dfrac{y^2}{b^2} = 1$.*

Solution: Let PCP' and QCQ' be two equi-conjugate diameters of the given ellipse. Then we have $CP = CQ$. Also we have

$$CP^2 + CQ^2 = a^2 + b^2 \implies 2CP^2 = a^2 + b^2 \tag{6.23}$$

Let ϕ be the eccentric angle of P, i.e., the coordinates of P be $(a\cos\phi, b\sin\phi)$. Then $CP^2 = a^2 \cos^2 \phi + b^2 \sin^2 \phi$.

So we get form (6.23),

$$2a^2 \cos^2 \phi + 2b^2 \sin^2 \phi = a^2 + b^2$$
or, $a^2(2\cos^2 \phi - 1) = b^2(1 - 2\sin^2 \phi)$
or, $a^2 \cos 2\phi = b^2 \cos 2\phi$ or, $(a^2 - b^2)\cos 2\phi = 0$
or, $\cos 2\phi = 0$ ($\because a \neq b$) $\therefore 2\phi = \dfrac{\pi}{2}, 3\dfrac{\pi}{2}$.

So, $\phi = \dfrac{\pi}{4}, 3\dfrac{\pi}{4}$, i.e., $45°, 135°$. □

Example 6.8.10 *If the angle between two equi-conjugate diameters of an ellipse is $60°$, find its eccentricity.*

Solution: The equations of two equi-conjugate diameters of the ellipse $\dfrac{x^2}{a^2} + \dfrac{y^2}{b^2} = 1$ are $y = \dfrac{b}{a} x$ and $y = -\dfrac{b}{a} x$ [See § **6.7** equation (6.13) and (6.14)].

Since the angle between them is given as $60°$, we have

$$\tan 60° = \dfrac{\dfrac{b}{a} - \left(-\dfrac{b}{a}\right)}{1 - \dfrac{b^2}{a^2}} = \dfrac{2ab}{a^2 - b^2} \implies \sqrt{3} = \dfrac{2ab}{a^2 - b^2} \tag{6.24}$$

Diameters & Conjugate Diameters

Let the length of each diameter be $2d$. Then by using the properties of § 6.5 we get $2d^2 = a^2 + b^2$ and $2 \cdot \frac{1}{2} d^2 \sin 60° = ab$ [vide. **Example 6.8.7**].

i.e., we get
$$2d^2 = a^2 + b^2 \tag{6.25}$$
and
$$\sqrt{3} d^2 = 2ab \tag{6.26}$$

Dividing (6.26) by (6.25) we get

$$\frac{\sqrt{3}}{2} = \frac{2ab}{a^2 + b^2} \tag{6.27}$$

From (6.24) and (6.27) we get

$$\frac{ab}{a^2 - b^2} = \frac{2ab}{a^2 + b^2} \quad \text{or,} \quad 2a^2 - 2b^2 = a^2 + b^2$$

$$a^2 = 3b^2 \quad \Rightarrow \quad \frac{b^2}{a^2} = \frac{1}{3}.$$

So, if e denotes the eccentricity of the ellipse, then

$$e^2 = 1 - \frac{b^2}{a^2} = 1 - \frac{1}{3} = \frac{2}{3}, \quad \text{i.e.,} \quad e = \frac{\sqrt{2}}{\sqrt{3}} = \frac{1}{3}\sqrt{6}. \qquad \square$$

Example 6.8.11 *Find the locus of the point Q when the normal at a variable point P on the ellipse $\frac{x^2}{a^2} + \frac{y^2}{b^2} = 1$ cuts the diameter CD conjugate of CP at Q.*

Solution: If $(a \cos \theta, b \sin \theta)$ be the coordinates of the point P then the coordinates of D will be $(-a \sin \theta, b \cos \theta)$.

The normal at P is $ax \sec \theta - by \csc \theta = a^2 - b^2$.

The equation to CD is $bx \cos \theta + ay \sin \theta = 0$.

Let Q be the point (α, β), then

$$a\alpha \sec \theta - b\beta \csc \theta = a^2 - b^2 \tag{6.28}$$
and
$$b\alpha \cos \theta + a\beta \sin \theta = 0 \tag{6.29}$$

From (6.29), we get
$$\frac{\sin \theta}{-b\alpha} = \frac{\cos \theta}{a\beta} = \frac{1}{\sqrt{b^2 \alpha^2 + a^2 \beta^2}}$$

$$\therefore \quad \sin \theta = -\frac{b\alpha}{\sqrt{b^2 \alpha^2 + a^2 \beta^2}} \quad \text{and} \quad \cos \theta = \frac{a\beta}{\sqrt{b^2 \alpha^2 + a^2 \beta^2}}.$$

Eliminating θ between (6.28) and (6.29), we get

$$\left(\frac{\alpha}{\beta}+\frac{\beta}{\alpha}\right)\sqrt{b^2\alpha^2+a^2\beta^2}=a^2-b^2$$

or, $\dfrac{(\alpha^2+\beta^2)^2}{\alpha^2\beta^2}(\alpha^2 b^2+\beta^2 a^2)=(a^2-b^2)^2$ or, $\dfrac{a^2}{\alpha^2}+\dfrac{b^2}{\beta^2}=\left(\dfrac{a^2-b^2}{\alpha^2+\beta^2}\right)^2$.

Hence the locus of Q is $\dfrac{a^2}{x^2}+\dfrac{b^2}{y^2}=\left(\dfrac{a^2-b^2}{x^2+y^2}\right)^2$. □

Example 6.8.12 *PSQ is a focal chord of the ellipse $\dfrac{x^2}{a^2}+\dfrac{y^2}{b^2}=1$ where S is a focus and CP, CQ are a pair of semi-conjugate diameters of the curve. If ϕ be the eccentric angle of P and e is the eccentricity of the curve, then show that $PQ = a$ and $e(\sin\phi+\cos\phi)=1$.*

Solution: The coordinates of P and Q can be taken respectively as $(a\cos\phi, b\sin\phi)$ and $(a\sin\phi, -b\cos\phi)$. Hence the equation of the line PQ is

$$\frac{y-b\sin\phi}{b\sin\phi+b\cos\phi}=\frac{x-a\cos\phi}{a\cos\phi-a\sin\phi} \tag{6.30}$$

Now the coordinates of the focus S is given by $(ae, 0)$ and this point $S(ae, 0)$

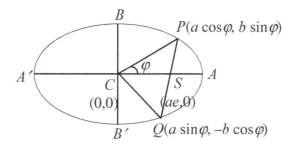

lies on (6.30)

$$\therefore\quad \frac{-\sin\phi}{\sin\phi+\cos\phi}=\frac{e-\cos\phi}{\cos\phi-\sin\phi}$$

or, $-\sin\phi\cos\phi+\sin^2\phi=e(\sin\phi+\cos\phi)-\sin\phi\cos\phi-\cos^2\phi$

or, $e(\sin\phi+\cos\phi)=1 \tag{6.31}$

Now, $PQ^2 = (a\cos\phi - a\sin\phi)^2 + (b\sin\phi + b\cos\phi)^2$
$= a^2(\cos\phi - \sin\phi)^2 + b^2(\sin\phi + \cos\phi)^2$
$= a^2(\cos\phi - \sin\phi)^2 + a^2(1 - e^2)(\sin\phi + \cos\phi)^2$
$\qquad\qquad\qquad\qquad\qquad\qquad [\because b^2 = a^2(1 - e^2)]$
$= a^2[(\cos\phi - \sin\phi)^2 + (\sin\phi + \cos\phi)^2] - a^2 e^2(\sin\phi + \cos\phi)^2$
$= a^2 \cdot 2(\cos^2\phi + \sin^2\phi) - a^2$ [using (6.31)]
$= 2a^2 - a^2 = a^2$

$\therefore\ PQ = a$. $\qquad\square$

Example 6.8.13 *Lines are drawn through the foci of an ellipse perpendicular to a pair of conjugate diameters and intersect at Q. Show that the locus of Q is a concentric ellipse.*

Solution: Let the two extremities of a pair of conjugate diameters of the ellipse $\dfrac{x^2}{a^2} + \dfrac{y^2}{b^2} = 1$ be $P(a\cos\theta, b\sin\theta)$ and $D(a\sin\theta, -b\cos\theta)$. If C and S be respectively the centre and focus of the ellipse then the coordinates of C and S are respectively given as $(0, 0)$ and $(ae, 0)$ where $b^2 = a^2(1 - e^2)$.

The slope of the diameter CP is $\dfrac{b\sin\theta}{a\cos\theta}$ and that of CD is $\dfrac{b\cos\theta}{-a\sin\theta}$.

The equation of the line passing through $S(ae, 0)$ and perpendicular to CP is
$$y = \dfrac{a\cos\theta}{-b\sin\theta}(x - ae)$$
and that of the straight line through the point $(-ae, 0)$ and perpendicular to CD is
$$y = \dfrac{a\sin\theta}{b\cos\theta}(x + ae).$$
If these two straight lines meet at $Q(\alpha, \beta)$, then we have
$$\cot\theta = \dfrac{b\beta}{a(ae - \alpha)} = \dfrac{a(\alpha + ae)}{b\beta}.$$

Eliminating θ, we get,
$$b^2\beta^2 = a^2(a^2 e^2 - \alpha^2)\ \text{or,}\ a^2\alpha^2 + b^2\beta^2 = a^4 e^2 = a^2(a^2 - b^2)$$
or, $\dfrac{\alpha^2}{b^2} + \dfrac{\beta^2}{a^2} = \dfrac{a^2 - b^2}{b^2}$.

Hence, the locus of Q is $\dfrac{x^2}{b^2} + \dfrac{y^2}{a^2} = \dfrac{a^2 - b^2}{b^2}$ which is an ellipse concentric with the ellipse $\dfrac{x^2}{a^2} + \dfrac{y^2}{b^2} = 1$. $\qquad\square$

Example 6.8.14 *If PCP' and DCD' be two mutually perpendicular diameters of the ellipse $\dfrac{x^2}{a^2} + \dfrac{y^2}{b^2} = 1$, prove that $\dfrac{1}{CP^2} + \dfrac{1}{CD^2} = \dfrac{1}{a^2} + \dfrac{1}{b^2}$.*

Solution: Let $(a\cos\phi, b\sin\phi)$ and $(a\cos\phi', b\sin\phi')$ be the coordinates of the points P and D respectively. The coordinates of the centre C of the given ellipse are $(0,0)$.

The slope of CP is $\dfrac{b}{a}\tan\phi$ and that of CD is $\dfrac{b}{a}\tan\phi'$.

Since CP and CD are perpendicular to each other so

$$\frac{b}{a}\tan\phi \cdot \frac{b}{a}\tan\phi' = -1 \quad \text{or,} \quad \tan\phi \cdot \tan\phi' = -\frac{a^2}{b^2} \qquad (6.32)$$

Now
$$\frac{1}{CP^2} + \frac{1}{CD^2} = \frac{1}{a^2\cos^2\phi + b^2\sin^2\phi} + \frac{1}{a^2\cos^2\phi' + b^2\sin^2\phi'}$$

$$= \frac{\sec^2\phi}{a^2 + b^2\tan^2\phi} + \frac{\sec^2\phi'}{a^2 + b^2\tan^2\phi'}$$

$$= \frac{1 + \tan^2\phi}{a^2 + b^2\tan^2\phi} + \frac{1 + \tan^2\phi'}{a^2 + b^2\tan^2\phi'}$$

$$= \frac{1 + \tan^2\phi}{a^2 + b^2\tan^2\phi} + \frac{1 + \frac{a^4}{b^4\tan^2\phi}}{a^2 + b^2\frac{a^4}{b^4\tan^2\phi}} \quad \text{[using (6.32)]}$$

$$= \frac{1 + \tan^2\phi}{a^2 + b^2\tan^2\phi} + \frac{a^4 + b^4\tan^2\phi}{a^4 b^2 + a^2 b^4 \tan^2\phi}$$

$$= \frac{a^2 b^2(1 + \tan^2\phi) + a^4 + b^4\tan^2\phi}{a^2 b^2(a^2 + b^2\tan^2\phi)}$$

$$= \frac{a^2(a^2 + b^2\tan^2\phi) + b^2(a^2 + b^2\tan^2\phi)}{a^2 b^2(a^2 + b^2\tan^2\phi)}$$

$$= \frac{(a^2 + b^2)(a^2 + b^2\tan^2\phi)}{(a^2 b^2)(a^2 + b^2\tan^2\phi)} = \frac{a^2 + b^2}{a^2 b^2} = \frac{1}{a^2} + \frac{1}{b^2}.$$

Hence $\dfrac{1}{CP^2} + \dfrac{1}{CD^2} = \dfrac{1}{a^2} + \dfrac{1}{b^2}$. □

Example 6.8.15 *Show that the product of the focal distances of a point on an ellipse is equal to the square of the length of the semi-diameter parallel to the tangent at this point.*

Solution: See § **6.5 Property 4**. □

Diameters & Conjugate Diameters

Example 6.8.16 *Prove that the locus of the middle points of chords of contact of tangents to the hyperbola $x^2 - y^2 = a^2$ from points on the auxiliary circle is the curve $(x^2 - y^2)^2 = a^2(x^2 + y^2)$.*

Solution: The auxiliary circle is $x^2 + y^2 = a^2$.

If $(a\cos\phi, a\sin\phi)$ be a point on it, then the chord of contact of this point with respect to the hyperbola $x^2 - y^2 = a^2$ is

$$xa\cos\phi - ya\sin\phi = a^2 \quad \text{or,} \quad x\cos\phi - y\sin\phi = a \tag{6.33}$$

If (x_1, y_1) is the middle point of this chord, then the equation (6.33) is identical with

$$xx_1 - yy_1 = x_1^2 - y_1^2 \tag{6.34}$$

Comparing (6.33) and (6.34) we get

$$\frac{\cos\phi}{x_1} = \frac{\sin\phi}{y_1} = \frac{a}{x_1^2 - y_1^2} \Rightarrow \cos\phi = \frac{ax_1}{x_1^2 - y_1^2}, \ \sin\phi = \frac{ay_1}{x_1^2 - y_1^2}$$

$$\text{or,} \quad 1 = \cos^2\phi + \sin^2\phi = \frac{a^2(x_1^2 + y_1^2)}{(x_1^2 - y_1^2)^2} \quad \text{or,} \ (x_1^2 - y_1^2)^2 = a^2(x_1^2 + y_1^2)$$

Therefore the required locus is $(x^2 - y^2)^2 = a^2(x^2 + y^2)$. $\square$

Example 6.8.17 *Any tangent to an ellipse with centre C meets the director circle in P and D. Prove that CP and CD are in the directions of the conjugate diameters of the ellipse.*

Solution: The straight line

$$y = mx + \sqrt{a^2m^2 + b^2} \tag{6.35}$$

is always a tangent to the ellipse $\dfrac{x^2}{a^2} + \dfrac{y^2}{b^2} = 1$.

The equation of the director circle of the ellipse is

$$x^2 + y^2 = a^2 + b^2 \tag{6.36}$$

The combined equation of CP and CD is obtained by making equation (6.36) homogeneous with the help of (6.35), which is

$$(a^2m^2 + b^2)(x^2 + y^2) = (a^2 + b^2)(y - mx)^2$$

$$\left[\because \text{form (6.35),} \ \frac{(y-mx)^2}{a^2m^2 + b^2} = 1\right]$$

$$\text{or,} \quad (b^2 - b^2m^2)x^2 + (a^2m^2 - a^2)y^2 + 2(a^2 + b^2)mxy = 0$$

$$\text{or,} \quad b^2(1 - m^2)x^2 + 2(a^2 + b^2)mxy + a^2(m^2 - 1)y^2 = 0 \tag{6.37}$$

If m_1 and m_1 be the slopes of CP and CD respectively, then from (6.37) we have
$$m_1 m_2 = \frac{b^2(1-m^2)}{a^2(m^2-1)} = -\frac{b^2}{a^2}.$$

Thus CP and CD are in the directions of the conjugate diameters of the ellipse. □

Example 6.8.18 *Prove that the straight lines joining the centre of the ellipse $\frac{x^2}{a^2} + \frac{y^2}{b^2} = 1$ to its points of intersection with the straight line $y = mx + \sqrt{\frac{a^2m^2+b^2}{2}}$ are conjugate diameters of the curve.*

Solution: The equations of the lines joining the centre of the ellipse to its points of intersection with the line $y = mx + \sqrt{\frac{a^2m^2+b^2}{2}}$ can be written as

$$\frac{x^2}{a^2} + \frac{y^2}{b^2} - \frac{(y-mx)^2}{\frac{a^2m^2+b^2}{2}} = 0 \qquad (6.38)$$

[This is obtained by making homogeneous the equation of the curve with the help of the equation of the line.]

or, $\left[\frac{1}{a^2} - \frac{m^2}{\left(\frac{a^2m^2+b^2}{2}\right)}\right]x^2 + \left[\frac{1}{b^2} - \frac{1}{\left(\frac{a^2m^2+b^2}{2}\right)}\right]y^2 + \frac{2m}{\frac{a^2m^2+b^2}{2}}xy = 0$

or, $(y - m_1 x)(y - m_2 x) = 0$

where $m_1 m_2 = \dfrac{\frac{1}{a^2} - \frac{m^2}{\left(\frac{a^2m^2+b^2}{2}\right)}}{\frac{1}{b^2} - \frac{1}{\left(\frac{a^2m^2+b^2}{2}\right)}} = \dfrac{\left(\frac{a^2m^2+b^2}{2} - a^2m^2\right)b^2}{\left(\frac{a^2m^2+b^2}{2} - b^2\right)a^2}$

$= \dfrac{(b^2 - a^2m^2)b^2}{(a^2m^2 - b^2)a^2} = -\dfrac{b^2}{a^2}.$

Therefore the lines $y - m_1 x = 0$ and $y - m_2 x = 0$, i.e., the lines represented by (6.38) are conjugate diameters of the ellipse. □

Example 6.8.19 *Show that the locus of the middle points of the normal chords of the rectangular hyperbola $x^2 - y^2 = a^2$ is $(y^2 - x^2)^3 = 4a^2 x^2 y^2$.*

Diameters & Conjugate Diameters

Solution: Let (x_1, y_1) be the middle point of a chord of the rectangular hyperbola

$$x^2 - y^2 = a^2 \qquad (6.39)$$

The equation of the chord is

$$xx_1 - yy_1 = x_1^2 - y_1^2 \qquad (6.40)$$

The equation of the normal at the point $(a \sec \phi, b \tan \phi)$ to the hyperbola is

$$x \cos \phi + y \cot \phi = 2a \qquad (6.41)$$

Comparing (6.40) and (6.41) we get

$$\frac{\cos \phi}{x_1} = \frac{\cot \phi}{-y_1} = \frac{2a}{x_1^2 - y_1^2} \quad \therefore \cos \phi = \frac{2ax_1}{x_1^2 - y_1^2} \text{ and } \sin \phi = -\frac{x_1}{y_1}.$$

Squaring and adding these we get

$$1 = \cos^2 \phi + \sin^2 \phi = \frac{4a^2 x_1^2}{(x_1^2 - y_1^2)^2} + \frac{x_1^2}{y_1^2}$$

or, $\dfrac{4a^2 x_1^2}{(x_1^2 - y_1^2)^2} + \dfrac{x_1^2}{y_1^2} - 1 = 0$ or, $\dfrac{4a^2 x_1^2}{(x_1^2 - y_1^2)^2} + \dfrac{x_1^2 - y_1^2}{y_1^2} = 0$

or, $4a^2 x_1^2 y_1^2 + (x_1^2 - y_1^2)^3 = 0$ or, $(y_1^2 - x_1^2)^3 = 4a^2 x_1^2 y_1^2$.

Therefore the required locus is $(y^2 - x^2)^3 = 4a^2 x^2 y^2$. □

Example 6.8.20 *Show that the locus of the poles of line joining the extremities of two conjugate diameters of the ellipse* $\dfrac{x^2}{a^2} + \dfrac{y^2}{b^2} = 1$ *is* $\dfrac{x^2}{a^2} + \dfrac{y^2}{b^2} = 2$.

Solution: Let $(a \cos \theta, b \sin \theta)$ and $(-a \sin \theta, b \cos \theta)$ be the ends of the conjugate diameters of the given ellipse $\dfrac{x^2}{a^2} + \dfrac{y^2}{b^2} = 1$.

The equation of the tangents at these points are

$$\frac{x}{a} \cos \theta + \frac{y}{b} \sin \theta = 1 \qquad (6.42)$$

and $\dfrac{x}{a} \sin \theta - \dfrac{y}{b} \cos \theta = -1 \qquad (6.43)$

The point of intersection of these tangents is the pole of the line joining the ends of the conjugate diameters.

Eliminating θ from (6.42) and (6.43) we get the required locus. Squaring (6.42) and (6.43) and then adding we get

$$\frac{x^2}{a^2}(\cos^2\theta + \sin^2\theta) + \frac{y^2}{b^2}(\cos^2\theta + \sin^2\theta) = 1^2 + (-1)^2$$

i.e., $\quad \dfrac{x^2}{a^2} + \dfrac{y^2}{b^2} = 2.$

□

Diameters & Conjugate Diameters 297

6.9 Exercises

Section A: Objective Type Questions

1. Show that the locus of the middle point of a chord of a parabola which passes through a fixed point $(a, 0)$ is the parabola $y^2 = 2a(x - a)$.

2. Show that for the ellipse $x^2 + 4y^2 = 36$, the diameters $y = 4x$ and $x + 16y = 0$ are conjugate diameters.

3. If the line $\dfrac{l}{a}x + \dfrac{m}{b}y = n$ cuts the ellipse $\dfrac{x^2}{a^2} + \dfrac{y^2}{b^2} = 1$ at the ends of conjugate diameters of the ellipse, prove that $l^2 + m^2 = 2n^2$.

4. Find the equations of two conjugate diameters of the hyperbola $\dfrac{x^2}{5} - \dfrac{y^2}{4} = 1$ if one of them passes through the point $(10, 1)$.

5. Show that the locus of the middle points of the chords of the ellipse $\dfrac{x^2}{a^2} + \dfrac{y^2}{b^2} = 1$ which pass through a fixed point (α, β) is given by $b^2 x(x - \alpha) + a^2 y(y - \beta) = 0$.

6. Find the equation of the chord of the parabola $y^2 = 8x$ which is bisected at the point $(2, -3)$.

7. Find the equation of the diameter of the ellipse $3x^2 + 4y^2 = 5$ conjugate to the diameter $y + 3x = 0$.

8. For the hyperbola $16x^2 - 9y^2 = 144$ find the equation of the diameter which is conjugate to the diameter $x = 2y$.

9. Find the condition that the pair of straight lines $Ax^2 + 2Hxy + By^2 = 0$ may be conjugate diameters of the ellipse $\dfrac{x^2}{a^2} + \dfrac{y^2}{b^2} = 1$.

10. If the points of intersection of the two ellipses $\dfrac{x^2}{a^2} + \dfrac{y^2}{b^2} = 1$ and $\dfrac{x^2}{\alpha^2} + \dfrac{y^2}{\beta^2} = 1$ be the ends of the conjugate diameters of the former then prove that $\dfrac{a^2}{\alpha^2} + \dfrac{b^2}{\beta^2} = 2$.

Section B: Broad Answer Type Questions

1. Find the middle point of the chord of the parabola $x^2 = 8y$, whose equation is $3x + 4y + 1 = 0$.

2. Find the locus of the middle points of the chords of the circle $x^2 + y^2 =$

13 which are parallel to the straight line $3x + 4y = 5$.

3. Prove that the chord $x + y = 0$ of the parabola $y^2 = 4x$ is bisected at the point $(2, -2)$.

4. Find the equation of the locus of the mid-points of the chords of the parabola $y^2 = 4x$ which have a gradient $\dfrac{1}{2}$.

5. CP and CQ are conjugate semi-diameters of the ellipse $2x^2 + 7y^2 = 30$. If the coordinates of P are $(1, 2)$, find the length of CQ.

6. P and Q are the extremities of conjugate diameters of the ellipse $2x^2 + 3y^2 = 24$. Find the locus of the point of intersection of the tangents at P and Q.

7. Show that the diameters whose equations are $y + 3x = 0$ and $4y - x = 0$ are conjugate diameters of the ellipse $3x^2 + 4y^2 = 5$.

8. If $P(x_1, y_1)$ and $D(x_2, y_2)$ be the extremities of the conjugate diameters CP and CD of an ellipse, prove that $\dfrac{x_1 x_2}{a^2} + \dfrac{y_1 y_2}{b^2} = 0$.

Hence deduce $x_1 y_2 - y_1 x_2 = \pm ab$.

9. Show that the diameters whose equations are $y + 3x = 0$ and $4y - x = 0$ are the conjugate diameters of the ellipse $3x^2 + 4y^2 = 5$.

10. CP and CD are conjugate diameters of $\dfrac{x^2}{a^2} + \dfrac{y^2}{b^2} = 1$. Show that the ortho-centre of the triangle $\triangle CPD$ is

$$2(b^2 y^2 + a^2 x^2) = (a^2 - b^2)^2 (b^2 y^2 - a^2 x^2)^2.$$

11. If P and D are the ends of conjugate diameters find the locus of

    (i)   the middle point of PD.

    (ii) the point of intersection of the tangents at P and D.

    (iii) the foot of the perpendicular form the centre upon PD.

and (iv) the point of intersection of the normals at P and D.

12. A circle passes through the ends of a diameter of the ellipse $\dfrac{x^2}{a^2} + \dfrac{y^2}{b^2} = 1$ and also touches the curve. Prove that the centre of the circle lie on the ellipse $4a^2 x^2 + 4b^2 y^2 = (a^2 - b^2)^2$.

13. The normal to the rectangular hyperbola $xy = c^2$, at a point P on it meets the curve again at Q and touches the conjugate hyperbola. Show that $(PQ)^2 = 512 c^4$.

14. Show that $4x - 3y + 4 = 0$ and $x + 3y - 7 = 0$ are parallel to the conjugate diameters of the ellipse $4x^2 + 9y^2 = 36$.

Diameters & Conjugate Diameters

15. If the line $lx + my = 1$, passes through the extremities of a pair of conjugate diameters of the ellipse $\dfrac{x^2}{a^2} + \dfrac{y^2}{b^2} = 1$, then $a^2l^2 + b^2m^2 = 2$.

16. Show that the conjugate diameters of $\dfrac{x^2}{a^2} + \dfrac{y^2}{b^2} = 1$ which are equally inclined to each of the axes are $bx + ay = 0$ and $bx - ay = 0$.

17. Prove that the equation to the equi-conjugate diameters of the conic $ax^2 + 2hxy + by^2 = 1$ is

$$\frac{ax^2 + 2hxy + by^2}{ab - h^2} = \frac{2(x^2 + y^2)}{a + b}.$$

18. Show that the pair of lines

$$(Ha - hA)x^2 + (aB - Ab)xy + (hB - Hb)y^2 = 0$$

are conjugate diameters of both the conics $ax^2 + 2hxy + by^2 = 0$ and $Ax^2 + 2Hxy + By^2 = 0$.

19. Show that the tangents at the ends of the diameters of an ellipse are parallel to each other and to the conjugate diameters.

20. If O is the centre of the hyperbola $\dfrac{x^2}{a^2} - \dfrac{y^2}{b^2} = 1$, find two conjugate diameters OD and OD_1 of the curve such that OD bisects the angle between the positive x-axis and OD_1.

21. If $P(x_1, y_1)$ and $Q(x_2, y_2)$ are the extremities of a pair of conjugate diameters of the ellipse $\dfrac{x^2}{a^2} + \dfrac{y^2}{b^2} = 1$, then show that

(i) $x_1^2 + x_2^2 = a^2$, (ii) $y_1^2 + y_2^2 = b^2$, (iii) $x_1y_1 + x_2y_2 = 0$

and (iv) the equation of the chord PQ is $\dfrac{x_1 + x_2}{a^2}x + \dfrac{y_1 + y_2}{b^2}y = 1$.

22. If CP and CQ are conjugate semi-diameters of the ellipse $\dfrac{x^2}{a^2} + \dfrac{y^2}{b^2} = 1$, show that the line joining the middle points of P and Q touches the conic $\dfrac{x^2}{a^2} - \dfrac{y^2}{b^2} = \dfrac{1}{8}$.

23. Find the angle between the equi-conjugate diameter of the ellipse $x^2 + 3y^2 = 3$.

24. Show that the condition that the lines $Ax^2 + 2Hxy + By^2 = 0$ should be conjugate diameters of the conic

$$ax^2 + 2hxy + by^2 = 1 \text{ is } aB + bA - 2hH = 0.$$

ANSWERS

Section A: 4. $y = \frac{1}{10}x$, $y = 8x$; **6.** $4x + 3y + 1 = 0$; **7.** $x = 4y$; **8.** $32x = 9y$; **9.** $Aa^2 + Bb^2 = 0$.

Section B: 1. $(-3, 2)$; **2.** $4x = 3y$; **4.** $y = 4$; **5** $\frac{10}{\sqrt{7}}$; **6.** $\frac{x^2}{12} + \frac{y^2}{8} = 2$; **11. (i)** $\frac{x^2}{a^2} + \frac{y^2}{b^2} = \frac{1}{2}$; **(ii)** $\frac{x^2}{a^2} + \frac{y^2}{b^2} = 2$; **(iii)** $2(x^2 + y^2)^2 = a^2x^2 + b^2y^2$; **(iv)** $2(a^2x^2 + b^2y^2) = (a^2 - b^2)^2(a^2x^2 - b^2y^2)^2$; **20.** Equation of OD is $y = mx$ and that of OD_1 is $y = m_1 x$ where $m = \frac{\pm b}{\sqrt{2a^2 + b^2}}$, $m_1 = \frac{b^2}{a^2 m}$; **23.** $60°$.

Bibliography

[1] Vyvyan T.G., *Elementary Analytic Geometry*, Deighton, Bell and Company, 1867.

[2] Loomis E., *The Elements of Analytic Geometry*, Harper, Princeton University Press, 1881.

[3] Loney S.L., *The Elements of Coordinate Geometry*, Macmillan and Co., Harvard, 1896.

[4] Askwith E.H., *The Analytical Geometry of the Conic Sections*, Adam and Charles Black, London, 1908.

[5] Bohannan R.D., *Plain Analytic Geometry*, R.G. Adams & Company, 1913.

[6] Coleman P., *Coordinate Geometry: An Elementary Course*, Clarendon Press, Oxford, 1914.

[7] Gibson G.A. and Pinkerton P., *Elements of Analytical Geometry*, Macmillan and Company Limited, Harvard, 1919.

[8] Poor V.C., *Analytic Geometry*, Wiley, New York, 1934.

[9] Kar J.M., *Analytical Geometry of the Conic Sections*, The Globe Library, Calcutta (Kolkata), 1960.

[10] Bohuslov R.L., *Analytical Geometry: A Pre-Calculus Approach*, Macmillan and Pennsylvania State University, Pennsylvania, 1970.

[11] Khanna M.L., *Coordinate Geometry*, Jai Prakash Nath Publication, Meerut, 1983.

[12] Das B., *Analytical Geometry with Vector Analysis*, Orient Book Company, Calcutta (Kolkata), 1990.

[13] Chaki M.C., *A Textbook of Analytic Geometry (Vol. II)*, Calcutta Publishers, Calcutta (Kolkata), 1991.

[14] Jain P.K. and Ahmad K., *Textbook of Analytical Geometry of Two Dimensions*, New Age International Pvt. Ltd., New Delhi, 1996.

[15] Dutta N. and Jana R.N., *Analytical Geometry of Two and Three Dimensions*, Shreedhar Prakashani, Kolkata, 1999.

[16] Kar B.K., *Advanced Analytic Geometry and Vector Analysis*, Books & Allied Pvt. Ltd., Kolkata, 2000.

[17] Robson A., *An Introduction to Analytic Geometry*, Cambridge University Press, Cambridge, 2003.

[18] Sengupta S.B., *Coordinate Geometry and Vector Analysis*, Joydurga Library, Kolkata, 2005.

[19] Chakravorty J.G. and Ghosh P.R., *Advanced Analytical Geometry*, U. N. Dhur & Sons Pvt. Ltd., Kolkata, 2009.

[20] Khan R.M., *Analytical Geometry of Two and Three Dimensions and Vector Analysis*, NCBA, Kolkata, 2012.

[21] Das A.N., *Analytical Geometry of Two and Three Dimensions*, NCBA, Kolkata, 2013.

[22] Pal K.C., *Analytical Geometry (Two and Three Dimensions) and Vector Analysis*, Books & Allied Pvt. Ltd., Kolkata, 2013.

[23] Vittal P.R., *Analytical Geometry 2D and 3D*, Pearson, New Delhi, 2013.

[24] Mondal A.K., *A Textbook of Advanced Analytic Geometry*, Sreetara Prakashani, Kolkata, 2015.

[25] Mukherjee A. and Bej N.K., *Analytical Geometry of Two and Three Dimensions (Advanced Level)*, Books & Allied Pvt. Ltd., Kolkata, 2016.

Index

Angle between a pair of straight lines, 174
Angle-bisectors of a pair of lines, 175
Area
    of a polygon, 5
    of a triangle, 5, 104
Asymptote, 120

Canonical form, 27, 61
Central curve, 56
Centre, 56
Centroid, 6
Chord of circle, 13
Chord of contact, 16, 117
Circle, 12, 111
Circumcentre, 7
Co-normal Points, 218
Concurrent lines, 11
Conic, 111
Conjugate
    diameters, 19, 225, 276
    lines, 237
    points, 237
Coordinate, 3

Degenerate curve, 69
Diameter of a conic, 225, 273
Directed line segment, 1
Directed lines, 1
Director circle, 222
Distance between two points

    in Cartesian coordinates, 4
    in Polar coordinates, 103

Ellipse, 15, 111
Equi-conjugate diameters, 283
Ex-centre, 7
Excribed circle, 7

Family of straight lines, 12
Fixed point
    under rigid motion, 28
    under rotation, 28
    under translation, 27

General orthogonal transformation of axes, 26
General second degree equation, 27

Half line, 3
Harmonic conjugate, 227
Homogeneous equation, 166
Hyperbola, 16, 111

Imaginary ellipse, 63
Incentre, 7
Initial line, 101
Invariants, 27

Length of chord, 16
Location of a point, 3
Locus, 3

Matrix notation
    of orthogonal transformation, 25
    of rigid motion, 25
    of rotation, 25
    of translation, 24

Non-central curve, 56
Non-degenerate curve, 69
Non-singular curve, 69
Normal, 16, 216
Normal form, 27, 66

Ordinate of a diameter, 274
Orthocentre, 8
Orthogonal
    circle, 13
    projection, 3
    transformation, 24

Pair of
    coincident straight lines, 69
    imaginary straight lines, 172
    intersecting straight lines, 69
    parallel straight lines, 69
    straight lines, 70, 111, 165
    tangents, 219
Parabola, 14, 111
Parallel chord, 273
Point ellipse, 69
Polar, 226
Polar
    axis, 101
    coordinates, 101
Polar equation of
    a chord of a conic, 114
    a chord of contact, 117
    a circle, 108
    a conic, 111
    a line in normal form, 107
    a normal to a conic, 116
    a straight line, 104
    a tangent to a conic, 115
    an asymptote, 120
Pole, 101, 226
Projection
    of a directed line segment, 3
    of a point, 2
Proper conics, 70

Radical axis, 13
Radius vector, 101
Rank of a second order curve, 69
Rigid motion, 24
Rotation, 22

Second order curve, 55
Section formula, 4
Self-conjugate, 238
Self-polar triangle, 238
Semi-conjugate diameters, 278
Simultaneous equation, 60
Singular curve, 69
Standard conics, 27
Standard form, 66
Straight line, 9

Tangent, 16, 209
Transformation of axes, 21
Translation, 21

Vectorial angle, 101